Paige G. Andrew
Mary Lynette Larsgaard
Editors

Maps and Related Cartographic Materials: Cataloging, Classification, and Bibliographic Control

Maps and Related Cartographic Materials: Cataloging, Classification, and Bibliographic Control has been co-published simultaneously as *Cataloging & Classification Quarterly,* Volume 27, Numbers 1/2 and 3/4 1999.

Pre-publication
REVIEWS,
COMMENTARIES,
EVALUATIONS . . .

"**M**ap cataloging has been a difficult aspect of library organization since . . . even before libraries had to deal with printed books. Today, with the digital 'revolution,' libraries have even more reason to be concerned. The answer is this new book. . . . It begins with the basics of . . . map cataloging and classification; moves on to . . . different formats, such as sheet maps, globes, atlases, aerial photography, and digital cartography; and concludes with retrospective conversion and archived materials. . . . The section devoted to the cataloging of aerial photographs and other remote-sensing materials . . . make this a very up-to-date volume. . . . This collected work should be in any library. . . ."

David C. McQuillan, MA, ML
Map Librarian
University of South Carolina
Chairman
IFLA Section of Geography
and Map Libraries

Maps and Related Cartographic Materials: Cataloging, Classification, and Bibliographic Control

Maps and Related Cartographic Materials: Cataloging, Classification, and Bibliographic Control has been co-published simultaneously as *Cataloging & Classification Quarterly*, Volume 27, Numbers 1/2 and 3/4 1999.

The *Cataloging & Classification Quarterly* Monographic "Separates"

Below is a list of "separates," which in serials librarianship means a special issue simultaneously published as a special journal issue or double-issue *and* as a "separate" hardbound monograph. (This is a format which we also call a "DocuSerial.")

"Separates" are published because specialized libraries or professionals may wish to purchase a specific thematic issue by itself in a format which can be separately cataloged and shelved, as opposed to purchasing the journal on an on-going basis. Faculty members may also more easily consider a "separate" for classroom adoption.

"Separates" are carefully classified separately with the major book jobbers so that the journal tie-in can be noted on new book order slips to avoid duplicate purchasing.

You may wish to visit Haworth's Website at . . .

http://www.haworthpressinc.com

. . . to search our online catalog for complete tables of contents of these separates and related publications.

You may also call 1-800-HAWORTH (outside US/Canada: 607-722-5857), or Fax 1-800-895-0582 (outside US/Canada: 607-771-0012), or e-mail at:

getinfo@haworthpressinc.com

Maps and Related Cartographic Materials: Cataloging, Classification, and Bibliographic Control, edited by Paige G. Andrew, BA, MLS and Mary Lynette Larsgaard, BA, MA (Vol. 27, No. 1/2/3/4, 1999) *Discover how to catalog the major formats of cartographic materials, including sheet maps, early and contemporary atlases, remote-sensed images (i.e., aerial photographs and satellite images), globes, geologic sections, digital material, and items on CD-ROM.*

Portraits in Cataloging and Classification: Theorists, Educators, and Practitioners of the Late Twentieth Century, edited by Carolynne Myall, MS, CAS, and Ruth C. Carter, PhD. (Vol. 25, No. 2/3/4, 1998) *"This delightful tome introduces us to a side of our profession that we rarely see: the human beings behind the philosophy, rules, and interpretations that have guided our professional lives over the past half century. No collection on cataloging would be complete without a copy of this work." (Walter M. High, PhD, Automation Librarian, North Carolina Supreme Court Library; Assistant Law Librarian for Technical Services, North Carolina University, Chapel Hill)*

Cataloging and Classification: Trends, Transformations, Teaching, and Training, edited by James R. Shearer, MA, ALA and Alan R. Thomas, MA, FLA (Vol. 24, No. 1/2, 1997) *"Offers a comprehensive retrospective and innovative projection for the future." (The Catholic Library Association)*

Electric Resources: Selection and Bibliographic Control, edited by Ling-yuh W. (Miko) Pattie, MSLS, and Bonnie Jean Cox, MSLS (Vol. 22, No. 3/4, 1996) *"Recommended for any reader who is searching for a thorough, well-rounded, inclusive compendium on the subject." (The Journal of Academic Librarianship)*

Cataloging and Classification Standards and Rules, edited by John J. Reimer, MLS (Vol. 21, No. 3/4, 1996). *"Includes chapters by a number of experts on many of our best loved library standards. . . . Recommended to those who want to understand the history and development of our library standards and to understand the issues at play in the development of new standards." (LASIE)*

Classification: Options and Opportunities, edited by Alan R. Thomas, MA, FLA (Vol. 19, No. 3/4, 1995). *"There is much new and valuable insight to be found in all the chapters. . . . Timely in refreshing our confidence in the value of well-designed and applied classification in providing the best of service to the end-users." (Catalogue and Index)*

Cataloging Government Publications Online, edited by Carolyn C. Sherayko, MLS (Vol. 18, No. 3/4, 1994). *"Presents a wealth of detailed information in a clear and digestible form, and reveals many of the practicalities involved in getting government publications collections onto online cataloging systems." (The Law Librarian)*

Cooperative Cataloging: Past, Present and Future, edited by Barry B. Baker, MLS (Vol. 17, No. 3/4, 1994). *"The value of this collection lies in its historical perspective and analysis of past and present approaches to shared cataloging. . . . Recommended to library schools and large general collections needing materials on the history of library and information science." (Library Journal)*

Languages of the World: Cataloging Issues and Problems, edited by Martin D. Joachim (Vol. 17, No. 1/2, 1994). *"An excellent introduction to the problems libraries must face when cataloging materials not written in English. . . . should be read by every cataloger having to work with international materials, and it is recommended for all library schools. Nicely indexed." (Academic Library Book Review)*

Retrospective Conversion Now in Paperback: History, Approaches, Considerations, edited by Brian Schottlaender, MLS (Vol. 17, No. 1/2, 1992). *"Fascinating insight into the ways and means of converting and updating manual catalogs to machine-readable format." (Library Association Record)*

Enhancing Access to Information: Designing Catalogs for the 21st Century, edited by David A. Tyckoson (Vol. 13, No. 3/4, 1992 *"Its down-to-earth, nontechnical orientation should appeal to practitioners including administrators and public service librarians." (Library Resources & Technical Services)*

Describing Archival Materials: The Use of the MARC AMC Format, edited by Richard P. Smiraglia, MLS (Vol. 11, No. 3/4, 1991). *"A valuable introduction to the use of the MARC AMC format and the principles of archival cataloging itself." (Library Resources & Technical Services)*

Subject Control in Online Catalogs, edited by Robert P. Holley, PhD, MLS (Vol. 10, No. 1/2, 1990). *"The authors demonstrate the reasons underlying some of the problems and how solutions may be sought. . . . Also included are some fine research studies where the researches have sought to test the interaction of users with the catalogue, as well as looking at use by library practitioners." (Library Association Record)*

Library of Congress Subject Headings: Philosophy, Practice, and Prospects, edited by William E. Studwell, MSLS (Supp. #2, 1990). *"Plays an important role in any debate on subject cataloging and succeeds in focusing the reader on the possibilities and problems of using Library of Congress Subject Headings and of subject cataloging in the future." (Australian Academic & Research Libraries)*

Authority Control in the Online Enviroment: Considerations and Practices, edited by Barbara B. Tillett, PhD (Vol. 9, No. 3, 1989). *"Marks an excellent addition to the field. . . .[It] is intended, as stated in the introduction, to 'offer background and inspiration for future thinking.' In achieving this goal, it has certainly succeeded." (Information Technology & Libraries)*

National and International Bibliographic Databases: Trends and Prospects, edited by Michael Carpenter, PhD, MBA, MLS (Vol. 8, No. 3/4, 1988). *"A fascinating work, containing much of concern both to the general cataloger and to the language or area specialist as well. It is also highly recommended reading for all those interested in bibliographic databases, their development, or their history." (Library Resources & Technical Services)*

Cataloging Sound Recordings: A Manual with Examples, Deanne Holzberlein, PhD, MLS (Supp. #1, 1988). *"A valuable, easy to read working tool which should be part of the standard equipment of all catalogers who handle sound recordings." (ALR)*

Education and Training for Catalogers and Classifiers, edited by Ruth C. Carter, PhD (Vol. 7, No. 4, 1987). *"Recommended for all students and members of the profession who possess an interest in cataloging." (RQ-Reference and Adult Services Division)*

The United States Newspaper Program: Cataloging Aspects, edited by Ruth C. Carter, PhD (Vol. 6, No. 4, 1986). *"Required reading for all who use newspapers for research (historians and librarians in particular), newspaper cataloguers, administrators of newspaper collections, and–most important–those who control the preservation pursestrings." (Australian Academic & Research Libraries)*

Computer Software Cataloging: Techniques and Examples, edited by Deanne Holzberlein, PhD, MLS (Vol. 6, No. 2, 1986). *"Detailed explanations of each of the essential fields in a cataloging record. Will help any librarian who is grappling with the complicated responsibility of cataloging computer software." (Public Libraries)*

The Future of the Union Catalogue: Proceedings of the International Symposium on the Future of the Union Catalogue, edited by C. Donald Cook (Vol. 2, No. 1/2, 1982). *Experts explore the current concepts and future prospects of the union catalogue.*

Published by

The Haworth Information Press, 10 Alice Street, Binghamton, NY 13904-1580 USA

The Haworth Information Press is an imprint of The Haworth Press, Inc., 10 Alice Street, Binghamton, NY 13904-1580 USA.

Maps and Related Cartographic Materials: Cataloging, Classification, and Bibliographic Control has been co-published simultaneously as *Cataloging & Classification Quarterly,* Volume 27, Numbers 1/2 and 3/4 1999.

Cover design by Thomas J. Mayshock Jr.

Library of Congress Cataloging-in-Publication Data

Maps and related cartographic materials : cataloging, classification, and bibliographic control / Paige G. Andrew, Mary Lynette Larsgaard, editors.
    p. cm.
    Co-published simultaneously as v. 27, no. 1/2 and 3/4, 1999 of Cataloging & classification quarterly.
    Includes bibliographical references and index.
    ISBN 0-7890-0778-9 (alk. paper) – ISBN 0-7890-0813-0 (pbk. : alk. paper)
    1. Cataloging of maps–United States. I. Andrew, Paige G. II. Larsgaard, Mary Lynette, 1946-

Z695.6 .M377 1999
025.3'46–dc21
                                                    99-051487

Maps and Related Cartographic Materials: Cataloging, Classification, and Bibliographic Control

Paige G. Andrew, BA, MLS
Mary Lynette Larsgaard, BA, MA
Editors

Maps and Related Cartographic Materials: Cataloging, Classification, and Bibliographic Control has been co-published simultaneously as *Cataloging & Classification Quarterly*, Volume 27, Numbers 1/2 and 3/4 1999.

The Haworth Information Press
An Imprint of
The Haworth Press, Inc.
New York • London • Oxford

INDEXING & ABSTRACTING

Contributions to this publication are selectively indexed or abstracted in print, electronic, online, or CD-ROM version(s) of the reference tools and information services listed below. This list is current as of the copyright date of this publication. See the end of this section for additional notes.

- *BUBL Information Service. An Internet-based Information Service for the UK higher education community <URL: http://bubl.ac.uk/>*
- *CNPIEC Reference Guide: Chinese National Directory of Foreign Periodicals*
- *Computing Reviews*
- *Current Awareness Abstracts of Library & Information Management Literature, ASLIB (UK)*
- *IBZ International Bibliography of Periodical Literature*
- *Index to Periodical Articles Related to Law*
- *Information Science Abstracts*
- *Informed Librarian, The*
- *INSPEC*
- *Journal of Academic Librarianship: Guide to Professional Literature, The*
- *Konyvtari Figyelo-Library Review*
- *Library & Information Science Abstracts (LISA)*
- *Library & Information Science Annual (LISCA)*
- *Library Literature*
- *National Clearinghouse for Primary Care Information (NCPCI)*
- *Newsletter of Library and Information Services*
- *PASCAL*
- *Periodica Islamica*
- *Referativnyi Zhurnal (Abstracts Journal of the All-Russian Institute of Scientific and Technical Information)*

(continued)

Special Bibliographic Notes related to special journal issues (separates) and indexing/abstracting:

- indexing/abstracting services in this list will also cover material in any "separate" that is co-published simultaneously with Haworth's special thematic journal issue or DocuSerial. Indexing/abstracting usually covers material at the article/chapter level.
- monographic co-editions are intended for either non-subscribers or libraries which intend to purchase a second copy for their circulating collections.
- monographic co-editions are reported to all jobbers/wholesalers/approval plans. The source journal is listed as the "series" to assist the prevention of duplicate purchasing in the same manner utilized for books-in-series.
- to facilitate user/access services all indexing/abstracting services are encouraged to utilize the co-indexing entry note indicated at the bottom of the first page of each article/chapter/contribution.
- this is intended to assist a library user of any reference tool (whether print, electronic, online, or CD-ROM) to locate the monographic version if the library has purchased this version but not a subscription to the source journal.
- individual articles/chapters in any Haworth publication are also available through the Haworth Document Delivery Service (HDDS).

Maps and Related Cartographic Materials: Cataloging, Classification, and Bibliographic Control

CONTENTS

ABOUT THE EDITORS

Paige Andrew, BA, MLS, is the Maps Cataloger and Selector for Geography at the Pennsylvania State University Libraries at the University Park campus. He also holds the rank of Assistant Professor at the University. Mr. Andrew is a long-time active member, and past Chair, of the Geography and Map Division of the Special Libraries Association. He currently serves as Captain of the North East Map Organization, and is a member of and Secretary to the North American Cartographic Information Society. Mr. Andrew has written and co-authored several articles relating to map cataloging and map librarianship topics and has presented papers at the Special Libraries Association Annual Conference, the Annual Meeting of the North American Cartographic Information Society, and the American Library Association Annual Conference. In addition, he is also a member of the Western Association of Map Libraries and the Pennsylvania Mapping and Geographic Information Consortium, and is a representative to the Anglo-American Cataloguing Committee for Cartographic Materials.

Mary Lynette Larsgaard, BA, MA, is Assistant Head of the Map and Imagery Library of the Davidson Library at the University of California in Santa Barbara. She is the author of numerous journal articles and several books, including three editions of *Map Librarianship: An Introduction* (1998, Libraries Unlimited). Ms. Larsgaard is recognized as the leader of cartographic cataloging in the field of map librarianship and she remains very active in the Map and Geography Round Table of the American Library Association (ALA), the Western Association of Map Libraries and the Anglo-American Cataloguing Committee for Cartographic Materials.

Foreword

These papers represent an important contribution to map librarianship. They help fill a long-standing need for background information and guidance as well as sound, practical advice in most aspects of cartographic materials classification and cataloging. Many of us have been asked by map catalogers (or those contemplating doing such work) about the topics raised in these papers, particularly as they apply to remote-sensing imagery and digital data. Answers, when they existed, often were only in the minds of a few expert practitioners and were rarely dealt with at any length in the literature. As is the case with issues not committed to paper, differing and dynamic ways of how to handle or view particular kinds of cartographic materials cataloging emerged. On occasion, the advice that was given to the inquirer was to be patient since a variety of agencies and committees were working on this issue. For regular map conference attendees, information, incomplete and fragmented as might be, could sometimes be obtained by identifying the experts in the field and pumping them for information, or perhaps attending a good workshop. Some help was also available on an ad hoc basis via e-mail discussion groups like MAPS-L. For the most part, however, such information was largely unavailable, dispersed, and generally difficult to come by.

This collection of papers is one more step on a rather remarkable path of progress in cartographic materials cataloging, particularly over the last thirty years. Among the milestones developed during that period are the MARC Map format from the 1960s, the International Standard Bibliographic Description for Cartographic Materials (ISBD-CM), the much-expanded 4th edition of the Library of Congress's G-classification schedule and the first appearance of map records in the bibliographic utilities in the '70s, as well as AACR2 and *Cartographic Materials: A Manual of Interpretation for AACR2* in the '80s. These tools, and quite a few others, collectively helped to create the impetus and the means for the cataloging of cartographic materials by more and more libraries, and the switching of existing idiosyncratic cataloging practices to newer, standardized ones. All of these tools, many now taken pretty much for

[Haworth co-indexing entry note]: "Foreword." Hoehn, Philip. Co-published simultaneously in *Cataloging & Classification Quarterly* (The Haworth Information Press, an imprint of The Haworth Press, Inc.) Vol. 27, No. 1/2, 1999, pp. xv-xvi; and: *Maps and Related Cartographic Materials: Cataloging, Classification, and Bibliographic Control* (ed: Paige G. Andrew, and Mary Lynette Larsgaard) The Haworth Information Press, an imprint of The Haworth Press, Inc., 1999, pp. xiii-xiv. Single or multiple copies of this article are available for a fee from The Haworth Document Delivery Service [1-800-342-9678, 9:00 a.m. - 5:00 p.m. (EST). E-mail address: getinfo@haworthpressinc.com].

xiii

granted, were of course created by thousands of hours (often volunteered) of hard work by dozens of practitioners and institutions in the field.

In recent years cartographic materials specialists have been forced to add to their already full job descriptions the management of digital data, Geographic Information Systems (GIS), and Internet resources–frequently while experiencing staff reductions. This speed-up shows no signs of abating. About the only way out of this dilemma, short of greatly curtailing services or hours, is to get users and non-map library staffers to become more self-reliant in their searches for cartographic materials. While Webpage guides and instruction can go a long way toward that end, having records in the institution's catalog for cartographic materials is perhaps the real key. Some library leaders have suggested (or even implemented) centralized reference services within an institution, or have envisioned subject specialists who would provide reference services to more than one institution from a centralized telephone- and email-accessible workstation. Successful implementation of such notions rests upon having online access to catalog records (sometimes linked to online images of the material itself). Cost-saving automated circulation systems also require catalog records. For these reasons and others library administrators in growing numbers seem to recognize the need to provide catalog access to cartographic materials, particularly since large amounts of copy cataloging is now economically available from LC, GPO, and the bibliographic utilities. Given these developments, this collection of papers has made its appearance at an opportune time.

I believe that every library department doing (or contemplating) geospatial data cataloging will need this work. In addition, catalogers specializing in this field will probably want a reference copy at their desks. Ruth Carter, Paige Andrew and Mary Larsgaard and, of course, all of the contributing authors, are to be commended and congratulated for envisioning this work and on seeing it through to publication.

Philip Hoehn
Branner Earth Sciences Library & Map Collection
Stanford University

Introduction

Paige G. Andrew
Mary Lynette Larsgaard

SUMMARY. This introduction to a body of literature on cataloging cartographic materials gives an overview of the arrangement and contents of the volume. Additionally, it includes a list of article topics for which no writer could be found. *[Article copies available for a fee from The Haworth Document Delivery Service: 1-800-342-9678. E-mail address: getinfo@haworth pressinc.com <Website: http://www.haworthpressinc.com>]*

KEYWORDS. Cartographic materials, cataloging, digital geospatial data, remote-sensing images, geologic sections, maps, globes, early maps, atlases, map series, classification, subject headings, retrospective cataloging, OCLC Enhance for maps, metadata, cartographic archives

Welcome to this re-issue, in book form, of two double issues of *Cataloging & Classification Quarterly*, devoted to the cataloging and classification of

Mr. Paige G. Andrew, Assistant Professor, is the faculty Maps Cataloger at the Pennsylvania State University Libraries, as well as Selector for the discipline of Geography. Formerly he was the Maps Cataloger, from November 1986-January 1995, for the University of Georgia Libraries, Athens, Georgia. He earned his bachelors degree in 1983 at Western Washington University, Bellingham, Washington, majoring in Geography. He earned his MLS in 1986 at the University of Washington, Seattle, Washington and is a native Seattleite. He can be reached via email at: pga@psulias.psu.edu

Mary Lynette Larsgaard has, since 1988, been Assistant Head of the Map and Imagery Laboratory, Davidson Library, University of California, Santa Barbara. She has a BA in geology, an MA in library science, and an MA in geography. She is the author of *Map Librarianship, An Introduction* (Third edition, 1998, Libraries Unlimited). She can be reached via email at: mary@sdc.ucsb.edu

[Haworth co-indexing entry note]: "Introduction." Andrew, Paige G., and Mary Lynette Larsgaard. Co-published simultaneously in *Cataloging & Classification Quarterly* (The Haworth Information Press, an imprint of The Haworth Press, Inc.) Vol. 27, No. 1/2, 1999, pp. xv-xviii; and: *Maps and Related Cartographic Materials: Cataloging, Classification, and Bibliographic Control* (ed: Paige G. Andrew, and Mary Lynette Larsgaard) The Haworth Information Press, an imprint of The Haworth Press, Inc., 1999, pp. xv-xviii. Single or multiple copies of this article are available for a fee from The Haworth Document Delivery Service [1-800-342-9678, 9:00 a.m. - 5:00 p.m. (EST). E-mail address: getinfo@haworthpressinc.com].

xv

cartographic materials, or if you prefer an alternate phrase, geospatial data. This opening sentence sets the tone for the current situation in what might be called the "cataloging of worlds"–changing times, adding of new genres and physical forms to what was already a demanding format, and rules that are not just in flux but also in the process of being created! We are fortunate indeed to have had such a positive response to the need for authors of the various articles, each of whom has lent their knowledge, know-how, and expertise to this collection.

In this volume we present a format-focused guide for what is, for the sake of brevity, called the map cataloger. It includes articles for each major type of "map" cataloging with a general procedure to be followed, an itemization of problem areas in the bibliographic record and–where we know of them–answers to those problems. In addition, there are articles that help us view the larger, and ever-changing, picture of the map cataloger's world and what s/he might soon be tackling, and also look at how traditional maps are controlled when they reside in the realm of an archives.

ORGANIZATION

To lend some perspective and set the tone toward the whole area of cartographic cataloging the first section starts off with an article that looks at the current status of map cataloging, focusing on the "who," "what," and "where" of the task.

From there the next series of articles in the first section could be labeled the "how to" of cataloging the myriad formats that cartographic materials are packaged in, from the basic paper sheet map, to globes, to geologic sections. Next is an article on cataloging twentieth-century atlases, followed by an article that closely inspects some of our most recently collected cartographic items, remotely-sensed images, including aerial photographs and satellite images. Concluding section one are two articles on rare, or early, cartographic materials, one each for maps and atlases, and finally an article that looks at how to handle early maps within printed publications.

Moving from the paper-based or "hard copy" world the next set of articles, in section two of the volume, tackles cartographic materials in digital form. It includes a look at metadata standards on a worldwide scale, the use of the latest fields added to MARC 21 for describing specific information found in digital forms of cartographic material, and concludes with an article outlining the creation of bibliographic records for those materials on CD-ROM.

The next two articles examine how to classify catrographic materials as well as how to apply subject headings to these items. Following these are two articles looking at retrospective conversion projects and one on quality control of the bibliographic record via OCLC's Enhance Program. We close the volume by taking a different viewpoint on access to cartographic materials–

how they are described, organized, and controlled within the setting of an archives.

THOSE THAT GOT AWAY

A sure way to find out more than you ever wanted to know about any editor's woes is to ask about the topics that the editor knew needed to be addressed but could not find any volunteers to write the articles! We certainly had that same problem; fortunately the list of "ones that got away" is a short one. We do hope that these topics will be elucidated upon in the near future within the literature of map librarianship or cataloging or both. The list of topics *not covered* in this issue includes:

1. *Original versus copy cataloging of sheet maps.*

Library administrators, map collection curators, cataloging department heads, and yes, even map catalogers often wonder about the differences between creating an original bibliographic record for an item versus "borrowing" a record from OCLC or elsewhere and tweaking it to fit the local collection. Issues include: levels of staff expertise needed to do the work; quality of existing versus to-be-created bibliographic records; and naturally, difference in the time that it takes to create an original record as opposed to using copy available from a bibliographic utility or even locally created records from an earlier time.

2. *Linking techniques used in the bibliographic record to indicate relationships such as: parent/child; other edition/original edition; and original form/ other physical form of material.*

A mammoth percentage of all geospatial data, perhaps as much as 80 percent, has some sort of relationship with its cohorts. The most common, both for maps and for remote-sensing images, is that of parent/child. Another common relationship for maps is that of other edition; many topographic sheets, for example, are issued in more than one edition. Increasingly more obvious these days is the relationship between the original hard-copy item and its digital form. How are these relationships most accurately and clearly presented in the bibliographic record and ultimately to our users? What about those circumstances where all of the above relationships apply or could apply?

3. *Creating a bibliographic record from a surrogate for a sheet map.*

Those of us in institutions that centrally catalog for library collections physically some distance from a main campus or in another building on campus sometimes deal with the dreaded "surrogate" issue. Usually the collection that wants to have a map cataloged mails the map cataloger a photocopy of the panel of a folded map or the area of a flat map that contains the title, not realizing how much critical information is unavailable for use in the description. If the map cataloger is fortunate s/he can track down copy based on the title alone, but often this is not the case.

Some of the issues involved include: should we ever catalog from a surrogate? If we agree to catalog in this manner what is the minimum level of information we need to have in hand to locate a copy of a bibliographic record? What about the minimum level of information needed to create an original record?

4. *Cataloging of maps on microform, facsimile maps, and maps published in photocopy form.*

Each of these types of maps do indeed occur, and fairly often. Questions to be dealt with range from, "How can I catalog this item based on the original when all I have is a microfiche and there is no previously existing catalog record or citation in a published cartobibliography?", to, "When do I use MARC 21 field 533 and when do I use 534?", with a good many other perplexing queries in-between.

There are many other knotty topics that need to be written about so we encourage you to share the knowledge you have gained with the rest of the map cataloging community. While in many cases there are rules scattered throughout AACR and other guides on how to handle some of these situations, it is much easier to read one coherent guide that pulls together in one place all the rules needed to catalog one given type of cartographic material.

CONCLUSION

As our fast-paced world of cartographic information continues to evolve and change, we are aware that some of the information to be gleaned from the papers contained in this volume may become dated, if not outright obsolete. Additionally, new forms of cartographic information will require new ways of handling and describing these items and, therefore, we again encourage publication of articles describing methodologies and techniques in this or other map librarianship and cataloging literatures.

The co-editors believe very strongly that this compilation meets a substantial demand for cataloging information that is prescriptive in terms of describing and organizing all forms of cartographic materials. We thank the contributing authors for sharing their expertise–doing so with patience and understanding–and Ruth Carter for first inviting us to tackle this much needed publication, and then for assisting us in many ways as we worked toward completion of our editorial tasks.

Table of Acronyms
Found Throughout This Volume

AACCCM – Anglo-American Cataloguing Committee for Cartographic Materials. Published as *Cartographic Materials: A Manual of Interpretation for AACR2*. Chicago: American Library Association, 1982. (Revision in progress).

AACR2 – *Anglo-American Cataloguing Rules, Second Edition*. Chicago: American Library Association, 1980.

AACR2R – *Anglo-American Cataloguing Rules, Second Edition, 1988 Revision*. Chicago: American Library Association, 1988, with amendments.

CM – *Cartographic Materials: A Manual of Interpretation for AACR2*. (See AACCCM)

CONSER – Cooperative Online Serials Program.

DCRB – *Descriptive Cataloging of Rare Books*.

DLC – OCLC's symbol for the Library of Congress.

GPO – Government Printing Office.

ISBD(A) – *International Standard Bibliographic Description for Older Monographic Publications (Antiquarian)*.

ISBD(CM) – *International Standard Bibliographic Description for Cartographic Materials*.

LC G&M – Library of Congress Geography and Map Division.

LOCIS – Library of Congress Information System. telnet: locis.loc.gov

MARC – Machine-Readable Cataloging.

MARC 21 – Machine-Readable Cataloging (U.S. Standard, formerly known as USMARC).

[Haworth co-indexing entry note]: "Table of Acronyms Found Throughout This Volume." Co-published simultaneously in *Cataloging & Classification Quarterly* (The Haworth Information Press, an imprint of The Haworth Press, Inc.) Vol. 27, No. 1/2, 1999, pp. 1-2; and: *Maps and Related Cartographic Materials: Cataloging, Classification, and Bibliographic Control* (ed: Paige G. Andrew, and Mary Lynette Larsgaard) The Haworth Information Press, an imprint of The Haworth Press, Inc., 1999, pp. 1-2. Single or multiple copies of this article are available for a fee from The Haworth Document Delivery Service [1-800-342-9678, 9:00 a.m. - 5:00 p.m. (EST). E-mail address: getinfo@haworthpressinc.com].

1

MARCIVE – MARCIVE, Inc., P.O. Box 47508, San Antonio, TX 78265-7508

MCM – *Map Cataloging Manual.* Prepared by the Map and Geography Division, Library of Congress.

NUC-CM – *National Union Catalog: Cartographic Materials.* Washington, D.C.: Library of Congress, 1993- .

OCLC – (formerly Online Computer Library Center, Inc.). Bibliographic utility. 6565 Frantz Rd., Dublin, OH 43017-3395

RLIN – Research Libraries Information Network. Bibliographic utility like OCLC.

USGS – United States Geological Survey.

WLN – (formerly Washington Library Network). Bibliographic utility like OCLC.

MARC 21/OCLC/RLIN
Fixed Field Crosswalk Table for Maps in Alphabetical Order by OCLC Field Name

MARC 21	OCLC	RLIN	DEFINITION
Leader/07	BL	BLT (2nd position)	Bibliographic level
008/22-24	BME	PRJ and PRM	Map projection + prime meridian
008/25	CM/TYP	GRP	Type of cartographic material
008/06	D/CODE	PC	Type of date/ publication status
Leader/18	DCF	DCF	Descriptive cataloging form
008/07-10	DT/1	PD (1st part)	Date 1/Beginning date of publication
008/11-14	DT/2	PD (2nd part)	Date2/Ending date of publication
Leader/17	E/L	EL	Encoding level

[Haworth co-indexing entry note]: "MARC 21/OCLC/RLIN Fixed Field Crosswalk Table for Maps in Alphabetical Order by OCLC Field Name." Co-published simultaneously in *Cataloging & Classification Quarterly* (The Haworth Information Press, an imprint of The Haworth Press, Inc.) Vol. 27, No. 1/2, 1999, pp. 3-4; and: *Maps and Related Cartographic Materials: Cataloging, Classification, and Bibliographic Control* (ed: Paige G. Andrew, and Mary Lynette Larsgaard) The Haworth Information Press, an imprint of The Haworth Press, Inc., 1999, pp. 3-4. Single or multiple copies of this article are available for a fee from The Haworth Document Delivery Service [1-800-342-9678, 9:00 a.m. - 5:00 p.m. (EST). E-mail address: getinfo@haworthpressinc.com].

MARC 21	OCLC	RLIN	DEFINITION
007/00	GMD	COM	Category of material
008/28	GOVT	GPC	Government publication
008/31	INDX	II	Index
008/35-37	LANG	L	Language
008/38	MOD	MOD	Modified record
008/15-17	PLACE	CP	Place of publication/ production/ execution
	POL	PL	Positive/negative aspect (Maps)
008/18-21	RELIEF	RLF	Relief
	REPRO	TREP	Type of reproduction
008/33-34	SP/FORM	FMT	Special format characteristics
008/39	SRC	CSC	Cataloging source

MARC Tags for Cataloging Cartographic Materials

Velma Parker

SUMMARY. This is a table of those MARC fields most frequently used when cataloging cartographic materials. The table gives fields both for monographs and for serials. *[Article copies available for a fee from The Haworth Document Delivery Service: 1-800-342-9678. E-mail address: getinfo@haworth pressinc.com <Website: http://www.haworthpressinc.com>]*

KEYWORDS. MARC, CONSER, cataloging, cartographic materials

It is very nearly impossible to explain "map" cataloguing without mentioning MARC tags, and the authors of this volume have borne this out. Therefore the following MARC field table is provided. In addition, the table includes tags which may be used for bibliographic records of cartographic materials, whether monograph or serial; the second column contains a code indicating which fields should be filled out to create a core serial record for CONSER. According to the CONSER documentation, minimal level records

Miss Parker is a Standards Officer in the Visual and Sound Archives Division, National Archives of Canada, 395 Wellington Street, Ottawa, Ontario KIA 0N3, Canada. She is a longtime member of the Bibliographic Control Committee of the Association of Canadian Map Libraries and Archives. She received her Honours BA in Geography from the University of Western Ontario, London, Ontario, her Masters of Library Science from the University of Western Ontario, London, Ontario, and her Dip. in Education, Type A high school teaching certificate from Althouse College, University of Western Ontario, London, Ontario.

[Haworth co-indexing entry note]: "MARC Tags for Cataloging Cartographic Materials." Parker, Velma. Co-published simultaneously in *Cataloging & Classification Quarterly* (The Haworth Information Press, an imprint of The Haworth Press, Inc.) Vol. 27, No. 1/2, 1999, pp. 5-8; and: *Maps and Related Cartographic Materials: Cataloging, Classification, and Bibliographic Control* (ed: Paige G. Andrew, and Mary Lynette Larsgaard) The Haworth Information Press, an imprint of The Haworth Press, Inc., 1999, pp. 5-8. Single or multiple copies of this article are available for a fee from The Haworth Document Delivery Service [1-800-342-9678, 9:00 a.m. - 5:00 p.m. (EST). E-mail address: getinfo@haworthpressinc.com].

are the same as the core records except that subject analysis is optional in the minimal level record.

An attempt has been made to include cartographic materials in all forms (manuscript through print, microform and computer files) and formats (i.e., serials). The main listing is for the eye-readable formats. If there are variations for microform, computer files or serials these are noted.

Note that the 008 fixed field to be used is that for cartographic material. If the item, etc., is issued on microform or as computer data, then use the 008 for cartographic materials and, in addition, the 006 for the form (i.e., microform or computer files). If the item catalogued is a serial, also add the 006 for serials.

Codes for the following table: M–mandatory, MA–mandatory if applicable, O–optional, R–required if available, blank–not on CONSER list.

CARTOGRAPHIC MATERIALS MARC TAGS	
FIXED LENGTH FIELDS	CORE SERIALCONSER RECORDS
Record status (Leader/5)	
Type of record (Leader/6)	M
Bibliographic level (Leader/7)	M
Encoding level (Leader/17)	M
Descriptive cat. form (Leader/18)	M
Type of date/Publication status (008/6)	M
Date 1 and Date 2 (008/7-14)	M
Place of publication, etc. (008/15-17)	M
Relief (CM 008/18-21)	R
Map projection (CM 008/22-23)	
Type of cartographic material (CM 008/25)	M
Government publication (CM 008/28)	
Index (CM 008/31)	M
Language (008/35-37)	M
Modified record (008/38)	M

Cataloguing source (008/39)	M
Form of material (SE 006/00 = s for serial publication; also CF 006/00 = m for computer file).	M
Frequency (SE 006/01; CF 006/01)	M
Regularity (SE 006/02; CF 006/02)	M
Type of serial (SE 006/04)	M
Form of original item (SE 006/05)	M
Form of item (SE 006/06 for cartographic serials on microform)	M
Type of computer file (CF 006/9)	MA
Conference publication (SE 006/12)	
Successive/latest entry (SE 006/17)	M
007 Physical description fixed fields (Computer file, Map, Microform)	M for microforms and computer files
CONTROL FIELDS–0XX	
010 LC control no	M
012 CONSER fixed length field ($i–NST code)	M
022 ISSN	R
034 Coded cartographic mathematical data	R
040 Cataloguing source	
041 Language code	O
042 Authentication code	M
043 Geographic area code	O
05X, 09X, etc. Call no.	
074 GPO item no.	R
086 Gov't doc. class. no.	R
VARIABLE FIELDS–1XX-9XX	
1XX Main entry	MA
240 Uniform title	MA
245 Title Statement (including $h)	M

246 Varying form of title	MA
250 Edition statement	MA
255 Cartographic math. data ($a–scale, $b–projection)	R
256 Computer file characteristics	
260 Publication, etc. (Imprint)	M
300 Physical description	M; MA for microforms; not applicable to direct access computer files
310 Current publication frequency	O
362 Dates of pub., vol. designation	MA
4XX Series statement	MA
500 Source of title, DBO note	MA
533 Reproduction note	MA for microforms
538 System details	M
5XX Notes	O
6XX Subject added entries	MA
700-730 Name/title added entries	MA
776 Additional physical form entry	MA microforms only
780/785 Preceding/Succeeding entry	MA
7XX Other linking entries	O
850 Holding institution	M
8XX Series added entries	MA
936 CONSER variable length field	MA

Codes: M–mandatory, MA–mandatory if applicable, O–optional, R–required if available, blank–not on CONSER list.

OVERVIEW

Map Cataloging and Classification: The Basic *Who*, *What*, and *Where*

Harry O. Davis
James S. Chervinko

SUMMARY. This study explores map cataloging in academic and research libraries, analyzed by four size ranges of map collections. Topics include the locus of map cataloging, the involvement of professionals, paraprofessionals, and students in map cataloging, and perceived adequacy of staffing. Additional topics include the extent and level of MARC records in map catalogs and bibliographic utilities, and classification systems used for map cataloging. Finally, there is an

Harry O. Davis (BA, MSc, MA) is Map and Assistant Science Librarian and James S. Chervinko (BA, MA, MS) is Head of Original and Special Materials Cataloging, both at Morris Library, Southern Illinois University, Carbondale, at Carbondale, IL 62901-6632. Their e-mail addresses are hdavis@lib.siu.edu and jchervin@ lib.siu.edu.

The authors wish to acknowledge research support for this paper provided by the Research and Publication Committee of Library Affairs at Southern Illinois University, Carbondale.

[Haworth co-indexing entry note]: "Map Cataloging and Classification: The Basic *Who*, *What*, and *Where*." Davis, Harry O., and James S. Chervinko. Co-published simultaneously in *Cataloging & Classification Quarterly* (The Haworth Information Press, an imprint of The Haworth Press, Inc.) Vol. 27, No. 1/2, 1999, pp. 9-37; and: *Maps and Related Cartographic Materials: Cataloging, Classification, and Bibliographic Control* (ed: Paige G. Andrew, and Mary Lynette Larsgaard) The Haworth Information Press, an imprint of The Haworth Press, Inc., 1999, pp. 9-37. Single or multiple copies of this article are available for a fee from The Haworth Document Delivery Service [1-800-342-9678, 9:00 a.m. - 5:00 p.m. (EST). E-mail address: getinfo@haworthpressinc.com].

examination of the predominant attributes of map cataloging and the adequacy and quality of current statistics to allow for understanding and improvement of map cataloging. *[Article copies available for a fee from The Haworth Document Delivery Service: 1-800-342-9678. E-mail address: getinfo@haworthpressinc.com <Website: http://www.haworthpressinc.com>]*

KEYWORDS. Maps, cataloging, catalogers, MARC, collections, statistics, classification, staffing, locus, location

PREFACE

This article originates from data collected by the authors for a larger research project still underway which explores how academic and research library staff catalog separate maps accompanying monographs or serial volumes and how they catalog separate text accompanying maps. Data from this larger project, however, also contributes to understanding a group of narrower topics, that being who is responsible for map cataloging, where the activity occurs, and under what conditions.

Data were acquired in two surveys, one to heads of cataloging and one to map librarians[1] at 119 member libraries of the Association of Research Libraries (ARL) and at thirty non-ARL libraries chosen for the size or significance of their map collections. Eighty-four surveys were returned from the heads of cataloging group at these libraries and 54 (a 36.2% return rate) from map librarians. (The numbers represent two kinds of situations, those where the head of cataloging and the map librarian at a particular institution *both* responded, each to their particular survey, and those where *only one or the other* at a given institution responded.) Of the 54, one library indicated that no current map cataloging is being done and that none of the 130,000 maps in its collection is cataloged. This library is excluded in data reported below, thus yielding a total useable sample of 53 map-cataloging libraries.

For more efficient and understandable tabulation and analysis of the survey results, the responding map-cataloging libraries have been divided in four ranges by collection size. *Range 1* encompasses 14 libraries, which reported map collections between 20,000 and 140,000 maps. *Range 2* consists of 11 libraries, with between 150,000 and 190,000 maps. In *Range 3* there are 13 libraries reporting between 200,000 and 280,000 maps. Finally, *Range 4* embraces the remaining 15 map libraries with very large collections of between 300,000 and 4,800,000 maps.[2] These ranges are separated at natural breakpoints in the data and allow uniformity in statistical analysis because of similar sample sizes.

For all size ranges, 18.9% of the survey respondents in the United States are in the Northeast; 11.3% are in the Southeast; 20.8% are in the Midwest;

and 31.2% are in the West. The remaining 16.9% are located in Canada. Of the total sample population, 71.7% of the respondents are ARL libraries. Among the academic institutions in the total sample, four have fewer than 10,000 students; six have between 10,000 and 14,000 students; twenty have between 15,000 and 24,999 students; nine have between 25,000 and 34,999 students; and seven have over 35,000 students. The remaining seven are non-academic institutions.

The data presented in this paper are for the four ranges of map collection size and for the total sample, with all sizes combined. Cross-tabulations are presented for the separate size ranges or for the total sample. The intent is to allow readers to relate the data to their own institution and size of collection so as to understand better how their situation compares to that of others. An additional intent is to show similarities and contrasts at the various collection size levels. In summary, it is hoped that the data will allow map librarians and administrators to assess local strengths and deficiencies and to effect planning for improved map cataloging.

The survey results in this paper do not permit an examination of all aspects of *who, what,* and *where.* The data do allow, however, a basic understanding of where map cataloging occurs, of what kind of map cataloging occurs, and by whom. It is, essentially, an overview of the extent to which map cataloging is occurring in variously sized academic and research libraries. The findings reflect the data provided by map libraries returning completed surveys. The authors are satisfied that the data can be considered sufficiently representative of American and Canadian map libraries. Furthermore, the size cohorts used are close enough in sample size to allow a measure of confidence in making comparative analyses between size ranges.

It should be noted that many survey respondents attached comments to their survey answers. Some of these comments have helped the authors of this paper to understand the intent of a given answer, while others provide worthy elucidation and extension to the responses. Consequently, we have tried to incorporate at least some of the flavor of selected comments, a number of which reflect confusion or concern in the ranks of map librarians, especially the concern of inadequacy, either individual or institutional. Finally, percentages shown in the graphs and tables may not total 100% for each cohort because of missing or spurious data. In other instances, percentages may total more than 100% because of multiple data combinations.

PRIOR RESEARCH

The authors have identified no prior research directly similar to that reported in this paper.[3] There is a variety of papers found in the literature which consider library staff workload, the adequacy of the number of staff members, or the map cataloging process, but there seems to be a lack of previous literature reporting on a combination of these topics. Just as the *who* and *what* questions have not been explored jointly, the *where* question has been ignored. A considerable number of papers in the literature examine issues such as levels of staffing in cataloging and the use of paraprofessionals or others instead of professional librarians in cataloging. Several explore the general issue of workload in library operations, and the future of cataloging, but none have been found that relate directly to map cataloging. The 1997 bibliography on map cataloging edited by Hughes and Demetracopoulos confirms the general lack of literature on who catalogs maps and where.[4,5]

Important correlative reading includes the standards for university map collections promulgated in 1987 by the Committee on Standards, Geography and Map Division, Special Libraries Association.[6] Standard A.2, and the Commentary thereon, set forth expectations for map cataloging and classification, which allow assessment of certain findings reported in the present paper. Similarly, Standard C.1 and its Commentary set forth personnel expectations, albeit in non-specific form with respect to map cataloging.

Data are generally lacking in the literature to provide insight and guidance for map library staffing and operational capacity, including that for map cataloging. One interesting example with very basic data is in a report by Franz Wawrik,[7] which includes a table showing the number of map curators in comparison to the number of map collections in each country reported. One must assume that those countries with the weakest ratio between curators and collections also would be the weakest in terms of numbers of map catalogers.

Although not specific for either map librarians or catalogers, a 1996 report by Kyrillidou[8] presenting statistical ratios among ARL libraries may give some insight into what one might expect (or hope for) in the staffing of map libraries. The bulk of this report gives data ratios that are not germane to the present research, but Table 4.a, "Ratio of Support to Professional Staff Sorted by Ratio Values in 1995" can be compared to ARL map library situations. This table shows ARL member ratios for 1995 varying between 0.5 support staff for each professional to 4.6, with a mean of 2.0. This data can be updated using values reported in the annual *ARL Statistics,*[9] as can the size of collections of cartographic materials, but again, unfortunately, there is a lack of data to help in understanding support levels provided to cataloging.

An earlier report providing some statistical insight is by McGarry in 1981 for the Ad Hoc Committee on Cataloging of the Geography and Map Division, Special Library Association.[10] This report indicates levels of use of OCLC and AACR2 at that time for map cataloging, as well as the differentiation in roles for cataloging professionals and paraprofessionals. Rieke and Steele (1984) report on "A Survey of Major Map Collections in Texas," with limited data for map cataloging.[11] A paper by Hiller provides statistical data and discussion of map cataloging as of 1987.[12] None of these reports, however, provides direct cross-tabulation of statistical variables for map libraries.

A good example of the presentation and use of comparative statistics for map libraries is provided in the 1992 paper by Kaye who uses such data to determine the adequacy of, and improvements to, the Yale University map collection.[13] Even in this paper, however, there is not a true sorting-out of the *who, what,* and *where* questions with respect to map cataloging. Nonetheless, it offers insight into how comparative data can be used in assessing the status and adequacy of a map library operation.

Several other papers provide additional perspective. Zögner's 1983 survey of German map libraries (updating Kramm's 1959 survey) provides some data similar to that provided by American researchers in the 1980s.[14] David Cobb's 1985 overview of map librarianship includes statistical data for map libraries. His Table 4 on map cataloging and classification activity allows limited comparison with the data reported below, as does his Table 3 concerned with staffing levels.[15] Another study published in 1985 by Robar recognizes the value of providing comparative statistics for university map collections.[16] The principal similarity to the present research is in the use of collection size ranges to present comparative data. The comparability with the research reported in this paper is limited to staffing by collection size, and that comparison is impaired by the fact that Robar used quite different collection size ranges from those incorporated in this paper.

A 1993 report by Charles Seavey summarized the status of his 1991 survey of ARL cartographic resources.[17] This followed a 1992 publication that assessed methodology for ranking and evaluating map collections, allowing a better understanding of the findings by size ranges in the present paper in the context of expectations for "small" *versus* "large" map collections.[18] The 1993 presentation includes a copy of the questionnaire used by Seavey in his 1991 survey and displays selected survey results for map libraries in the western United States. He also presents a useful discussion of the difficulty in collecting accurate data that yield unambiguous conclusions. Seavey shows a preference to develop comparative data and ratios similar to the research reported in the present paper. Unfortunately, the full results of the 1991 survey were never published, and the intended annual follow-up surveys did not transpire.

In sum, the existing literature provides only very limited data or perspective to augment that collected and used for this study. There seems to be a dearth of prior literature about *who* catalogs maps and *where*. There are somewhat more data for *what* is cataloged, but these data are not correlated with the other two factors. It is hoped that the present paper will correct part of the gap and give encouragement to additional research and, just as importantly, to more regular and systematic data collection for map library staffing and functionality, including cataloging.

THE **WHERE** *OF MAP CATALOGING*

Physical Location

Of all the respondents, 43.3% reported that map cataloging occurs only in the map library (with some other locations but not including the cataloging department), while 28.3% stated that it is carried out only in the cataloging department (with some other locations but not including the map library). In 11.3% of other responding libraries it occurs in both the map library and the cataloging department (both with some other locations). In some cases map cataloging occurs only in the cataloging department because the necessary computerized workstations are located only there, although in some of these cases map library personnel do the actual cataloging. Another 11.3% of the respondents stated that all map cataloging is done only in a different location from the map library or the cataloging department, such as a branch library or government documents unit.

Figure 1 illustrates the wide variation in responses summarized by size ranges. Clearly the map library predominates in *Ranges 3 and 4,* while in *Range 2* it occurs equally in each department. But in the smaller *Range 1* libraries the map cataloging is done primarily in the cataloging department.

Organizational Locus

Of all the respondents, 37.7% indicated that map cataloging occurred only under the map library's jurisdiction, whether or not this activity was carried out in the map library. Among 32.1% map cataloging was only under the cataloging department's jurisdiction. In 24.5% of the libraries map cataloging was done by catalogers or other staff from both jurisdictions.

Figure 2 indicates, as did Figure 1, the diversity of data in the four collection size ranges. Here, as in Figure 1, the cataloging department predominates in *Range 1*. In general one might expect smaller map libraries with smaller staff resources to be less able or likely to achieve cataloging on their own, and the data support this hypothesis. But, as the collections become larger, the map library takes over as both the predominant organizer of and the primary physical location for map cataloging, as evidenced by the much higher percentages for *Range 2* and *3* in Figures 1 and 2. A different phenomenon occurs in *Range 4,* wherein the cataloging department begins to predominate again as the organizer of map cataloging, and the percentage approaches that of the *Range 1* libraries (see Figure 2). But, in *Range 4* the physical location

FIGURE 1. Physical Location of Map Cataloging

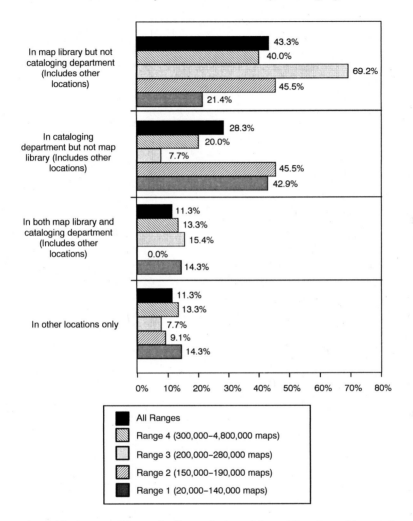

remains with the map library in the majority of these libraries with very large map collections (see Figure 1).

THE WHO OF MAP CATALOGING AND CLASSIFICATION

In 49% of the responding libraries either the map librarian or a professional map cataloger from the map library, or a combination of both, catalog

FIGURE 2. Organizational Locus of Map Cataloging

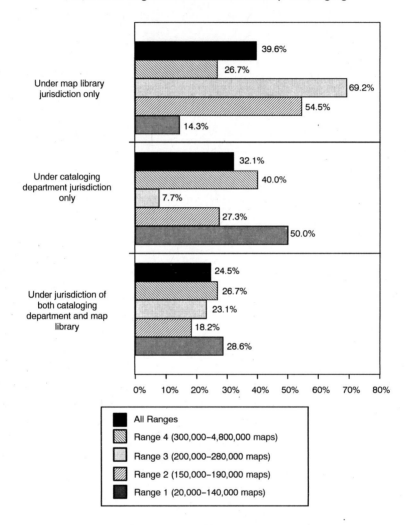

maps. In 30.2% of the libraries map cataloging is undertaken by map library paraprofessionals either exclusively or in some cases assisting a professional cataloger, but only 5.7% of the libraries indicated that students in the map library assist with cataloging.

Among 45.3% of all the respondents, professional staff members in the cataloging department do the map cataloging. In many of the cataloging departments there is a part-time map cataloger who also catalogs other formats such as serials, electronic resources, or special formats. In 32.1% of the libraries, paraprofessionals in the cataloging department perform map cataloging, but no libraries use student assistants to catalog maps in the cataloging department.

Figure 3 shows that the involvement of professional librarians in map cataloging follows the same response pattern as Figure 2 indicated for organizational locus. Cataloging department professionals are the predominant map catalogers in *Ranges 1* and *4,* and map library professionals predominate in *Ranges 2* and *3.* Furthermore, cataloging department paraprofessionals follow the same pattern of involvement as cataloging department professionals, higher in *Ranges 1* and *4* and lower in *Ranges 2* and *3.* Interestingly, the highest levels of involvement by both map library professionals and paraprofessionals are in *Range 3.* Likewise, this range encompasses the lowest level for both cataloging department professionals and paraprofessionals.

Adequacy of the Number of Staff Members

One of our respondents is the Library of Congress, which presents a unique situation in terms of map library staffing. It employs a far greater number of personnel than any of the other respondents. Thus, it is more meaningful to report statistics on adequacy of the number of staff members without including data for the Library of Congress. Consequently, the average number of map library professionals in the remaining 52 map-cataloging libraries is 1.1 FTE. The average number of paraprofessionals is 0.9 FTE and of student assistants 1.7 FTE. Staff personnel described as "other" has an average of 0.3 FTE. Therefore, the average total number of FTE employees of the above four types is 4 FTE. The average total number of FTE personnel in the map libraries who are able to create MARC records is substantially less, 0.8 FTE or 20% of the average total number of FTE employees. This value is also lower than both the professional FTE value (1.1) and the paraprofessional FTE value (0.9).

Figure 4 shows that the percentages of map libraries with staff able to create MARC records for maps are both surprisingly low and nearly consistent through the four size ranges. To compensate for this and to generate an acceptable amount of map cataloging, cataloging department personnel assume a significant role in map cataloging. Figure 3 indicates that among all the respondents, the percentage of cataloging department professional (45.3%) and paraprofessional (32.1%) personnel cataloging maps almost equals the percentage of map library professional (49.0%) and paraprofessional (30.2%) personnel doing the same. The predominant reasons given for limited cataloging of maps in the map library are lack of time, lack of staff, and the difficulty of creating MARC records for maps, especially since format integration was introduced.

FIGURE 3. Who Catalogs Maps

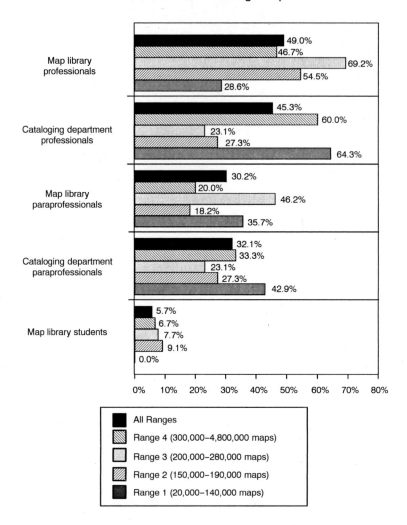

Table 1 indicates that the total FTE staff in *Range 1* and *Range 2* map libraries is essentially the same at 3.3 FTE staff and not much different in *Range 3* at 3.7 FTE. Then there is a significant increase at the *Range 4* level, where there is an average of 5.5 FTE. In correlating this with the ability of map library staff to create MARC records, it can be seen that the data likeness is maintained at *Ranges 1* and *3* at an average of 0.7 FTE and in *Range 4* at 1.2 FTE. The *Range 2* data, however, indicates only 0.5 FTE able to create

FIGURE 4. Map Library Staff Able to Create MARC Records

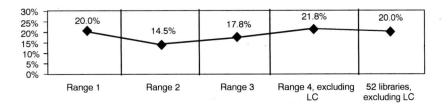

TABLE 1. FTE Staff in Map Libraries (excluding LC)

	Professional	Paraprofessional	Student	Other	Total FTE (Columns 1-4)	Ratio of paraprofessionals to professionals	Can create MARC records
Range 1	0.6	0.7	1.7	0.3	3.3	1.2	0.7
Range 2	0.7	0.7	1.9	0.0	3.3	1	0.5
Range 3	1.0	0.7	1.5	0.5	3.7	0.8	0.7
Range 4	2.1	1.4	1.8	0.2	5.5	0.6	1.2
All Ranges	1.1	0.9	1.7	0.3	4.0	0.8	0.8

MARC records, making map libraries at this size disproportionately weaker in map cataloging ability.

If one combines the data for map library professionals and paraprofessionals, who are the staff members most likely to do map cataloging, as part of the map library staff (since students are used relatively little for this purpose), then the combined averages for FTEs for the four ranges are as follows: *Range 1* has 1.3 FTE; *Range 2* has 1.4 FTE; *Range 3* has 1.7 FTE; and *Range 4* has 3.5 FTE. As can be seen, as the combined FTE for professionals and paraprofessionals increases, the ability to create MARC records decreases from *Range 1* to *Range 2* and then increases from *Range 2* to *Range 4*. A similar pattern in the ability to create MARC records occurs as the paraprofessional to professional ratio decreases.

In contrast, and again somewhat surprisingly, 45.3% of the responding map-cataloging libraries (see Figure 5) indicated that the number of staff members in the map library is adequate to maintain reasonable levels of processing and service. In the survey, map librarians were asked to rank, more specifically, the need for additional staffing in 7 categories of map library activities: direct reference service to patrons, basic processing, map cataloging, collection development, collection maintenance, GIS/electronic

FIGURE 5. Percentage of Map Libraries Which Consider Total Number of Staff to Be Adequate

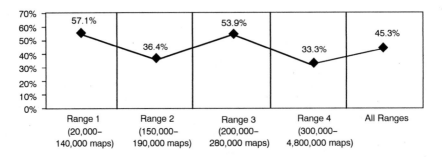

data services, and other services. Among all the responding libraries 41.5% regard the need for more map cataloging staff as the top priority (see Table 2).

Table 2 shows the percentage of map libraries in each size range that ranked the need for additional staff to catalog maps as first priority, second priority, and so forth, to sixth priority (none ranked it as seventh priority). Also shown are the libraries that did not provide any ranking. The consequence of this non-ranking situation is that the percentages in any combination of the first through sixth priority columns could be increased by all or part of the values shown in the "No Response" column. Stated differently, 35.7% of the *Range 1* map libraries said that additional map cataloging staff was their number one need, but 28.6% gave no response. If responses were provided by the libraries represented by the 28.6%, it seems likely that these libraries would have ranked their need priority similar to the majority of *Range 1* respondents; that is, the need for map cataloging staff would have been ranked first or second in priority. The same type of rationale can be applied to the other cohorts.

Even with the complication of the no response situation, it can be seen in Table 2 that 41.5% of all map libraries ranked map cataloging as first priority for additional staffing, and 20.6% ranked this as the second priority. Combined, these priorities total 62.1% of all libraries, and to this one could add another potential 18.9% from non-respondents, for a potential total sum of 81%. Clearly, the need for staff to catalog maps is a very real need for map libraries. The emphasis on this need is found at all size ranges, although less clearly as a first priority in *Range 2*. Of particular interest is the response by *Range 4* libraries; of these largest map libraries, 60% gave map cataloging as their number one staffing need, and the value increases to 87% if one adds the second priority and no-response data. None of the *Range 4* libraries ranked the need for map cataloging staff as a third, fourth, or fifth priority, but 13.3% ranked this as their sixth priority in staffing need. This seems to indicate

TABLE 2. Perceived Need for Additional Staff to Catalog Maps

	Ranked as first priority	Ranked as second priority	Ranked as third priority	Ranked as fourth priority	Ranked as fifth priority	Ranked as sixth priority	No ranking (question not answered)
Range 1	35.7%	21.4%	7.1%	0%	7.1%	0%	28.6%
Range 2	18.2%	45.5%	9.1%	9.1%	0%	0%	18.2%
Range 3	46.2%	15.4%	15.4%	7.7%	7.7%	0%	7.7%
Range 4 (including LC)	60%	6.7%	0%	0%	0%	13.3%	20.0%
All Ranges (including LC)	41.5%	20.8%	7.6%	3.8%	3.8%	3.8%	18.9%

rather clearly that a definite minority of the largest map libraries is content with its cataloging capability. An overwhelming majority, on the other hand, feels that either it is unable to cope with the volume of their cataloging need (including backlogs and retrospective conversion needs), or the majority sense an unmet responsibility as large "leader" map libraries.

As the range data in Table 2 demonstrate, the percentage of libraries ranking staffing need for map cataloging first priority increases both as the collection size grows and as the physical location (see Figure 1) and organizational locus (see Figure 2) of map cataloging move to the map library. When the organizational locus is only in the map library, 38.1% of this group ranked staff need for map cataloging as first priority, but, when the locus is only in the cataloging department, only 23.5% ranked this need as first priority. The highest percentage (69.2%) ranked this need as first priority when the locus was in both locations. This seems to suggest the greatest satisfaction regarding adequacy of staffing occurs only with a cataloging department locus.

Another way to examine map library staffing is to relate staffing ratios to need, adequacy of the number of staff members, and capability data. The survey data provide FTE equivalencies for professional and paraprofessional staffing in map libraries and thus permits a ratio for the two. The ratio, in turn, can be used to indicate if ratios correlate with stronger or weaker value for a given attribute.

Table 1 indicates, based on a 52-library sample, that, on average, map libraries have 0.8 paraprofessionals to each professional. The inverse statement is that the average map library has 1.3 professionals for each paraprofessional. Corresponding data are provided for each size range. *Range 1* libraries have, on average, slightly more paraprofessionals than professionals in a ratio of 1.2:1; *Range 2* libraries have an even 1:1 ratio; the *Range 3* libraries match the total sample average; and *Range 4* libraries fall below the average at 0.6 paraprofessionals to each professional.

Each of these ratios can then be compared to the attributes in Figures 4 and 5 and in Table 2. In terms of overall adequacy of the number of staff members in the map library, those libraries with more paraprofessionals than professionals (i.e., *Range 1* libraries) rate themselves as most adequately staffed. Nevertheless, there is a wide gap between a value of 57% adequacy and 100% or ideal adequacy and certainly between the overall average of 45.3% and 100%. The data for *Ranges 2* and *4* indicated a very weak sense of adequacy in staffing even as the professional component increases.

With the exception of the datum for *Range 2,* the perception of need for map cataloging staff increases dramatically as the paraprofessional component decreases in the staffing ratio. The two findings may, of course, be unrelated, with the perception of map cataloging as the number one need related instead to variables such as acquisition rates, uncataloged backlogs, or demands imposed by cataloging standards at the larger map libraries. On the surface, however, it would appear that map cataloging becomes a more urgent priority as paraprofessional staffing decreases relative to professional staffing. One explanation might be that in such situations the professional becomes more encumbered with what might otherwise be assigned as paraprofessional responsibilities, thus leaving less time for professional (original) cataloging.

The majority of values in Figures 4 and 5 falls below 50%, indicating, in general, that map libraries are inadequately staffed to achieve optimal map cataloging capability and production. The data in Table 2, indicating map cataloging as a first priority staffing need, is more problematic in that the lower values (especially in *Range 2*) could suggest satisfaction with cataloging activity. The alternative rationale is that map cataloging is needed, but not so much as patron service staffing needs (including the increasing need to have GIS capability).

THE **WHAT** *OF MAP CATALOGING AND CLASSIFICATION*

Of all the responding map-cataloging libraries, 88.7% indicated that MARC records reflect 5% to 100% of the individual cataloged map collections, but such records represented only an average of 59.8% of the combined cataloged map collections. Of these libraries, 56.6% reported that their map catalogs consisted of at least 70% MARC records.

Figure 6 clearly shows that as the map collections get larger the percentages of MARC records representing the cataloged maps decrease. This also indicates that in large collections significant portions of the map catalogs have not been retrospectively converted to machine-readable form.

FIGURE 6. MARC Records in Map Catalog

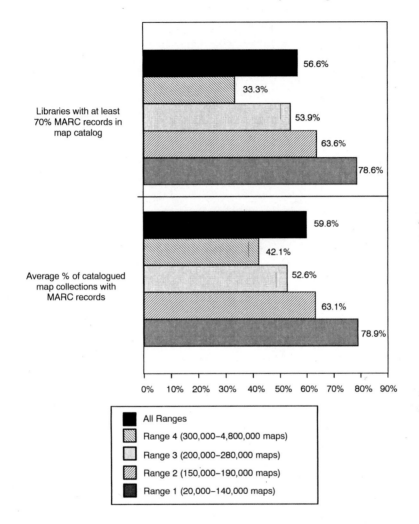

Bibliographic Utilities

Of the respondents, 86.8% use OCLC for cataloging maps, but only 41.5% use OCLC exclusively. Another 30.2% use OCLC in conjunction with another system, either RLIN and/or a different system such as AG-Canada or a local system, or they acquire records via online catalogs accessible through a Z39.50 server. A far smaller group of libraries, 24.5%, use RLIN, but only one library uses RLIN exclusively. Only 9% use another utility and not OCLC or RLIN. In addition, MARCIVE provides map records to 22.6% of all the responding map-cataloging libraries. A small number of the responding libraries, only 7.6%, outsource any of their map cataloging.

Figure 7 vividly demonstrates the vast difference between percentages of libraries that use only one utility for acquiring machine-readable cataloging data and those that employ more than one utility and/or purchase records

FIGURE 7. Use of Bibliographic Utilities

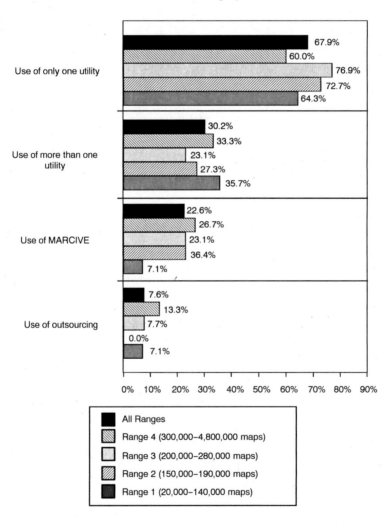

from MARCIVE. Only three libraries reported using a combination of MARCIVE and more than one bibliographic utility. Outsourcing is not reported by *Range 2* libraries in the survey and is most significant (at 13.3%) in the largest (*Range 4*) map libraries, in which the largest percentage of map libraries (60%) ranked the need for additional staff for map cataloging as first priority.

Classification Systems

Eighty-three percent of all the responding map-cataloging libraries classify maps with the Library of Congress Classification System (LCC), while only 5.7% use the Dewey Decimal Classification System (DDC). The Superintendent of Documents Classification System (SuDocs) is used at 16.0% of the libraries, while 20.8% use another system, usually locally devised. Some local systems are modifications of the LCC system.

In terms of size, 64.3% of the *Range 1* libraries use only LCC, while 21.4% use LCC in conjunction with SuDocs. Only 7.1% use SuDocs and a local system, while another 7.1% use only a local system. In *Range 2*, 63.6% of the libraries use only LCC, while 18.2% use LCC in conjunction with SuDocs. More libraries (27.3%) use LCC in conjunction with a local system. In addition, one library in this group uses SuDocs as a third system. In *Range 3*, 53.9% of the libraries use only LCC, while 7.7% use only DDC and 15.4% use only local systems, but 23.1% use LCC in conjunction with one or more other of these systems. In *Range 4*, 66.7% of the libraries use only LCC, while only 6.7% use LCC in conjunction with SuDocs. Twenty percent use only a local system.

Figure 8 indicates a predominant use of a single classification system for maps throughout the four size ranges.

Original Cataloging

Of all the responding map cataloging libraries, 69.8% indicated that either all or some maps originally cataloged are done at the full level. But, only 26.4% of the respondents provide full level original cataloging for at least 70% of the maps needing original cataloging.

The reporting of serial maps was somewhat problematic, since there was some confusion about what kinds of maps should be regarded as serials. Maps in monographic series were most often regarded as, and reported as, monographs rather than as serials. In fact, most responding libraries were unable to provide data on serial maps. For purposes of tabulation, therefore, true serial maps were not included.

Minimal level original cataloging is far less common. Only 26.4% of all

FIGURE 8. Percentage of Libraries Using Only One Map Classification System

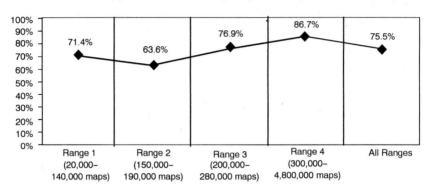

the responding map-cataloging libraries indicated that some maps are originally cataloged at the minimal level. In only one case does 100% of such
cataloging occur. In addition, in most responding libraries, minimal level
copy records are upgraded to full level.

Figure 9 demonstrates that full-level original cataloging across all size
ranges occurs at a consistently low level for *Ranges 1, 2,* and *4.* But *Range 3*
libraries, which previously reported (see Figure 3) the highest level of map
library professional and paraprofessional involvement in cataloging, show
the highest percentage in this category.

SUMMARY OF FINDINGS

It can be instructive to see which characteristics are reported by more than
50% of the responding libraries and which by less than 50%. Figure 10 arrays
the attributes reported in this paper from those most commonly shared to
those least often reported as library attributes related to map cataloging. As
can be seen, only three attributes are common to more than 50% of the map
libraries: the use of a single classification system for map collections, the use
of a single utility in map cataloging, and the presence of map catalogs with at
least 70% MARC records. A fourth attribute, the use of map library professionals to catalog maps, is close enough at 49% to be grouped with the
majority. By contrast, all the remaining 17 attributes fall into the minority,
that is, fewer than 49% of the map libraries report each of the 17 attributes.
This seems to indicate rather unequivocally that there is little consistency in
the characteristics of map libraries and of map cataloging functionality. The
data may indicate not only inconsistency in map library operations but also
confusion about optimization of operations and ability to set and achieve

FIGURE 9. Percentage of Libraries Creating Full Level Original Records for at Least 70% of Maps Needing Original Cataloging

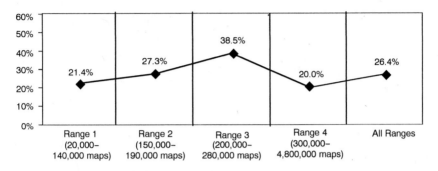

goals. The data probably indicate uneven guidance and support for map libraries.

The authors wanted to use the same approach to see what attributes would prevail as the majority characteristics in each of the four map-collection-size ranges. For at least two of the cohorts, however, this would have meant a somewhat abbreviated array of commonality–that is, attributes shared by more than 50% of the libraries in the size range. Rather than using 50% as a majority-minority breakpoint in the data, it was decided that the cohort data would be examined in the sense of those attributes common to two-thirds of the libraries in each range and that array would be considered as the "majority" or predominating characteristics of the range. In actuality, 32% was chosen as the breakpoint to determine inclusion in Figures 11 to 14. This percentage is slightly below the two-thirds to one-third division at 33.3%, but it approximates the median of the percentages as displayed in Figure 10.

Figure 11 shows that in the majority of libraries in *Range 1* (those with 20,000 to 140,000 maps) policies and practices differ considerably from those listed above the median in Figure 10. Map cataloging is organizationally and physically located in the cataloging department rather than in the map library, and the burden of map cataloging is borne primarily by cataloging department professionals with cataloging department and map library paraprofessionals participating in the cataloging effort to a lesser extent. Map library professionals in this range do not catalog maps in the majority of libraries, and the need for more map catalogers is ranked first priority. Nonetheless, the highest percentage of libraries in this group reported that MARC records represent at least 70% of the cataloged map collections, and more than one bibliographic utility is used to search for map records.

In *Range 2* libraries with 150,000 to 190,000 maps (see Figure 12), map cataloging is organizationally located in the map library, but the physical location is evenly divided between the map library and the cataloging depart-

FIGURE 10. Ordered Attributes of Map Cataloging

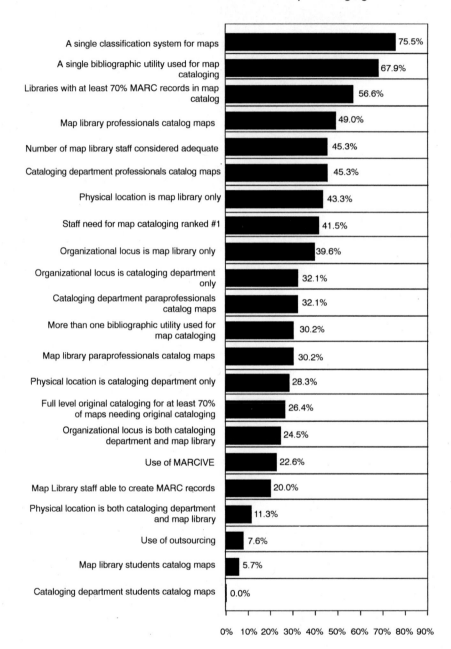

FIGURE 11. Attributes Common to Majority of Libraries in Range 1 (20,000 to 140,000 maps)

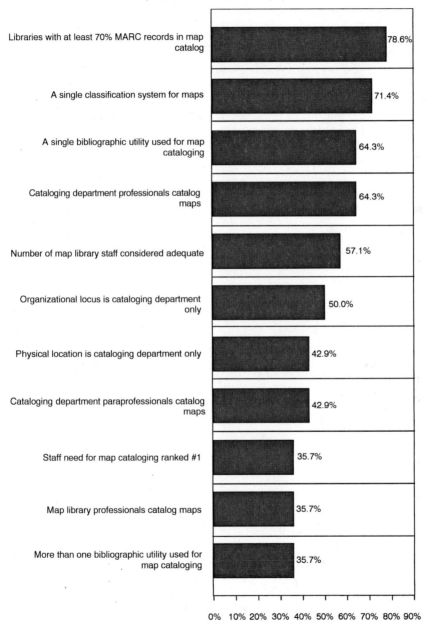

Attribute	Percentage
Libraries with at least 70% MARC records in map catalog	78.6%
A single classification system for maps	71.4%
A single bibliographic utility used for map cataloging	64.3%
Cataloging department professionals catalog maps	64.3%
Number of map library staff considered adequate	57.1%
Organizational locus is cataloging department only	50.0%
Physical location is cataloging department only	42.9%
Cataloging department paraprofessionals catalog maps	42.9%
Staff need for map cataloging ranked #1	35.7%
Map library professionals catalog maps	35.7%
More than one bibliographic utility used for map cataloging	35.7%

0% 10% 20% 30% 40% 50% 60% 70% 80% 90%

ment. In contrast to professionals in *Range 1* libraries, map library professionals in *Range 2* are much more involved in map cataloging than are cataloging department personnel, but, surprisingly, even map library paraprofessionals are far less involved in map cataloging. The number of map library staff members is considered adequate, and staff need for map cataloging is ranked less than first priority. In fact, staff for GIS/electronic data services was most commonly regarded as first priority. Again, MARC records represent at least 70% of the cataloged map collection in the vast majority of these libraries, but MARCIVE provides a portion of these records.

Among the majority of the libraries in *Range 3* with 200,000 to 280,000 maps (see Figure 13), map cataloging is organizationally and physically located in the map library at higher percentages than in *Range 2,* and map library professionals and paraprofessionals do the vast majority of map cataloging. In this group the number of map library staff is regarded as adequate, but staff need for map cataloging is ranked first priority. Nonetheless, again the cataloged map collections contain at least 70% MARC records but to a lesser degree than in *Range 2.* Surprisingly, in spite of the larger size of these collections, full level original cataloging is done for at least 70% of the maps needing original cataloging.

Finally, among the majority of the very large *Range 4* libraries (see Figure 14), the organizational locus for map cataloging shifts to the cataloging department, although this cataloging is performed by both cataloging department and map library professionals, with far less involvement by cataloging department paraprofessionals and none from map library paraprofessionals. Consequently, and quite predictably, the percentage of libraries reporting that the size of the map library staff is adequate is far lower than in *Range 3.* Also, since the libraries in *Range 4* have very large map collections and undoubtedly high rates of map acquisitions, a very large percentage ranked staff need for map cataloging as first priority. Nonetheless, the cataloged map collections contain at least 70% MARC records, but this occurs in a far lower percentage of libraries than in the other three ranges, although more than one bibliographic utility is used to search for MARC records.

DISCUSSION OF FINDINGS

The discontinuity found so often between size-adjacent cohorts, especially at the *Range 2* interval, may be perplexing but, nevertheless, may represent aspects of the data not yet understood. The data discontinuities may reflect current and historical developments in attention to, and support for, map librarianship, at the four collection levels. The discontinuities could also, of course, result from skewing in the size-range samples if caused by an inadequate sample population or inappropriate range intervals, and this could

FIGURE 12. Attributes Common to Majority of Libraries in Range 2
(150,000 to 190,000 maps)

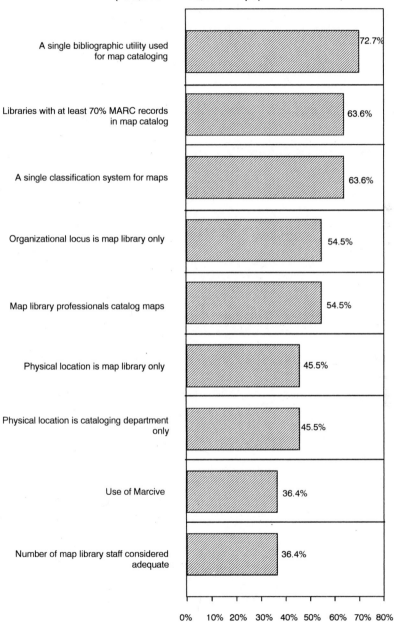

A single bibliographic utility used for map cataloging	72.7%
Libraries with at least 70% MARC records in map catalog	63.6%
A single classification system for maps	63.6%
Organizational locus is map library only	54.5%
Map library professionals catalog maps	54.5%
Physical location is map library only	45.5%
Physical location is cataloging department only	45.5%
Use of Marcive	36.4%
Number of map library staff considered adequate	36.4%

0% 10% 20% 30% 40% 50% 60% 70% 80%

FIGURE 13. Attributes Common to Majority of Libraries in Range 3
(200,000 to 280,000 maps)

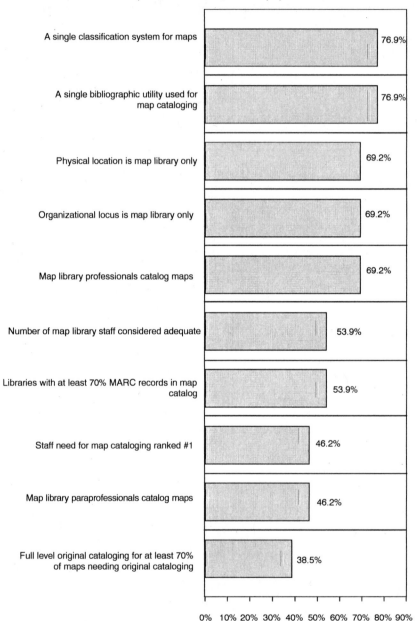

FIGURE 14. Attributes Common to Majority of Libraries in Range 4
(300,000 to 4,800,000 maps)

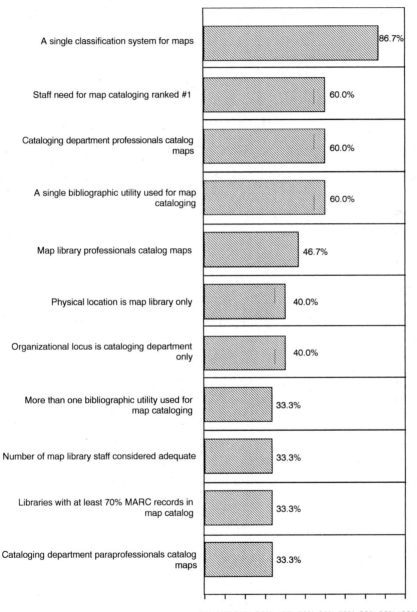

provide a caution to the use and interpretation of data in individual ranges. As often as not, however, data from which a cohort average as a percentage is derived, do not vary greatly within their range. Consequently, it is likely that additional data would not alter the percentages sufficiently to remove the discontinuities between ranges.

The purpose of the present paper is to report survey data and to provide basic analysis and interpretation, but not to attempt complete understanding of the data that result among cohorts or to rationalize seeming discrepancies in terms of probable expectations for the data. If the data match a reader's situation at his or her map library, then the data correlate with a real situation, and the comparative data can be understood in that context. If the data do not match, the challenge is to try to understand why and to see if that understanding provides guidance for the local library situation.

A larger and more relevant view is to recognize the lack of pluralities in the data. Figure 10 illustrates so forcefully the lack of shared attributes in map cataloging at even a 50% level and, thus, what must, in so many instances, be recognized as perceivable weaknesses in map librarianship. As heartening as it may be to see what does exist for map cataloging (e.g., that nearly 57% of all respondents have at least 70% of their catalog records in MARC format), it is discouraging to find that nearly 55% of respondents find the number of their map library staff members to be inadequate and that 42% rank map cataloging as the function in their map library in greatest need of attention (ahead of attention needed for GIS, we might add). Another alerting finding is that, at best, no more than 20% of all map library staff are able to create MARC records (see Figure 4). One would hope for a much stronger capability among map library staff, even if the map cataloging is done outside the map library.

Another alert in the findings is the data reported for map library staffing. Kyrillidou indicates that the ARL average for total library staffing is two paraprofessionals for each professional, but the map library average is only 0.8 paraprofessionals per each professional.[19] This presents a serious question about staffing equity for map libraries and appropriate division of functionality within map library operations. The ARL statistics do not detail staffing ratios for cataloging operations, but there seems to have been a clear tendency in the past decade or so to decrease professional components and increase the paraprofessional to professional ratio in cataloging. If so, then map libraries are not sharing equitably in any benefits in the trend to free professionals for more advanced cataloging duties. Furthermore, the strength of any paraprofessional ratio is only as strong as the adequacy of the professional component itself and the survey results (see Table 1) suggest that this is also a weak attribute in map libraries.

The rather extreme weakness in the ability of map library staff to create

MARC records is also alarming (see Figure 4 and Table 1). Only about one-fifth of those in map libraries have this capability, and this situation can only weaken map libraries in the organizational and functional structure of host libraries. It suggests an increasing likelihood for map cataloging to shift away from map libraries to cataloging departments and for map libraries to have less influence in points of access for their patrons. Arguments can be made both for the locus of map cataloging to be in the map library or in the cataloging department, but, either way, it can certainly be argued that map library staff should understand the map cataloging process sufficiently to participate in map cataloging policy and the adoption of procedures that serve map library patrons to best advantage. Indeed, persuasive reasoning can be presented to support greater sharing in the locus of map cataloging, but the survey results indicate that this is not that common (see Figure 2).

BASIC CONCLUSIONS SUGGESTED BY THIS STUDY

- Map cataloging location is less critical than the organizational locus for map cataloging. Locus is important in terms of the conjoint relationship of cataloging ability and map and spatial resource understanding.
- Map cataloging ability is inordinately weak among the majority of map library personnel; therefore, map libraries must depend more on sources from outside the map library for obtaining quality map cataloging records.
- The percentage of map-cataloging libraries creating full-level original MARC records is, overall, quite low, with only 26.4% of libraries providing this level of cataloging for 70% to 100% of their maps needing original cataloging. The corollary is that 73.6% of libraries create full-level original records for less than 70% of their maps needing original cataloging. The survey results indicate that it is, in fact, often much less than 70% and actually as low as zero percent. This means that the bulk of full-level original map cataloging rests with a disproportionately small number of libraries and that most libraries depend on the relatively few (the 26.4%) to provide the majority of full-level original map cataloging.
- The number of staff members available to catalog maps is quite inadequate. Either this is true or there is an adequate number of staff members, but these employees are either inadequately qualified or are underproductive. The more logical conclusion is that the total number of staff members remains inadequate in most map libraries and the workload spread is too great to allow adequate attention to map cataloging. Impor-

tantly, this conclusion does not become that much more positive even when the cataloging responsibility has locus outside the map library.

RECOMMENDATIONS

Two basic sets of recommendations result from this study. The first is to use available data to influence more positive outcomes. The second is to promote more regular collection of data germane to understanding map library goals and capabilities, including map cataloging, whether performed in the map library or elsewhere. The corollary is to collect and use more comparative data, which means going beyond the single attribute type of data most often solicited and published by ARL.

The professional organizations representing map librarianship can assist by re-examining the adequacy of map library statistics and their potential for meaningful and constructive use. They can go beyond such assessment to providing enablement for the collection of adequate statistics, either by their own collection activities or by persuading other organizations, such as the Association of Research Libraries, to do so. This effort need not rest with the professional map librarianship organizations alone, however. Any organization representing professional catalogers can realize the merit of map cataloging and a professional responsibility to assist in support for map cataloging. One would hope that all catalogers will lend their professional influence to improvements in map and spatial data cataloging.

Ultimately, if map cataloging is to advance, both in quality and quantity, persuasive information and rationale must prevail. Relevant data must be communicated to library administrators in a professional and convincing manner. In short, map librarianship and map cataloging must achieve their own sense of primacy and importance and achieve their measure of equity in librarianship. The map library patron, after all, deserves as much in library effort and result as a patron using other collections elsewhere in the library.

It is hoped that the survey results presented in this paper provide both information and stimulus. The authors encourage others to undertake additional research to advance the cause of map and spatial data librarianship.

NOTES AND REFERENCES

1. The terms "head of cataloging," "map librarian," "map cataloger," "cataloger," "map library," and "cataloging department," and similar terms are used in this paper in a generic sense, even though specific nomenclature varies from library to library. Similarly, "map" used with other terms is understood to incorporate all cartographic and spatial data forms.

2. Data for the Library of Congress is included in most findings reported in this paper. Exclusions are noted.

3. The primary search was for literature dated 1980 and later.

4. Glenda Jo Fox Hughes and Constance Demetracopoulos, ed., *Map Cataloging Bibliography: Selectively Annotated* (Washington, DC: Special Libraries Association, Geography and Map Division, 1997.) (Special Publication no. 4), 116.

5. *Ibid.*, p. 59. The following paper, not seen, is cited as including recommendations for map library staffing standards: Maureen Wilson, "Work Standards and Procedures for Map Libraries," *in* Association of Canadian Map Libraries, *Proceedings of the Annual Conference, 2nd, 1968, Department of Geography, University of Alberta, Edmonton, Alberta, June 17-19, 1968.* (Ottawa: The Association, 1969.), 75.

6. Special Libraries Association, Geography and Map Division, Committee on Standards, "Standards for University Map Collections," *Bulletin,* Geography and Map Division, Special Libraries Association, no. 148 (June 1987): 2-12.

7. Franz Wawrik, *Basic and Continued Training of Map Curators.* http://python.konbib.nl/kb/skd/liber/articles/wawrik.htm

8. Martha Kyrillidou, *Developing Indicators for Academic Library Performance–Ratios from the ARL Statistics, 1993-1994 and 1994-1995* (Washington, DC: Association of Research Libraries, 1996), 190, esp. p. 30.

9. *ARL Statistics* (Washington, D.C.: Association of Research Libraries, issued annually).

10. Dorothy McGarry, "Responses to a Map Cataloging Questionnaire, Report by the SLA Geography and Map Division Ad Hoc Committee on Cataloging," *Bulletin,* Geography and Map Division, Special Libraries Association, no. 125 (September 1981): 32-39.

11. Judith L. Rieke and Leslie D. Steele, "A Survey of Major Map Collections in Texas," *Texas Libraries* 45:3 (Fall 1984): 86-89.

12. Steve Hiller, "Mainstreaming of Map Libraries: Twenty Years of Integration within Academic Research Libraries," *Information Bulletin* (Western Association of Map Libraries) 19(3) (June 1988): 151-160.

13. Margit Kaye, "A Comparison of Major University Map Collections to Determine the Adequacy of and to Propose Improvements to the Yale University Map Collection," *Bulletin,* Geography and Map Division, Special Libraries Association, no. 168 (June 1992): 2-23.

14. Lothar Zögner, *Verzeichnis der Kartensammlungen in der Bundesrepublik Deutschland einschliesslich Berlin (West)* (Wiesbaden: Harrosowitz, 1983).

15. David Cobb, "Map Librarianship in the U.S.: An Overview," *Wilson Library Bulletin* 60:2 (October 1985): 14-16.

16. Terri J. Robar, "University Map Collections: Comparative Statistics," *Bulletin,* Geography and Map Division, Special Libraries Association, no. 139 (March 1985): 10-16.

17. Charles Seavey, "The 1991 ARL Cartographic Resources Survey," *Information Bulletin* (Western Association of Map Librarians) 24(3) (July 1993): 175-180.

18. Charles A. Seavey, " Ranking and Evaluating the ARL Library Map Collections," *College & Research Libraries* 53:1 (January 1992): 31-43.

19. Kyrillidou, 30.

CATALOGING SPECIFIC
MATERIAL TYPES

Problem Areas
in the Descriptive Cataloging
of Sheet Maps

Ken Rockwell

SUMMARY. This article discusses areas of the bibliographic record where differences from cataloging monographs are commonly encountered in the descriptive cataloging of flat or folded sheet maps. Major fields in the bibliographic record are treated, such as title proper, main entry, and scale, pointing out common misunderstandings and errors which those unfamiliar with cataloging maps may experience. Hints,

Ken Rockwell received his MLS at the University of California at Los Angeles, 1989. Ken has been the Map Cataloger at the University of Utah's Marriott Library since 1989. He also catalogs science books and provides science and map reference in the Science and Engineering Department, and shares in the responsibility of acquiring maps for the collection.

[Haworth co-indexing entry note]: "Problem Areas in the Descriptive Cataloging of Sheet Maps." Rockwell, Ken. Co-published simultaneously in *Cataloging & Classification Quarterly* (The Haworth Information Press, an imprint of The Haworth Press, Inc.) Vol. 27, No. 1/2, 1999, pp. 39-63; and: *Maps and Related Cartographic Materials: Cataloging, Classification, and Bibliographic Control* (ed: Paige G. Andrew, and Mary Lynette Larsgaard) The Haworth Information Press, an imprint of The Haworth Press, Inc., 1999, pp. 39-63. Single or multiple copies of this article are available for a fee from The Haworth Document Delivery Service [1-800-342-9678, 9:00 a.m. - 5:00 p.m. (EST). E-mail address: getinfo@haworth pressinc.com].

guidelines, illustrations, and examples for the resolution of these problems are given. *[Article copies available for a fee from The Haworth Document Delivery Service: 1-800-342-9678. E-mail address: getinfo@haworthpressinc.com <Website: http://www.haworthpressinc.com>]*

KEYWORDS. Map cataloging, descriptive cataloging, title proper, scale

INTRODUCTION

It doesn't take long to determine that, in cataloging circles, maps are one format that nobody wants to work on. A case in point: when this author arrived at the University of Utah's Marriott Library to fill the position of map cataloger, it had been vacant for four years–and the size of the backlog reflected this. Though nominally the responsibility of the Science cataloger that person had done little map cataloging. Books always seem to take priority among catalogers.

Maps have a reputation for being difficult to catalog. Part of it has to do with their format, usually individual flat sheets of paper as opposed to the familiar bound book; and maps also have unique features that require learning to apply additional fields in the bibliographic record. There are also details which maps share in common with books, but which, due to the nature of the map or the presentation by the publisher, make for added difficulty.

MAP TITLES

Whereas titles of books are usually evident from the title page maps quite often provide more than one title from which to choose. For instance, when receiving a new map for cataloging one may see a title on a panel, the portion of the sheet visible when the map is folded. Then, after unfolding the map, one may find a different title in the top margin of the sheet, and a third above the legend at the bottom right. Which one constitutes the map's title that is entered in MARC field 245, subfield "a" (and possibly also "b" if part of the title can be considered "other title information")?

A rule of thumb is to choose the title that most clearly reflects the contents of the cartographic work, first looking for the words in the largest typeface. The sheet may contain two maps of equal importance, such as the state of Nevada on one side and Utah on the other, as commonly occurs with road maps. The folded panel title may read "Nevada, Utah," and thus be more comprehensive than the titles of each map alone. This is the obvious choice for title proper, while the individual titles can be accessed through added title entries.

If the titles are fairly equal in size of typography, choose one of the titles on the face of the map. This follows the analogy that a folded panel title is like the cover of the book and the face of the map, especially near the legend, is like the title page of a book.

Sometimes, though, the cover is all the cataloger has. Certain European publishers attach the map to a cardboard cover in which the map is folded, and confine their maps' titles to the cover. These covers present a problem if the map is to be stored flat in a map case drawer. If the cover remains attached to the map, it can easily catch on other maps and lead to tearing. The library may therefore decide to remove and discard the cover. This was historically done at the Marriott Library prior to cataloging the maps. The problem, from the cataloging standpoint, was in identifying the map once all bibliographical data was discarded along with the cover. Even if the map were cataloged prior to the cover's removal, such practice hinders future identification of the map, including when a map of the same title is purchased. The cover details may differ, including the distributor of the map, but the map itself may in fact be the same.

In the case of the Marriott Library it was decided to leave the maps in the covers and to store them in vertical files rather than in the map cases. Filing in vertical files does have the drawback of providing a separate location for the maps with the possibility of being overlooked by patrons browsing for maps in the map drawers. Another option is to store the map itself unfolded (this is best in terms of preservation) and file the cover separately, as accompanying text, in standard book shelving or in "Princeton" boxes that can be shelved or kept on top of map cases.

Layout or size and presentation of the title may also present a problem. The title may begin with a place name in small print, followed by another place in much larger print. One must choose between entering the title parts in direct order or rearranging them to put the words that are given emphasis at the beginning. Anglo-American Cataloging Rules (AACR2R) clearly allow for such rearrangement, but judging from records in OCLC some catalogers seem reluctant to use this option. Cataloging records for U.S. National Imagery and Mapping Agency (NIMA), formerly the U.S. Defense Mapping Agency (DMA), nautical charts entered into OCLC by the Government Printing Office invariably use direct order, starting with the small-print larger place first followed by other small-print places, and finally by the large-print place on the individual map. This may be useful for indexing and listing online all of the maps, say, of the Western Pacific Ocean; but the patron is less likely to want an online catalog display filled with similar titles, some of which may truncate before the individual place can be displayed. It is a better practice with these maps to place the large-print place name first in the title proper.

In case of doubt, the best resource for determining which title to use as the primary title is the "APPLICATION" following Rule 1B8b in *Cartographic Materials: A Manual of Interpretation for AACR2 (Cartographic Materials),* which includes a table showing an order of preference for choosing based on the location of the title.

In any catalog, be it card or online, of course one is not limited to searching under only one form of the title. The cataloger should, therefore, make liberal use of added title (MARC 21 field 246) entries, based on the different ways in which someone can interpret the title and search for it. Therefore, catalog for different title structure and location situations, always basing decisions made on the appropriate cataloging rules and rule intepretations.

STATEMENTS OF RESPONSIBILITY
AND APPLYING MAIN ENTRY

Maps, like books, may be primarily the work of an individual. Quite commonly though they are the product of a group of persons working for the publishing government agency or commercial company and the individuals will not be named. When the second edition of Anglo-American Cataloging Rules (AACR2) was published in 1978, the rules for corporate body main entry in Rule 21.1B2 were changed to greatly limit the number of cases in which a corporate entity would be considered the main entry in the cataloging record. The rules excluded important map publishers, including Rand McNally and the United States Geological Survey (USGS). Recognizing this, the map librarian community effectively lobbied the American Library Association's Committee on Cataloging:Description and Access (CC:DA) to work with the Joint Steering Committee for Revision of AACR (JSC) to make room for corporate authorship of cartographic materials; and this change is reflected in the 1988 revision of AACR2 as Rule 21.1B2, item "f," rectifying this situation. After all, government agencies and companies do produce most of our maps!

Rule 21.1B2f states that corporate main entry is acceptable for "cartographic material emanating from a corporate body. . . ." The key word in this phrase is "emanating from," and both footnote number 2, "Consider a work to emanate from a corporate body if it is issued by that body *or* has been caused to be issued by that body *or* if it originated by that body" and the Rule Interpretation for 21.1B2f gives us much latitude in terms of using corporate main entry in the bibliographic record. Chapter 3 of AACR2R, Rule 3.1F1 says "Transcribe statements of responsibility relating to *persons or bodies* [author's emphasis] as instructed in 1.1F." Turning to 1.1F1, "Transcribe statements of responsibility *appearing prominently* [author's emphasis] in the item in the form in which they appear there." And Rule 0.8 in AACR2R defines "prominently" as meaning ". . . that a statement to which it applies must be a formal statement found in one of the prescribed sources of information (see 1.0A) for areas 1 and 2 for the class of material to which the item being catalogued belongs." For maps, the prescribed source of information is the chief source of information, which in Rule 3.0B3 is the "cartographic item itself . . ." and the "container . . . or case . . . ."

Although in the vast majority of cases maps will have been created or caused to be created by a corporate body this does not mean we ignore or eliminate personal authorship of these materials. The sole purpose of having Rule 21.1B2f is to make corporate body main entry acceptable in those situations where there is no personal author indicated. When a personal author *is* given, it should be used as main entry. Thus, if a person is given responsibility by phrasing such as "by . . ."; "cartography by . . ."; "drawn by . . ."; or other phrases such as those found in *Cartographic Materials* on pages 35-36, then personal main entry may be warranted. In fact, unlike in the world of monographs, even those maps "compiled by . . ." an individual can be given main entry status because the compiling of layers of information for a map is much more closely tied to the creation of the final product. Remember that if you have three persons or less given on the map they all are given in the statement of responsibility with the first used as main entry, if they qualify under Rule 21.6B1, and the other one or two given an added entry in the record, as indicated in Rule 21.6B2.

What is often confusing is when both a person(s) and a corporate body are involved in the creation of a map, and both are prominently stated, as is the case with many maps published as part of the U.S. Geological Survey's different map series. Using the list of adjectival phrases in *Cartographic Materials* as a guide, preference should be given to the individual even though that individual may be employed by the company or agency publishing the map (this is because the person(s) are usually more prominently displayed on the map as a part of the title whereas the USGS is located in a corner or given in smaller type). In such cases one should also give added entries to the company or agency involved for the simple reason that so much of cartographic material does "emanate from a corporate body" and thus our patrons typically search by those well-known map creating companies and agencies. In all cases each map or map set being cataloged must be taken on a case-by-case basis as wording, typography (i.e., size, bolding, color of lettering), layout, and other variable factors can lead to different decisions even amongst those titles published by the same company or agency.

Statements of responsibility and main entry status run the gamut for maps, from the most straightforward of situations where a single government

agency or company, or individual is named on the map and thus is given both
responsibility and main entry, to very complicated situations where multiple
corporate bodies and/or individuals are involved. Government agencies tend
to be more straightforward as to responsibility in the majority of cases as
opposed to private companies where the company may serve as map creator,
editor, publisher, and distributor or any combination of these.

An example of a "simple" statement of responsibility/main entry case is a
Rand McNally road map for the state of Utah as found in OCLC record
#36858009 (see Example 1). According to the record the only statement
found on the map is "Rand McNally," thus it is easy to see who is responsible for the map's creation. The name "Rand McNally" would be given in the
statement of responsibility area (245 |c) as well as becoming the main entry
under the correct form of the name's heading, "Rand McNally and Company."

EXAMPLE 1

12 110 2 Rand McNally and Company.
13 245 10 Utah state map : |b Logan, Ogden, Provo, Salt Lake city downtown, Salt Lake City &
vicinity, St. George / |c Rand McNally.
14 255   Scale [ca. 1:1,305,216]. 1 in. = approx. 20.6 miles |c (W 114^0--W 109^0/N 42^0--N 37^0).
15 260   Chicago : |b Rand McNally, |c c1996.
16 300   1 map : |b col. ; |c 56 x 43 cm., folded to 23 x 11 cm.

A less clear case might be that of geologic maps published by state or
federal geologic surveys. The geologist providing the data used for the map
created on a USGS topographic base is often prominently named though
sometimes his or her role is stated, sometimes not. What is often unclear is
whether the geologist, as an employee of the survey (or on occasion who has
been hired as an independent specialist), is the "author" of the map while the
survey is merely editor and publisher. Some cataloging agencies invariably
give the main entry to the survey, with the geologist given an added entry.
But, some institutions favor the individual over the survey in such cases even
though his/her role in the map's creation is not explicit.

An example similar to the situation above is found with the title "Geologic
map of the Bear Lake South quadrangle, Cache County, Utah," OCLC record
#36978138, (see Example 2) published by the Utah Geological Survey. In
this case the geologist, because of both wording ("by James C. Coogan") and
layout on the map, is listed first in the statement of responsibility and given
main entry status. In much smaller print and found in a less prominent
position on the map is the phrase "Lori J. Douglas, cartographer," indicating
that perhaps Ms. Douglas did some final cartographic editing work for the
publisher. [Note that the Utah Geological Survey is included as an added
entry in the record.]

EXAMPLE 2

9 100 1   Coogan, James C.
10 245 10 Geologic map of the Bear Lake South quadrangle, Cache [i.e. Rich] County, Utah / |c by James C. Coogan ; Lori J. Douglas, cartographer.
11 246 1 |i Title on envelope: |a Geologic map of the Bear Lake South quadrangle, Rich County, Utah
12 255   Scale 1:24,000 |c (W 111°22'30"--W 111°15'00"/N 42°00'00"--N 41°52'30").
13 260   Salt Lake City, Utah : |b Utah Geological Survey, |c 1997.
14 300   1 map : |b col. ; |c 58 x 44 cm. + |e 2 supplementary sheets (74 x 51 cm. and 79 x 69 cm.) and 1 pamphlet (maps ; 28 cm.)
...
24 710 2   Utah Geological Survey.

In extreme cases, multiple statements of responsibility make us wonder which person or body is primarily responsible for the map in hand. Take, for example, a map of the Santa Catalina Mountains of Arizona, found on OCLC record #30114754 (see Example 3). The cartography is "by the U.S.G.S. and Eber Glendening"; revisions were made "by the Southern Arizona Hiking Club, the Southern Arizona Rescue Association, and Rainbow Expeditions, with the cooperation of the National Forest Service," and the map was "edited and published by Rainbow Expeditions." Another detail on the map is a copyright statement naming the Southern Arizona Rescue Association. Such a statement can be a good indicator of the entity holding primary responsibility for a work; however, the presence of multiple entities may sometimes diminish the weight given to copyright.

EXAMPLE 3

8 110 2   Rainbow Expeditions (Firm)
9 245 10   Santa Catalina Mountains, Arizona : |b a trail and recreation map / |c edited and published by Rainbow Expeditions ; cartography by the U.S.G.S. and Eber Glendening ; revisions by the Southern Arizona Hiking Club, the Southern Arizona Rescue Association, and Rainbow Expeditions with the cooperation of the National Forest Service.
10 250   9th ed., 1993.

One may assume that the Geological Survey simply provided the base map and Eber Glendening did the cartography for it, therefore they are not the "authors" of this revised edition. Rainbow Expeditions served as final editor and publisher of the map and may or may not have been involved with the cartographic work that went into the 9th edition. Focusing on who worked on the revisions, which were made by three corporate bodies with the assistance of a fourth, it appears that one could name the first of these as main entry. In the end, a case such as this requires the cataloger to use some intuition as to which entity is the most prominently responsible, but, if the cataloger is undecided, one may choose to enter as main entry by title. It appears that the decisions made for the Statement of responsibility and main entry for this work may have been guided by Rule 21.6C1 for shared responsibility, although Rule 21.6C2 would allow for title main entry in this case as well.

For further guidance in determining the layout of the statement of responsibility and a possible main entry, see Rule 1F5 and its "APPLICATION" and Rule 1F6 and its "APPLICATION" in *Cartographic Materials.* It is wise to also be sure to give a person(s) or body(s) *not chosen as main entry* an added entry if the statement of responsibility on the map indicates that they were responsible for a significant contribution towards the creation of the map.

EDITION STATEMENTS

Maps can also be produced in specific editions, relating changes made to them over time. Rules 3.2B1-3.2B5 in AACR2R should be consulted when working with edition statements. Rule 3.2B1 says to "Transcribe a statement relating to an edition of a work . . . as instructed in 1.2B," which when consulted that area simply further describes in detail what and how edition statements should be transcribed.

ISSUES RELATING TO SCALE

Map records include certain fields of "mathematical data" not found in other formats, most important of which is scale. The scale of a map tells a user something about the relative size and detail of features on the map. It is expressed in the catalog record as a ratio of the form 1:X, where X is the denominator of a representative fraction. As an example, the scale 1:24,000 common to USGS topographic maps tells us that 1 unit of measure on the map represents 24,000 of the same unit on the ground. The larger the number following the colon, the smaller the scale. For smaller-scale maps, you will see more of the Earth's surface, but less individual detail, than on a larger-scale map.

Published maps often give some indication of scale, whether as a ratio, in the form of a line or bar (i.e., a bar scale), or as an equivalency statement (e.g., 1 inch equals approx. 60 miles), known as a verbal scale (see Figure 1 for an example of a bar scale and Figure 2 for both a bar scale and a verbal scale). When a bar scale is present use of a Map Scale Indicator, available from the Memorial University Cartographic Laboratory in St. Johns, Newfoundland, makes for easy calculating of the correct ratio, i.e., representative fraction or RF. There are exceptions: some publishers have had a habit of using odd numbers of feet for the endpoint of the bar scale, making it hard to translate into an approximate ratio. I have before me a map of Kerrville, Texas, done by Western Map and Publishing Company in 1988 (OCLC record #22111239). Its bar scale runs from 0 to 6400 feet, and is divided into five segments (1280 feet each), with the first segment also divided into two 640-ft. lengths. In such cases, it is easiest to estimate the location of an appropriate figure with which to work (e.g., 1000 feet), by marking it on the edge of a piece of scratch paper. Then, if necessary, use a ruler to mark out further to a point equivalent to 1 mile or other base-line point on which to use the Map Scale Indicator. The Map Scale Indicator works on lengths of 1 mile, 1 kilometer, 1000 feet, or 1 degree of latitude, four methods of measuring rolled into one item!

Equivalency or verbal scale statements in the form of 1 inch = x miles are not too difficult to work with using the standard of 1 inch to 1 mile, which is the ratio 1:63,360 (because there are 63,360 inches in a mile). One can then enter 63360 into a pocket calculator and simply multiply by the number of miles. The result becomes the denominator of the representative fraction. Other times, the statement is less straightforward (e.g., "1 inch = 0.76 miles") and requires a bit more math on the calculator to get to the form of 1 (unit) = x (larger unit), but multiplying 63360 by the number of miles given, even if not a whole number, will result in a correct RF value.

When no scale is indicated the presence of degree markers in the map margin, indicating degrees of latitude, may allow for an estimation by using the 1 degree of latitude scale on the Map Scale Indicator.

When the map lacks both scale depictions and degree marks it is honest enough to simply put "Scale not given" in the record. Sometimes an estimate can be made from artifacts of surveying. On small-scale maps that show state boundaries within the United States, for example, one may find two borders known to be 1 degree of latitude long, namely the north-south portion of the borders of Utah and Wyoming and of Colorado and Nebraska. (These work because latitude lines remain equidistant, with 60 statute miles to a degree; this trick won't work with longitude lines, which converge as they approach the Earth's poles.) And on larger-scale maps, one may sometimes observe areas outlined by roads, property lines, or changes of land use which fall into uniform square patterns. This, especially in the western U.S., reflects the Township and Range system of breaking the land into square-mile blocks and may allow one to estimate the length of a mile on a map. One should make sure, however, that the area being measured isn't actually a quarter-mile square. Familiarity with the area mapped aids in estimating the scale by this approach.

In cases where no such estimate is possible, there remain various cases that call for different wording in the scale statement. A straightforward map of apparently uniform scale receives the aforementioned statement, "Scale not given." But some kinds of maps, such as aerial views and those of non-uniform scale (e.g., pictorial and cartoon maps), are given the statement "Scale indeterminable." Maps with certain projections which vary from equator to pole take the statement "Scale varies," except when a statement is present giving, for example, the scale at the equator. Note that the statement "Scale varies" only applies to a single map where scale changes when moving from the center of the map outward, as outlined in Rule 3B3 and its "APPLICA-TION" as outlined in *Cartographic Materials.*

Records for cartographic works containing more than one map at different scales are given individual statements up to 2 in number; beyond that (3 differing scales or more) they receive one statement with the phrase, "Scales differ" given.

In addition to scale the mathematical data area may also include a projection statement and a statement of coordinates, i.e., the extent of latitude and longitude (and in the case of celestial charts, equinox). Projection is a simple matter to handle. If a projection is stated on the map it is the next piece of information given, transcribed into subfield "b" of the 255 field, if no projection is given the cataloger moves on to applying coordinates when present. The prescribed order of entry for coordinates is: westernmost longitude, easternmost longitude, northernmost latitude, southernmost latitude. These can be straightforward in the case of topographic quadrangles and maps using them as base maps–and in fact this is where they are most useful for patrons, who may have a locality and its coordinates from a gazetteer. If the online catalog allows for searching by coordinate values, one may easily locate the needed quadrangle, if it is available in the catalog.

Many maps, however, may show coordinates in the margins but not at the points needed, namely the corners of the map (see Figure 1). Various commercial trail map publishers who use USGS topographic quadrangles for base maps do this. Estimates are usually possible, but often not to the precision in which the marginal, or corner, coordinates display (i.e., to the nearest second). One should not attempt to give a more precise coordinate than the map merits. This technique is known as extrapolation.

In entering coordinates the rule is to have each coordinate reflect the precision of the most precise coordinate. That is, if the latitude of the top margin is 45 degrees, 7 minutes, 30 seconds North and the bottom margin is an even 45 degrees North, the latter should be cataloged to include the minutes and seconds, and likewise the longitudes, even if they only include even degrees or minutes. For example:

255 Scale 1:24,000 |c (W 104^{0}22'30"–W 104^{0}15'00"/N 45^{0}07'30"–N 45^{0}00'00").

"Unbalanced" entry of the coordinates, by leaving off the zero-value minutes and seconds, has been a frequent error in past OCLC records. In fact, errors in entry of coordinates is so common that it contributed to the decision by the Library of Congress in a rule interpretation that coordinates are optional. When there is no clear display of coordinates on a map, therefore, the cataloger should not go out of his or her way to calculate them. (However, for coordinates of states in the United States a "tip sheet," MAGERT's Open File Report No. 86-5 by David A. Cobb, "United States Coordinates," is available for purchase from the Map and Geography Round Table.)

PUBLISHER AND PLACE OF PUBLICATION

While most map publishers tend to advertise themselves somewhere on the map, there are occasional exceptions. Certain maps received regularly in government documents libraries display no word at all as to who is responsible for them, only an encoded identification number. The most obvious of these are of the variety of excellent base and thematic maps produced by that most circumspect of publishers, the Central Intelligence Agency. Fortunately, one learns quickly to recognize these by appearance.

Among commercial map publishers it is common to have more than one agency from which to choose. Certain large companies market maps whose name appears in a copyright statement. An example is a map of Lufkin and Nacogdoches, Texas, dated 1993, which displays "Rand McNally" prominently on the cover (OCLC record #38498981). Yet Rand McNally may not have produced the cartography for this map; the copyright statement in the legend identifies the probable cartographic producer as that of the International Aerial Mapping Company of San Antonio. In such cases one may argue that Rand McNally is acting simply as a distributor, while the originating agency is the publisher, but the prominence of a company's name on the cover suggests that in this case, Rand McNally serves as the publisher. MARC 21 field 260 would thus appear as follows:

260   [Chicago] : |b Rand McNally, |c 1993.

There are occasions, however, when the prominently named agency is not really the publisher, but only serving as a distributor. Decades ago it was common to receive free maps at gas stations that bore numerous logos of the oil company. Small print on the map, usually near the legend, would invariably identify the company that produced the actual cartography. Such companies as Rand McNally or H.M. Gousha had arrangements to produce AND publish the maps for different companies, customizing them to give prominent advertising to the oil company, which had no role in the actual production and printing. The 260 field would reflect this:

260   [San Jose, Calif.] : |b H.M. Gousha ; |a [S.l.] : |b Distributed by Shell Oil, |c 1961.

The question of which company published a particular map has become more complicated in recent years with the consolidation of the map publishing industry into fewer companies. Certain publishers have a unique cartographic style that an experienced map librarian will immediately recognize. An example is that of Champion Map Corporation, late of Daytona Beach, Florida. Now that Champion has been purchased by Rand McNally a cataloger must resist the temptation to enter their name on new editions as being published by Rand McNally, even though the cartographic style seems to say, "I'm a Champion Map."

Another source of irritation concerns the place of publication, which does not always appear on the map. For some common publishers, like Rand McNally, one comes to know their headquarters by memory; hence the entry of Chicago–or Skokie, Illinois depending on the date of the map–in brackets in the first example above. In some cases, such as with the Arrow Map Company, a history of frequent relocation can make such suppositions questionable, especially when (as with Arrow) the exact year of publication is also omitted. Use of the "place unknown" abbreviation "[S.l.]" or even [United States] is never wrong in such cases. In relatively simple cases, as with the U.S. Geological Survey, which moved from Washington, D.C. to Reston, Va. in 1974, one may insert the place in brackets when missing and use a question mark if the map was published in the transition year.

ISSUES RELATING TO DATE

The date of publication can sometimes be hard to determine on a map. Historically, commercial maps have often lacked any clear statement of date of publication. One reason may have been to disguise the map's obsolescence. Certain publishers devised alphanumeric codes for their own reference, particularly Rand McNally and H.M. Gousha. Because some such maps included the date, map catalogers have been able to decipher the date from these codes. See the following publications, and a Web site, for lists of codes and their meanings:

- [Dates on maps in the] Western Association of Map Libraries' *Information Bulletin.* 6: 15 (1975), 13: 340-41 (1982), and 16: 280-82 (1985)
- "Dating maps." 1990. U*N*A*B*A*S*H*E*D librarian. 74: 13-14.
- Hoehn, Phil. 1997. "Date codes on maps." Berkeley: UC Map Library, E-mail message of 22 April 1997, message-id: .OSF.4.90.970422130051. 10825A-100000@library.berkeley.edu
- Website: Road Map Collectors of America, http://falcon.cc.ukans.edu/ ~dschul/rmca/codes.html

If a map lacks a publication or copyright date, even in code form, there may still be clues that will allow for an estimate, at least to the correct decade. A road map with an index of places often includes population figures for the larger cities in a state, which may be compared to the figures in the decennial U.S. Census to get an estimate of the decade of publication. In early road maps the date of the census from which the population figures was derived is usually given, also allowing for an accurate inference of date of publication at the decade level. For example, because of the high volume of Utah maps received by the Marriott Library for its historical map collection in the Special Collections Department I keep on hand a list of Salt Lake City's population for each decennial census year. A map cataloger is well advised to keep a list of dates of changes in, e.g., political boundaries, such as the breakup of the former Yugoslavia, for just such situations.

Other clues leading to the establishment of a date may appear within illustrations, textual notes describing points of interest in the area covered, and advertisements. Illustrations in tourist notes or advertisements that may help include: portrayals of old models of cars, clothing styles, or cityscapes with changing skylines or landmarks familiar to the cataloger. Careful reading of descriptive text occasionally turns up notes such as ". . .constructed in 1971," or ". . . to be completed in 1975." When such predictions get down to the expected month, it's often a clue that the map was prepared for publication within a year of the aforementioned date. Finally, business advertisements may give a year of establishment along with a banner reading something like "50 years of professional service" so that by combining the year of establishment and the "50 years . . ." one can determine a possible date.

Sometimes, instead of lacking dates, one may have a number of different dates to choose from, e.g., the cover may have one date while the legend bears another. Give priority to the date in the legend or other dates on the map itself over a cover date. The cover date may reflect the latest printing, but the map may have been published before this date and is now being re-released without editing. Under the principle that a reprint does not necessarily constitute a new edition, one may want to check OCLC carefully to see if the bibliographic record for the map may be available after all under the date in the legend. One can always add the cover reprint date in a note.

What if the cataloger receives a map bearing the next year's date? Sometimes maps reach the point of publication prior to the date printed on the map. The cataloger may then "correct" the date; however, s/he should include the printed date and follow it with [i.e., (the earlier date)] for the sake of future record users.

This is all well and good for the library that purchases the map early. But if the same map arrives after the new year begins, the cataloger may not recognize it as the map bearing the corrected date. Since access to bibliographic records by date in OCLC's WorldCat is derived from the fixed field date, the subsequent user may find, for example, a 1997 record and believe the map in hand, which reads "1998," is a new edition. It is up to the first cataloger to have been as complete as possible, and for the later cataloger to check these "near-hits" and see if the two maps may actually be the same.

PHYSICAL DESCRIPTION: NUMBER OF MAPS

Books tend to be relatively straightforward in their description, especially if they have good pagination: the count of numbered pages is simply copied into the cataloging record. Maps often refuse to conform to such straightforward logic. They may come as single sheets or as sets, but the former may constitute more than one map, while the latter may constitute a single map.

In the case of a single map sheet, there is usually one map that clearly predominates, known as the "main map," when more than one map is included on the sheet. Other "maps" on the piece, although being cartographic items in their own right, that are not counted for the purpose of the physical description (MARC 21 field 300, subfield "a") are insets, area maps, location maps, and contiguous extensions of the main map at the same or at a different scale. But, in some commercial maps there appear to be two main maps, perhaps one on each side. H.M. Gousha commonly published maps in this manner, with separate titles on each side. An example is found on OCLC record #25207449 (see Example 4), in which the map bears the panel title "Oakland and East Bay cities citymap. . . ." On one side, the title above the legend reads "Street map of Oakland . . ."; the map on the other side bears the title "Street map of Alameda County southern section. . . ." These two maps are geographically contiguous, and might appear to be one map on both sides of the sheet. However, the maps also have slightly different scales, as well as different titles. The description then reads, as in the 300 field below, "2 maps on 1 sheet : |b both sides, col. ; |c . . .":

EXAMPLE 4

12 110 2 H.M. Gousha (Firm)

13 245 10 Oakland and East Bay cities citymap : |b including Alameda, Albany, Berkeley, El Cerrito, Emeryville, Piedmont, San Leandro and adjoining cities.

14 250   1989 ed.

15 255   Scale [ca. 1:36,749]. 1 in. = approx. 0.58 miles.

16 255   Scale [ca. 1:38,650]. 1 in. = approx. 0.61 miles.

17 260   San Jose, CA : |b H.M. Gousha, |c 1989, c1988.

18 300   2 maps on 1 sheet : |b both sides, col. ; |c 85 x 43 cm. and 104 x 36 cm., sheet 107 x 46 cm.

19 490 0 A Gousha travel publication

20 500   Panel title.

21 500   Includes indexes, ancillary map of "San Francisco Bay area," inset of northern Richmond, and col. ill.

22 505 0   Street map of Oakland : including all or portions of Alameda ... San Pablo -- Street map of Alameda County, southern section : including all or portions of Fremont ... Union City.

23 500   "M-10-ER-429-S. M-10-ER-1024-S."

24 651 0 Oakland (Calif.) |v Maps.

25 651 0 Oakland Metropolitan Area (Calif.) |v Maps.

In other cases, the two sides are clearly the same map, starting on one side and finishing on the other, even though the map legend title may vary slightly, such as "xxx County, northern section" and "xxx County, southern section." The description in this case would read:

300 1 map : |b both sides, col. ; |c . . .

Map sets consist of multiple map sheets tied together by an overall title, numbering scheme, and (usually) a common scale. (For a more detailed description on cataloging map sets see the article *Cataloguing Series and Serials* following this one.) Sets of topographic maps for a country or region come immediately to mind. Since each sheet often bears a distinctive title (usually the name of a place on the map), and can be useful by itself, it has been traditional to catalog them together as a set of maps without specifying the number, for example:

300 maps : |b col. ; |c . . .

then including the potential number of maps in the set in a note, for example:

500 Complete in 451 sheets.

OR

500 Geographic coverage complete in 451 sheets.

(Why not include the total number in the 300 field? Because the actual number of maps eventually produced in the set may be less; they may still be in the process of being published, or the series may have been interrupted during its publication cycle and cannot be expected to be completed. Additionally, there can be produced over time different versions of single sheets, which would also impact the total number of sheets in the set.)

In other cases, however, the physical set may be considered one map on several sheets. This is usually the case for smaller sets whose sheets don't bear separate identifying titles and whose numbering follows a straightforward enumeration (e.g., sheets 1-10) rather than a grid-based scheme (e.g., the alphanumeric system used in the old "International Map of the World"). A single map published on multiple sheets is produced in this manner because they often portray a small geographic area at a large scale. While it may appear to be a map set because of the number of sheets involved, this type of single map is usually relatively easy to identify due to the incomplete borders on each sheet, a single title given (on one sheet or strung across multiple sheets), and the map information being printed only on one side of each sheet involved. Laying out the sheets side-by-side reveals the context of the single map. This would be described as:

300   1 map on 4 sheets : |b col. ; |c . . .

where it is necessary to give the number of sheets involved because without this knowledge one would assume one map on one sheet.

PHYSICAL DESCRIPTION: OTHER PHYSICAL DETAILS

Another aspect which can be important to the patron is a brief description in field 300, subfield "b," of other map-related details such as whether it was printed in black and white or using multiple colors. Additionally, this is where the cataloger can let someone know that a map on a single sheet continues from one side of the sheet to the other. Or, in the case of two or more maps on a single sheet, that the maps have been printed on both sides of the sheet. The order of preference for giving information in this area is, as noted in Rules 3.5C1-3.5C5 in *AACR2R* and further clarified in *Cartographic Materials'* "APPLICATION" to Rule 5C2a, layout on recto and verso, manuscript, photocopy, color, material, and mounting.

The most common pieces of information found in this part of the bibliographic record are "both sides" and/or "col." (i.e., colored), separated by a comma if both terms are used. "Both sides" indicates either that a main map begins on one side of the sheet and finishes on the other or that two or more main maps are printed on both sides of the sheet. Note that the word "colored" has a very distinct meaning in cataloging maps, it means that two or more colors were used in the printing process. A map may be printed using the color brown for the ink but if brown is the only color used then one *may not* use the phrase "col." in the record.

Another important notation used in this area is the word "photocopy," which when combined with the 500 note "Blueline print" (sometimes given as "Blue-line print" or "Blue line print"), indicates that the map is not the original but has been copied by machine. The note becomes important in that it tells one what method of reproduction has been used. The form of the note "Blue-line print" seems to be the most commonly used in cataloging practice and thus is recommended over the other two forms.

A less often used notation is "mounted on . . ." since contemporary sheet maps typically are not produced and published on any other material but paper. However, if the cataloger does have an item that was produced in this manner, e.g., the paper map was glued to a cardboard backing, this is the place to indicate it, and one can use a 500 note to bring out further details if necessary.

PHYSICAL DESCRIPTION: DIMENSIONS

Sheet maps break the usual pattern of measurement used for other formats. Whereas books are measured by their maximum dimensions (particularly height of the cover) map dimensions consist of the extent of the map itself within the neat line when one is available, i.e., the geographic area displayed, rather than the size of the sheet. There may be several centimeters of margin which are disregarded in calculating the map size to be entered in the physical description.

Often the map size is straightforward, as with rectangular maps encompassed by a single straight line, called the "neat line." Most USGS topographic sheets, and the geologic maps that use them as base maps, display such a neat line. Note however, that in some high-latitude maps the pole-ward border is visibly shorter than the equator-ward border, and the east and west borders taper toward the pole. Thus one must pay attention to which edges one measures, choosing the sides with the maximum dimensions.

So what is the difference between a neat line, a border, and a margin? Using the glossary found in *Cartographic Materials* a neat line is defined as "A line, usually grid or graticule, which encloses the detail of the map." A border is defined as "The area of a map which lies between the neat line and the outermost line of the surrounding frame." And a margin is defined as "The area of a map which lies outside the border." One can see that in a case where all three of these exist on a map the neat line is the innermost line, the border is the space between the neat line and the next line framing the neat line, and the margin is the space from the border line to the edge of the physical sheet. Maps often are printed with only one line surrounding the geographic area and this automatically becomes the neat line.

Not all neat lines are as "neat" as the name implies. What of double lines and other more complex borders (see Figures 1 and 2 as an example)? As mentioned above, because we are measuring the extent of the map the guiding principle is to use the innermost of such multiple lines.

Many maps don't always fit within simple, geometric borders however. Certain maps of watershed areas show quite irregular shapes. See, for example, Hydrologic Investigations atlas map no. 704, "Approximate potentiometric surfaces for the aquifers of the Texas coastal uplands system." One finds three maps of an irregular-shaped portion of Texas on the one sheet. With such odd-shaped maps, it is a matter of careful measurement to obtain maximum extent. (The cataloger who entered a record for this map into OCLC chose to just give the measurements of the sheet instead, which is quite acceptable as noted in Rule 3.5D1 in AACR2R and further elucidated in Rule 5D1e in *Cartographic Materials*.)

In many other cases a map may not have a neat line to measure from. In this instance the geographic area has been printed to the edge of the sheet in

FIGURE 1. Portion of CIA Map of Austria

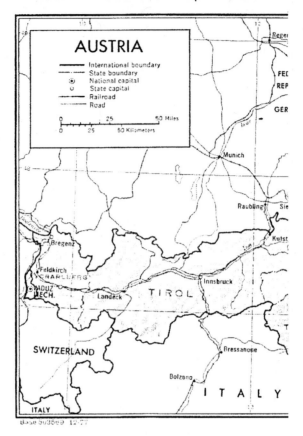

two or more directions. This is another situation in which Rule 5D1e of *Cartographic Materials* comes into play and one gives the sheet size as the dimensions of the map.

Maps on multiple sheets or on both sides of a single sheet present more complicated situations. Some commercial maps of an urban area have irregular (though usually linear) borders that do not line up with each other when turned over. They may also exhibit overlap between the parts. To correctly measure the maximum extent of the map one must determine the position of the "match line," or line of intersection between parts, and by how much the parts are offset. It is helpful when the publisher shows this match line on the map, but this isn't always provided.

FIGURE 2. Portion of Early Map of Hawaii

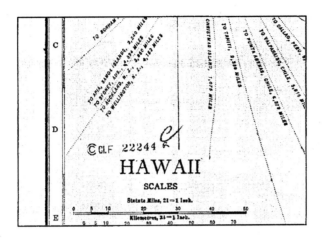

A rectangular grid used with an index for finding map features may also assist in measuring the map's extent. If, for example, the alphabetical guides along the east-to-west margin range from A to P on the north half of a map, and from Q to T on the south half, one may add the width of the north side to the measurement for sections Q through T on the south side to obtain the maximum left-to-right extent; and similarly for the top-to-bottom dimensions.

These calculations go hand-in-hand with determination of the number of maps, as discussed in the previous section. In addition, small insets may be considered part of the main map and need figuring into the calculation if they are at the same scale as, and contiguous to, the main map.

NOTES IN THE RECORD

As with a record for a book, the notes area in a map record can include all additional information that the cataloger thinks will help to identify the map for the patron. In the case of maps there are certain types of notes unique to the format as well as a prescribed order to giving them. (Keep in mind that the "patron" can also be another cataloger looking to find an existing record for a given map, so some notes that may uniquely identify a title would be of interest to the other cataloger and not necessarily to the library patron, such as a CIA code number.)

The notes prescribed to come first when the information is present are those pertaining to the nature and scope of the item. This includes identification of special map types, such as pictorial maps, bird's-eye views, and photomaps, which the cataloger needs to learn to identify.

Among the most important of these "nature and scope" notes for maps is for date of the data displayed, and next for describing relief, i.e., cartographic depiction of elevation and terrain. Recognition of relief features takes a little map-reading skill and patience. In some detailed maps one must scan a bit before locating such features. Among the most commonly found relief depictions are:

- Spot heights: a numerical value meant to identify elevation at a specific location, such as a hilltop, mountain peak or pass, or a town. When used to state the depth of a water body at a given point it is called a "sounding." (E.g., notice the numbers "1252," "1240," and "BM 708" in Figure 3.)
- Contour lines: lines that connect points of equal elevation. These are common on topographic maps and those maps that use topographic maps as their base, such as USGS geologic quadrangles. They must have numerical indications given at regular intervals to be called contours. (See Figure 4. The equivalent line used to portray depth in a body of water is called an isoline.)
- Gradient tints: zones of different color, usually identified by a key or diagram in the legend. The colors are typically separated by contour lines or formlines (similar visually to contours but not giving designated heights) or when used to depict depth, by isolines.
- Hachures: patterns consisting of short lines running up- and down-slope, often used on older maps to portray ridgelines in hilly or mountainous areas. Can also be used to show just the peaks of mountains or hills. In early maps these also took on the appearance of "woolly worms" or "caterpillars" due to the shape that they rendered. (See Figure 5.)
- Shading: a visual method of showing the shaded side of an elevated area such as a hill, ridge, mountain, or mountain range. The cartographer usually assumes the position of the sun as from the northwest, so the shaded areas typically are on the south and southeast side of the elevated area. (See Figure 1.)

Next in order are notes pertaining to the title, such as the source of the title that has been selected as the title proper. Formerly, this is where one would

FIGURE 3. Portion of 7.5 Minute Topo Quad Showing La Crosse and Winona County Borders

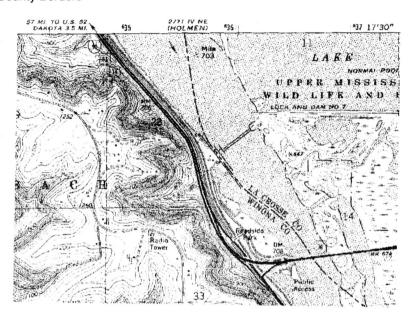

enter various notes on secondary titles, such as panel titles, but now these are included in the 246 field(s).

Then come notes pertaining to additional features such as indexes, ancillary maps, and marginal text. These notes typically begin with "Includes . . ." followed by a string of features. The order of listed features is not prescribed, but as a rule try to put them in relative order of importance as perceived.

Under this type of note are index maps, which take various forms. These range from a simple location map such as you find on many topographic sheets, to those that show the assemblage of sheets in a set, commonly known as an "adjoining sheets index."

Next is a note, in the case where the main map is only on one side of the sheet, known as the recto, listing types of information and "non-main" maps on the other side. The other side of the map is the verso and the prescribed order of the note is, e.g., "Text, key map, . . . and recreation table on verso."

Now come the bibliographic note (504) and contents note (505), if present. Bibliographical references often appear in the text of a thematic map, which may be found either in the map margin, on a separate sheet, or in an accompanying booklet. Contents notes are needed in the case of items containing

FIGURE 4. Portion of 7.5 Minute Topo Quad Showing Conestoga Woods/Conestoga Gardens

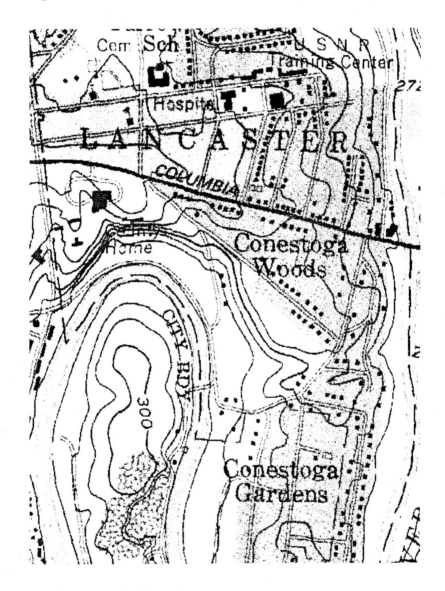

FIGURE 5. "Plan de l'Ile Borabora"

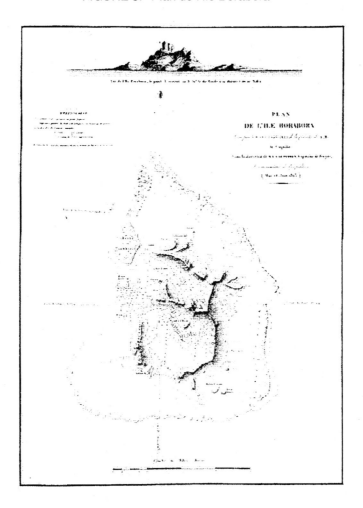

multiple maps with different titles when a collective title has been used as the primary title for the map, often a panel or cover title.

Then come notes for numbers and codes, such as publisher's numbers that may help to identify the map. These numbers are sometimes the sole source of the map's date, as with some Rand McNally and H.M. Gousha maps; or of the map's originating agency, as with Central Intelligence Agency maps (see the numerical code in the lower left corner of the map in Figure 1). It is important, primarily for other catalogers looking to conclusively identify a map or an edition of a map, to *always* record sets of numbers or alphanumeric designations as a quoted note.

Finally, notes regarding multiple languages are given, when applicable, in the 546 field.

CONCLUSION

The foregoing demonstrates some of the common issues which a new map cataloger is bound to encounter sooner or later as he or she begins working with this format of material. As one grows with experience, one will come to know just what to do in a given situation, and the best approach will seem intuitively obvious. I hope the hints and examples above help in the progress along the way to that point.

BIBLIOGRAPHY

Anglo-American Cataloguing Rules, Second Edition. Edited by Michael Gorman and Paul W. Winkler. Chicago: American Library Association, 1978.

Anglo-American Cataloguing Rules, Second Edition, 1988 Revision. Edited by Michael Gorman and Paul W. Winkler. Ottawa: Canadian Library Association, Chicago: American Library Association, c1988.

Cartographic Materials: A Manual of Interpretation for AACR2. Prepared by the Anglo-American Cataloguing Committee for Cartographic Materials, Hugo L. P. Stibbe, general editor, Vivien Cartmell and Velma Parker, eds. Chicago: American Library Association, 1982.

Map Cataloging Manual. Prepared by the Geography and Map Division, Library of Congress. Washington, D.C.: Library of Congress Cataloging Distribution Service, 1991. (Also available in electronic form as a part of the *Cataloger's Desktop* software.)

A "Natural Scale Indicator" is available from Dr. Clifford Wood at the Memorial University of Newfoundland (email address: chwood@kean.ucs.mun.ca). See the Website: http://www.mun.ca/geog/muncl.msi.htm

OTHER HELPFUL RESOURCES
FOR THE NEW MAP CATALOGER

Alpha, T. R. *Graphs Showing Linear to Fractional Scale Conversions for Maps.* (Miscellaneous Field Studies MF-I, 141). Reston, Va.: U.S. Geological Survey, 1979.

Bibliographic Formats and Standards. 2nd ed. Dublin, Ohio: OCLC, 1996.

Columbia Gazetteer of the World. Edited by Saul B. Cohen. New York, NY: Columbia University Press, 1998.

Geographic Names Information System (GNIS). http://mapping.usgs.gov/www/gnis
GEOnet Names Server. http://164.214.2.59/gns/html/index.html
ISBD(CM), International Standard Bibiliographic Description for Cartographic Materials. London: IFLA International Office, 1987.
Karrow, Robert W. *Manual for the Cataloging and Maintenance of the Cartographic Collections of the Newberry Library.* 2nd draft. Chicago: Newberry Library, 1977.
Larsgaard, Mary L. *Map Librarianship: An Introduction.* 3rd ed. Englewood, Colo.: Libraries Unlimited, 1998.
Saye, Jerry D. and Sherry L. Velluci. *Notes in the Catalog Record: Based on AACR2 and LC Rule Interpretations.* Chicago: American Library Association, 1989; OR Saliger, Florence A. and Eileen Zagon, *Notes for Catalogers: A Sourcebook for Use with AACR2.* White Plains, NY: Knowledge Industry Publications, 1984.
Technical Bulletin No. 212: Format Integration Phase 2, January 1996. Dublin, Ohio: OCLC, 1996.

Cataloguing Map Series and Serials

Velma Parker

SUMMARY. This article defines and outlines the characteristics of map series, map sets, map serials, maps in multiple editions and multi-sheet single maps. Brief instructions on sources of information and general methodology used in gathering information prior to creating the entry are presented. The different methods which may be used for cataloguing series and serials are explored. There is also a brief section on cataloguing bi- and multi-lingual works in a bilingual environment. For each relevant area of description, instructions and examples are given to illustrate problems. Sections on analysis (including multi-level cataloguing). *[Article copies available for a fee from The Haworth Document Delivery Service: 1-800-342-9678. E-mail address: getinfo@haworthpressinc.com <Website: http://www.haworthpressinc.com>]*

KEYWORDS. Map serials, map series, multi-sheet single maps, cataloging, bilingual works, multi-level cataloging, analysis

INTRODUCTION

Map series and serials are often confusing so the very first thing that one must do is to decide what you have. Differentiating among multi-sheet single

---

Miss Parker is a Standards Officer in the Visual and Sound Archives Division, National Archives of Canada, 395 Wellington Street, Ottawa, Ontario KIA 0N3, Canada. She is a longtime member of the Bibliographic Control Committee of the Association of Canadian Map Libraries and Archives. She receieved her Honours BA in Geography from the University of Western Ontario, London, Ontario, her Masters of Library Science from the University of Western Ontario, London, Ontario, and her Dip. in Education, Type A high school teaching certificate, from Althouse College, University of Western Ontario, London, Ontario.

[Haworth co-indexing entry note]: "Cataloguing Map Series and Serials." Parker, Velma. Co-published simultaneously in *Cataloging & Classification Quarterly* (The Haworth Information Press, an imprint of The Haworth Press, Inc.) Vol. 27, No. 1/2, 1999, pp. 65-101; and: *Maps and Related Cartographic Materials: Cataloging, Classification, and Bibliographic Control* (ed: Paige G. Andrew, and Mary Lynette Larsgaard) The Haworth Information Press, an imprint of The Haworth Press, Inc., 1999, pp. 65-101. Single or multiple copies of this article are available for a fee from The Haworth Document Delivery Service [1-800-342-9678, 9:00 a.m. - 5:00 p.m. (EST). E-mail address: getinfo@haworthpressinc.com].

maps, map series, map sets, map serials, and maps in multiple editions can pose problems as the differences are not always clear. In this article each will be defined and their characteristics outlined. As well, methods of cataloguing series and serials will be explored including brief explanations on analysis, access points and bilingual cataloguing.

Multisheet Single Maps

This section is included here solely to aid in identifying which sheets belong to a multisheet single map and which belong to a series/serial. A multisheet single map is considered to be a monograph which is issued in a number of separate physical parts; in other words, it is a multipart item. Thus, it will be catalogued as a monograph not as a serial or series.

Definitions

One complete map, printed in sections on several sheets of paper which may be assembled.[1]

Multipart item. A monograph complete, or intended to be completed, in a finite number of separate parts.[2]

Characteristics

- border complete only when sheets assembled
- title may only be fully seen when map is assembled
- other text (e.g., place names) may run across several sheets
- individual sheets may carry a number or other designation (e.g., north, south) to aid in assembly
- technical specifications exactly the same
- same publisher
- same date of publication/issuance
- revised simultaneously, or over a very short period of time
- assembly diagram for the whole map usually present on each sheet
- usually a small number of sheets
- sheets to be used together and not independently
- not usually available separately

Map Series

Map series form the bulk of most map collections and the bibliographic control of the series as a whole and of its various parts is extremely important. The more familiar series are the large topographic map series and the thematic series, notably the geological and hydrographic series.

Definitions

A number of related but physically separate and bibliographically distinct cartographic units intended by the producer(s) or issuing body(ies) to form a single group. For bibliographic treatment, the group is collectively identified by any commonly occurring unifying characteristic or combination of characteristics including: common designation (e.g., collective title, number, or a combination of both); sheet identification system (including successive or chronological numbering systems); scale; publisher; cartographic specifications; uniform format; etc.[3]

Another definition is as follows:

A map series . . . comprises a number of maps relating to a defined area of the world, published sequentially, based on a common grid or projection, drawn to a common scale, and using common cartographic symbols to convey information. The combined sheets of the series, possibly as many as several thousand sheets, theoretically form one large map, but each sheet is also a complete "self-sufficient" cartographic product. Because of adjustments related to the curvature of the earth, it may not be possible to actually join more than a few contiguous sheets.[4]

Types and Characteristics of Cartographic Series

Different types of cartographic series have been identified.[5] More complete information, with examples may be found in the sources listed in the endnote.

a. Contiguous-Area Map Series

This type is both the most common and the most familiar as every map collection has at least part of its own country's topographic map series in its collection. Most of these series are general geographic maps (normally topographic but some planimetric–usually the older ones). In addition, there are a few thematic contiguous-area map series as well (e.g., geology, land use). They are also the most abundant in terms of the number of sheets, which may range from two to hundreds of thousands.

- primary intent of the publisher is that each sheet may be used independently, thus each has characteristics of a single map (e.g., title, neat line, border, legend, margin notes, publication information); trimmed sheets can in theory be joined to form a larger map, but when a large number of maps are to be fitted together, matching presents some difficulties and the edges may not match perfectly (see second definition above)

- usually published by a single authority (e.g., the government agency responsible for mapping although its name may change over time); in a few cases several agencies may produce a series in co-operation (e.g., *International map of the world 1: 1 000 000*)
- may have the series title in addition to the sheet title on each sheet
- may have a sheet designation consisting of either a sheet numbering system (numeric and/or alphabetic, e.g., *3284; 76-A-6-III*) or a sheet titling/naming system (e.g., *Monteagle quadrangle, Tennessee*) or both (e.g., *31/G5: Ottawa; Sheet 1459: San Raphael*)
- usually long-term projects
- sheets are not usually all revised at same time (for some countries it may take many years to get out the first edition of all sheets; smaller countries may be able to initially produce and then revise all sheets at the same time or within a short time period)
- should have a graphic index usually issued and available separately, or may be on verso of each sheet
- index map on each sheet shows contiguous sheets, not all sheets in the series
- usually available separately
- common technical specifications (which may change over time for long-lived series) for sheet size, scale, projection, layout, typography, colour schemes, symbolization, etc., which make the individual sheets recognizable

Canada 1:50 000
Karta mira 1:2,500,000
National topographic series, 1:506 880 : [Canada]
Canada Land Inventory, land capability for wildlife–ungulates

b. Same-Area-Different-Themes Map Series

This type is rather rare. There may be some difficulty in deciding if it is an unbound atlas or a map series. The decision is based on publisher's intent as expressed on the item, or in publications or communiques issued by the publisher, and on whether it is bound or not.

- generally common technical specifications such as sheet size, scale, projection, typography, layout, etc.
- sheets may be published at the same time or over a specific period
- sheets may be used independently
- in some instances the sheets are not available separately

- single publisher

East Africa 1:4,000,000. D.O.S. misc. 299 A-G
(Sheets cover mineral resources, forest and game resources, vegetation, geology, soils, physical features)

c. Different-Areas-Common-Theme Map Series

This type, too, is rare. Again it is similar to an unbound atlas and the publisher's intent must be the deciding factor as to whether to catalogue it as an atlas or a series.

- may have different scales and projections
- non-contiguous areas
- sheets may be published at the same time or over a specified period
- sheets may be used independently
- in some instances the sheets are not available separately
- single publisher

Emergency Measures Organization, urban analysis series : [Canada]
(Includes Vancouver (32 sheets) and Toronto (32 sheets of each E & W)

d. Successively Numbered or Repetitively Titled Map Series

These groups of maps often cover a specific geographic area but the sheet lines are not regular, as are those in the contiguous-area series. Coverage at any given scale is spotty and there may be overlaps or gaps. When numbering is used, it has nothing to do with the geographic area covered as does the numbering for topographic series, but is the next available number, so that a sheet or group of sheets on Vancouver Island may be next to those for northern Ontario.

- maps monographic in nature
- high degree of independence of individual maps
- independent publishing history for updates, revisions and reprints
- sheets may be used independently
- normally single issuing body

Geological quadrangle map, GQ-
(Published by USGS)

Geological Survey of Canada. ["A" series maps]

Bartholomew world travel map

e. Part of Book Series

Some maps and atlases are published as part of a book series. These usually have the title of the series and the numbering of the book series.

- display bibliographic information similar to the books in the series

  (Special publication / State Bureau of Mines and Geology)
  (*Traced as:* Special publication (Montana. State Bureau of Mines and Geology))

  (Report / British Columbia Soil Survey; no. 44)
  (*Traced as:* Report (British Columbia Soil Survey) ; no. 44)

Map Sets

This is often considered a synonym for map series. Certainly, as far as cataloguing and classification is concerned they are normally catalogued as a unit as are map series. However, there are some materials which might be considered a set but not a series. Examples of this later case could be maps produced and published together in a portfolio to illustrate a theme such as transportation, or sets accompanying a report and consisting either of specially prepared maps, or of previously published maps gathered together from various sources. A good example of this latter case are the maps assembled for legal cases such as boundary disputes. These would be kept together as a set rather than dispersed through the collection.

Definitions

A set . . . is a group of distinctive maps, frequently published at the same time and having a common origin, a common theme, or a common [geographic] area. Such maps may be issued as a folio, or bound.[6]

[National energy transportation systems in the United States]
(*Set of 19 maps from the National atlas*)

Map Serials

Map serials are not as common, but every map collection has some. The map serials most of us are familiar with are the provincial or state road maps usually issued annually, the weather maps (daily, weekly, etc.) and some aeronautical charts. Some atlases are published as serials as well. Perhaps one of the better known of these is the *Rand McNally commercial atlas and marketing guide* with its road atlas supplement the *Road atlas of the United States, Canada and Mexico.*

Like text serial cataloguing, map serials present specific problems particularly as they are not often encountered. Some of these problems and proposed solutions will be presented in the next section. The examples will be primarily Canadian as these are what I have most frequently cataloged and are thus most familiar to me. A few will be borrowed from other venues.

Definitions

Cartographic materials: a manual of interpretation for AACR2 borrows the AACR2 definition of "serial" which reads as follows:

> A publication in any medium issued in successive parts bearing numerical or chronological designations and intended to be continued indefinitely.

Since few cartographic serials have other than what may be loosely called a chronological designation, the definition following may be more apt:

> A cartographic serial is a publication in any medium and which may be said to exist when the map, atlas, etc., is issued at regular intervals within a programme of continued publication.[7]

As mentioned above, most cartographic serials have a date which is somewhat analogous to the chronological designation mentioned in the definition. However, the presentation may not be in the more formal mode found in textual serials. Unless the serial uses words like "annual," "weekly" in the title or elsewhere on the sheets, there is often no indication that it is indeed a serial. In this latter case, a little detective work will usually find the answer. Usually this information is found in the catalogues or information sheets available from the publisher.

> FIR plotting chart : [Canada]
> Enroute low altitude, Canada and North Atlantic
> Perly's BJ map book Toronto and vicinity

Multiple Editions of Single Maps/Atlases

It is a matter of judgement when to consider a frequently updated map or atlas as a serial. Often annual road atlases or maps are considered as serials.

Larsgaard cites a case where material revised regularly but not annually is considered by some cataloguers to be a serial (e.g., U.S. Forest Service maps of the various national forests).[8] Even though the annual may be considered a serial, there may be instances where it will be preferable to catalogue each one individually, especially if the differences among issues is great requiring complex notes and many name and/or title added entries.

HOW TO SET ABOUT IT

In this section, we will be talking about the preliminary stages in acquiring the information necessary actually to create a catalogue record. Unlike books, one cannot sit down at the computer and simply input the information found on the item. The fact that the maps are often numerous and very large in size prevents this. Also, not all publishers put on the maps the necessary information, relying instead on their publicity and other information blurbs and, I frequently think, what they fondly (but mistakenly) hope is the intuitive knowledge of the users. All this to say that the preparatory work may take some time and may require research.

- Make an exploratory examination of the sheets (or a random sampling for very large series/serials) to identify a title and what body is primarily responsible for the series or serial.
- Do a bibliographic search for copy cataloguing and compare your information to that record.
- Even if you are going to use copy cataloguing, it is a good idea to go through all of the sheets, or as many as you possibly can, and make notes. I usually have a large, blank sheet of paper which is divided up into sections for variations in title, statements of responsibility, name of publisher, dates of publication, and also a section for notes on common features as well as exceptions. I include the date ranges for all variations in title, statement of responsibility and publisher names. Going through the series in this manner will also allow you to see if there are any experimental sheets, sub-series, or misfiled maps from other series or deviations which require either removal of the sheet from the series/serial or a note.
- For contiguous-area series, if you do not have a graphic index and have been unable to obtain one from the publisher or another map collection, then this could be made at the same time.
- Sort the maps, if required, by sheet designation, or, for serials, by date or any other logical manner.

Once this is done, you have the raw material to begin building a catalogue record.

Sources of Information

- The first source is the maps themselves (see above)
- Focus secondly on cataloguing copy.
- The most useful source is, of course, the publisher of the series/serial. If you can, archive the lists, catalogues and information sheets that are issued by the publisher, particularly those which give information about the serial itself and its publication history. Successive issues of such lists are often helpful in solving riddles.
- If you are fortunate enough to reside within economical phoning distance of the publisher, then a telephone call may answer the more difficult cases. I have had my share of playing telephone tag with government staff in unravelling mysteries but it has always paid off.

Chief Source of Information

Series and Sets

The following general definition of chief source for cartographic material will apply to map series and sets.

> The cartographic item itself; when the item is in a number of physical parts, treat all parts (including a title sheet) as the cartographic item itself.[9]

This means that all of the sheets in the series as a whole would be part of the chief source of information. Practically speaking, for large series which are not complete and will not be complete for many years to come, those sheets which comprise the series at the time of cataloguing will form the basis of the entry. Changes that occur over time may be recorded in notes, or, if there is time and resources, the whole entry may be revised, depending on the extent of the changes and how access would be affected.

Serials

For serials, we must look at several rules in AACR2R. Chapter 12 states that the chief source of information for textual materials is the title page of the first available issue (12.0B1) and for "nonprint"[10] serials, we are referred by rule 12.0B2 to the 0B rules of the relevant chapter. In addition, 12.1B8 states that "if the title proper of a serial changes, make a new description."[11] Again in 21.2C1 "if the title proper of a serial changes, make a separate main entry for each title."[12] According to 21.3B, a new entry is also made if there are changes in the name of the body responsible for the series or if the responsibility for the serial changes from one body or person to another body or person.

How do we apply all this to cartographic serials? We must remember that when the chief source for cartographic material was written, serial cartographic material was not considered. After discussions with serials cataloguers here and at the Library of Congress, it seems that the instructions in 12.1B1 and 21.2C1 apply no matter the chief source. Thus the first sheet (or the first available sheet) in the serial would be the chief source with a new entry every time the title changed significantly (see the section below on Successive entries, Serials) or when 21.3B applies.

Then, too, there are the atlases published in serial form. Since atlases have title pages, it is reasonable to assume that the same chief source which applies to books in serial form would also apply to atlases. Thus 12.0B1 would indeed apply to serials in atlas form making the title page the chief source of information.

12.0B1. Printed serials

Chief source of information. The chief source of information for printed serials is the title page (whether published with the issue or published later) or the title page substitute of the first issue of the serial. Failing this, the chief source of information is the title page of the first available issue. The title page substitute for an item lacking a title page is (in this order of preference) the analytical title page, cover, caption, masthead, editorial pages, colophon, other pages. Specify the source used as the title page substitute in a note . . . If information traditionally given on the title page is given on facing pages, with or without repetition, treat the two pages as the title page.

CATALOGUING METHOD

Monographic versus Cataloguing as a Unit

Most series, sets and serials are catalogued as a unit. However, this may not always be the most advantageous or efficient method for the researcher. For cartographic serials, more complex cases may be better served by treating them as monographs or monographic series. For example, where there are frequent variations in title, changes in the name of the body responsible, or ambiguous edition numbering, this may be the easiest way to get bibliographic control over the material. The decision must be weighed against the resources available to continue doing this over the life of the serial. On the other hand, the resources needed to revisit and update catalogue entries for serials catalogued as a unit must also be considered and the decision based on both sides of the issue. Certainly, if only a few issues are held and there are large gaps between issues, or if there are many variations in the form or order of the title elements, or if the publisher or the publisher's name changes over time (as often happens with government agencies), then cataloguing each one separately may be more useful. (A case in point is road maps, which certain institutions catalogue as monographs.) This will avoid making extensive notes and numerous added entries.

Series are usually catalogued as a unit. However, control of the sheets and their editions are of concern. In times past, this control has often taken the form of simply marking the sheets held on a graphic index. In a few instances, some collections went a step further by enlarging the graphic index so that the various editions could also be indicated. Increasingly, the individual sheets are also being catalogued and the records entered into a database which may be a separate database. Some institutions develop their own database using software such as Inmagic. Others find it more convenient to use packages developed elsewhere. An example of the latter is GEODEX (GEOgraphical inDEX system for map series) developed by Christopher Baruth of the American Geographical Society Collection (University of Wisconsin-Milwaukee) in 1988. When separate entries are created, no matter the system used, the record for each sheet must be linked in some manner to the description of the series as a whole. For a further discussion, see the section on Analysis below.

Sets are often catalogued as a unit. Because the maps may be from various sources, this is a case where more complete descriptions of each map may be advantageous. This may be done by using one of the analytical methods outlined in the section on Analysis below.

In general, catalogue as a single item:

- a single sheet or, in the case of an atlas, a single volume, of a serial if that is all that is held or will be held
- a looseleaf publication where new sheets are interfiled, or updated sheets are replaced, or outdated sheets are discarded (e.g., *Tübinger Atlas des Vorderen Orients* or the *Atlas of eastern and southeastern Europe*)
- individual issues which have separate titles (e.g., *JRO topic map : JTM*).

In general, catalogue as a unit most cartographic series, serials and sets with:

- continuous numeric designations (e.g., US *Daily weather maps. Weekly series*)
- same title and are identified by date designations only
- titles that imply continuing publication

- statement of frequency in title or elsewhere
- with numeric or chronological designation in the title or elsewhere on the map
- item intended to be published indefinitely with a frequency of triennial or more frequent
- ISSN
- new editions; frequently issued editions (e.g., annual, biennial, triennial); if the frequency is four or more years, treat as a monograph
- loose-leaf publications where sheets are added but not interfiled (if all pages are withdrawn in their entirety and new pages are added which completely replace them on an on-going basis, consider this a serial) (e.g., *Atlas Rzeczypospolitej Polskiej = Atlas of the Republic of Poland* ; *Maps on file*).

Consider as part of the unit special issues bearing only a designation that is part of the regular numbering for the serial.

Successive Entries

This may apply to both series and serial cartographic publications. In all cases, notes should be provided on all entries so the user of the catalogue may trace the history of the series or serial.

Series

As we have said above, AACR2R rule 12.1B8 says to create a separate entry if the title proper of a serial changes. Since cartographic series are not serials, technically this does not apply. But, are there cases when separate entries should be made? The answer is yes. Not all publishers assign titles to their series and if they do, they may change them slightly or significantly over time. This should not be the major or the only factor in deciding to create a new entry although it should not be completely ignored. Do some research to see what method is best for the researcher. However, consider creating a separate entry when there are changes of scale, numbering, or an increase in the geographic area covered.

Changes in Title

If the change in title is coupled with other changes such as that of scale and certainly of sheet numbering systems, then a new entry must be made and, if appropriate, a note added indicating the title, etc., of the continuing series. However, there are some series which carry different titles on some sheets yet they are part of the same series and are not to be considered as separate series nor as continuations. Certain of the topographic maps of Africa done by the Institut géographique national (France) are good examples of this. In their 1968 catalogue, the I.G.N. states that the *Carte de l'Afrique de l'ouest au 1:200 000* consists of both topographic and planimetric maps. In addition, the title on the six-colour planimetric map is *Fond planimétrique des régions désertiques* or *200 000 Sahariens* and that the title for the three-colour planimetric maps is *Croquis des régions Sahariennes.*[13] These variant titles should be included in the variant title note.

Changes in Scale

If one of the primary cartographic features of a map series is the scale, as it is for topographic map series, a change in scale is significant. In some cases, when the scale alone changes (e.g., from imperial 1:63 360 to metric 1:50 000) and the sheet designation system remains the same, the map producer may consider this as the same series and may continue the edition numbering through from the imperial system. For example, when the change from imperial to metric occurred for the Canadian topographic maps, some of the new sheets continued the edition numbering from the imperial sheets whereas other sheets in the same series began new edition numbering with the change in scale. Since scale is recorded in the catalogue record as a descriptive element and as part of the call number, it is usually more efficient to create a new entry for the new scale.

> Canada 1:506 800
> Canada 1:500 000

For other series, the scale may not be as significant. This is often the case for thematic series where the scale may vary considerably. In such cases, scale is not a factor in creating a new entry.

> Geophysical series (aeromagnetic) = Série des cartes geophysiques (aéromagnétiques)
> *Note:* Most maps before 1974 at scale 1:63 360; after 1974 at 1:50 000; some maps at scales 1:12 000, 1:24 000, 1:31 680, 1:125 000 and 1:126 720.

Numbering Changes

When a numeric or alphanumeric sheet designation system changes, then a separate entry should be made. The very obvious difficulty of interfiling will require a separate filing sequence. Initially, the new sheets will carry both the new and the old numbering systems but eventually this practice will be dropped. In the long run, it is more advantageous to create the new entry right away and keep the two filed separately. Thus, sheets with both new and old numbers would be filed in the new sequence.

> Geological Survey of Canada.
> [Publication number topographic map series]. 1908-1927.
> *Note:* Later editions with publication numbers part of: Geological Survey of Canada. [Topographic maps, "A" series].

> Geological Survey of Canada.
> [Topographic maps, "A" series]. 1910-1949.
> *Note:* Some sheets also have the publication numbers of the series: Geological Survey of Canada. [Publication number topographic map series].

Changes in Geographic Coverage Coupled with Changes in Title

When this happens, the coverage is usually expanded to include a larger area. This usually results in a new entry being created as there are several parts of the description which may be affected, namely the co-ordinates, the subject access points, call number, etc. For example the *Winnipeg navigation plotting chart* which covers the area between the middle of Alberta to the northwest corner of New Brunswick and only as far north as 59° is continued by *Canada plotting chart : AIR 1615,* which covers almost the whole country to 78°N.

Changes in Geographic Coverage with No Change in Title

In the cases that I have found where this happens, there has usually been a reduction in the area covered. Certain sheets are dropped. Such changes should be noted but no new entry created.

> Mercator plotting chart : [Canada]
> *(Five out of eleven areas are no longer produced; for the remaining, formerly single sheets are paired and printed back to back)*

> Maritime plotting chart
> *Note:* Series in 12 sheets. Atlantic area sheets 1-6 begun [1961]. Pacific area sheets 7-12 begun 1968. Only 8 sheets revised up until 1980. Since 1981 only sheets 7 and 8 rev.

Serials

Changes in Title

According to 21.2A, if there are changes in the first five words of, or the order of, the title proper (six words if the first word is an article) then a new entry is prepared. However, not all changes in the first five (six) words will lead to a new entry being prepared.

Do NOT count as changes:

- addition, deletion or change of an article, preposition or conjunction
- changes such as spelled out words for abbreviations, for signs or symbols, for numbers or dates or vice versa
- singular versus plural
- different spellings of a word including hyphenated words versus unhyphenated words, and one word versus two word compounds
- arabic numerals versus roman numerals
- initialisms and letters with separating punctuation versus without separating punctuation
- changes after the first five (six) words when meaning does not change or indicate different subject matter
- addition or deletion of words linking the title to the chronological or numeric designation (e.g., *for the year ending June 30*)
- addition or deletion of name of issuing body and any grammatical connection at end of title, or changes from initialism to spelled-out form, or from a longer to a shorter form
- addition, change or deletion of punctuation
- changes in numbering without a change in title

Changes in Statement of Responsibility

In AACR2R, 21.3B instructs us to create a new entry, even if the title proper remains the same in the following conditions:

- the corporate body in the main entry has changed its name and a new heading is required
- the person or corporate body is no longer responsible for the serial.

AREAS OF DESCRIPTION

Title and Statement of Responsibility Area

Series/Sets

Use a collective title and not sheet (volume) titles if such are present. Normally record the title based on the earliest available sheet (excluding experimental or preliminary issues). If, however, the title changes after the first few sheets and remains constant thereafter, then choose this more common title. Make a note explaining this and trace the title found on earlier or other issues.

In some cases, series/sets will be published without a collective title, the only titles being the sheet titles; if a title appears on the index map or in publisher's blurbs, use that. If the index is not part of the series itself but is a separate publication, then enclose the title in square brackets. Titles taken from publishers' sources are also enclosed in square brackets. If no title can be found, create one following the guidelines given in *Cartographic materials : a manual of interpretation for AACR2,* 1B7 Application 2. (These guidelines are summarized below under the heading "Supplied titles.") For sets of maps which are part of a larger work and which are catalogued separately from that larger unit, a title may be found in the report or may be supplied based on the report title.

Sheet Titles

When doing analytics for the sheets in a series or serial, there is special consideration in recording the title when there is both an alphanumeric designation and a sheet title. Both parts are considered as title proper. *Cartographic materials* (E.4B1) says "if a sheet has both a sheet title and a sheet number, record the sheet number first and separate [the rest of] the title from it by a colon, space.

> 31/G5: Ottawa
> 153: Oran
> ND-28-XIII: Dakar

Serials

Again, use the collective title and not the sheet title if such exists. AACR2R rule 12.1B7 states:

> If the title includes a date or numbering that varies from issue to issue, omit this date or numbering and replace it by the mark of omission, unless it occurs at the beginning of the title, in which case do not give the mark of omission.
>
> General highway map . . . Wyoming
> County map series (topographic) . . . Virginia

There is a proposal currently before the Joint Steering Committee for the Revision of AACR to expand this rule to include names as well as dates and numbers. If this is passed it will mean that examples such as the following will be covered by the rule:

> Land use of [name of] area, Nebraska

could be recorded as

> Land use of . . . area, Nebraska.

Supplied Titles

If the series does not have a title proper, supply one in square brackets (CM 1B7 app. 2, p. 23)

- from publisher's literature, or publisher's index, accompanying reports
- failing this, construct one including, preferably in this order, as many of the following as required to identify uniquely the series: area subject, scale (recorded as an RF and only if consistent on all sheets), series number if applicable, corporate body if required as a means of distinguishing it from others, edition, date of publication.

> [Publication number topographic map series]
> [Topographic maps, "A" series]
> [Canada, aeronautical chart, 1:500 000]
> [R.C.N. air plotting charts, northwest Atlantic]
> [Township plans of the Canadian West]
> [Aerial navigation charts for northern Canada]
> [Maps of east and west Florida]
> [Topographic map of Greece with British military overpint]

Statement of Responsibility Area

Series

Only formal statements of responsibility found on the majority of sheets are recorded. If the form of the name changes over time, or if the author body changes, then do not record this here. Such information may be recorded in a note, especially if added entries are required.

> Navigation plotting chart : [Canada] / prepared by the Surveys and Mapping Branch

Inventaire des terres du Canada, possibilités des terres pour la faune–ongulés / cartographie réalisée par l'Institut de recherche sur les sols, Direction de la recherche, Ministère de l'agriculture du Canada . . . = Canada Land Inventory, land capability for wildlife–ungulates / cartography by the Soil Research Institute, Research Branch, Canada Department of Agriculture . . .

Geological map of the circum-Pacific region / Circum-Pacific Map Project; map compilation coordinated by Warren O. Addicott and George Gryc.
Note: Cartography variously by Karen A. Johnson, Lori Garza and James L. Zimmermann.

Canada Land Inventory, critical capability areas / cartography by the Soil Research Institute, Research Branch, Agriculture Canada with the support of the Lands Directorate . . .
Note: Cartography on later sheets by Land Resources Research Institute.

In more complicated instances, when the title of a series which is produced by one agency varies over time and the form of the name of the agency also varies, then record the statement of responsibility found on the sheets bearing the title used as the title proper. If the form of the name varies for the title chosen as title proper, either choose the earliest one and record all of the other variations in a note or record them all in a note.

For series produced co-operatively, if there are three or fewer agencies, technically all three could be recorded here even though they do not all appear on any one sheet (see the definition of chief source of information). *Cartographic materials* Appendix E1B2 goes further to state that if there are four or more, that one may be recorded here and the rest in a note. However, I am not comfortable with this and so would tend to record all statements of responsibility in a note and trace them as appropriate.

For sets catalogued as a unit where the statement of responsibility is different for each title, these are recorded only in descriptions of the individual sheets. The individual sheets may be listed in a contents note or separately in analytical entries.

Serials

The information given above would apply to serials as well. In addition, there are rules in AACR2R which should be noted.

12.1F3. Do not record as statements of responsibility statements relating to persons that are editors of serials. If a statement relating to an editor is considered necessary by the cataloguing agency, give it in a note (see 12.7B6).

Edition Area

Edition statements are recorded only when it applies to the unit as a whole. Sheet editions would be recorded in the separate description for the sheet.

Gambia 1/50,000.–Provisional ed.

Ordnance Survey of England and Wales, 1/4 inch to one mile, third edition.–Civil air ed.

Mathematical Data Area

Scale

In cases where the scale changes slightly for later issues there are several options:

i) add an extra 255 MARC field for the second scale and make a note when this change occurred,

or

ii) just leave the original scale in 255 and make a note that the scale changed after a certain issue,

or

iii) catalogue each edition separately.

The first option is fine if there is only the one change in scale (but this is by no means certain for on-going serials). The second and third would be more useful choices if there are a number of small variations.

Projections

In some cases, the projection varies over time. For example, up to 1942 the Bonne projection was used for the *Carte d'Algérie au 1:50 000,* and later maps used the Lambert conformal conic projection.[14] This information is more comprehensibly expressed in a note.

Coordinates

For a contiguous-area series which is incomplete at the time of cataloguing, record the coordinates for the full extent of the intended complete coverage for the series. Use information provided by the producer such as an index sheet showing intended coverage.

For series covering non-contiguous areas (e.g., one showing various airports across the country) do not record coordinates in this area. If coordinates are deemed to be useful to the researcher they may be included in a contents or other note.

Numeric and/or Alphabetic, Chronological, or Other Designation Area

This area may be used, but I have found that in many instances it is somewhat repetitious of information already in the entry. Where I have found it to be useful is in recording edition information as instructed in AACR2R 12.2B2; the following examples are serials.

12.2B2. Give statements indicating volume numbering or designation, or chronological coverage (e.g., *1st ed., 1916 ed.*) in the numeric and/or alphabetic, chronological, or other designation area (see 12.3). Give statements indicating regular revision (e.g., *Rev. ed. issued every 6 months*) in the note area.

> 1st ed. (1976)-3d ed. (1978)
> *(Canada plotting chart : AIR 1614)*
>
> 1st ed. (1976)- .
> *(Canada plotting chart : AIR 1615)*
>
> 1st ed. (Jan. 1978)- .
> *(Section of Canada plotting chart : (AIR 1615)*
>
> Feb. 28-Mar. 6, 1983-Nov. 12-18, 1984.
> *(Daily weather maps. Weekly series. 1983-1984)*
>
> Aug. 19-25, 1996- .
> *(Daily weather maps. Weekly series. 1996-)*

I have not found any examples with numeric and or alphabetic designation, chronological designation, or alternate numbering systems.

Publication, Distribution, Etc. Area

When the publisher remains the same but the form of the name changes over time, record the earliest form of the name, or if you do not have the earliest editions, choose the form on the earliest available sheets. If you have not chosen the earliest title, then use the form of the publisher's name which goes with the title used. There may be instances where a different publisher takes over the series/serial. In such cases, record the earlier one here (if you have those sheets) and record all the names, with dates if possible, in the note area.

For series produced co-operatively, *Cartographic materials* (E.1B4b) says to record *[Various places : various publishers]* if there are four or more. This contradicts AACR2R's 1.4D5. If there are two or three, all places and names may be recorded here. There should be a note which explains that this is a co-operative effort along with any names of publishers not recorded here.

Physical Description

Extent and specific material designation. For many large series which are still in production, the extent will be open. For series which are obsolete (that are no longer being produced or revised) the total number of sheets, including all editions of all sheets, is recorded in the extent. Use the note area to give the number of sheets required to give complete coverage of the area, or the number of different themes in the series.

Dimensions. Normally record the largest dimensions of the map within the neatlines, or the sheet size.[15] However, there are some instances where discretion is to be exercised. There are some very prolific series which come in a variety of sizes. Some of these are folded and may be kept so, but even their folded size varies greatly. Many of the Canadian geophysical map series are examples of this. In my institution, we opted not to record any dimensions at all as it would have taken us days just to do the measuring.

Series Area

If parts of a series, set or serial are catalogued separately, the title of the series, set or serial may be recorded here. If the series, etc., uses identifiers such as alphanumerics or names, these may be given in the series number area.

(National atlas data base map series ; map no. NADM-2)

There are cases where the series/serial may itself belong to a series. The series title, etc., is then recorded here.

(National earth science series = Série nationale des sciences de la terre)

(*For series*: Magnetic anomaly map (residual total field) / Geological Survey of Canada)

Note Area

Notes should be made only when needed to explain parts of the descriptive entry, or to add information not carried in the entry which is necessary for understanding the entry or useful to the researcher. Most types of notes are listed below. In some instances, the heading for the note is followed only by examples and no explanation. In other instances, no examples were available.

Nature and Scope (Including Relief, Date of Situation, etc.)

Series contains no aeronautical information.
(*For a series entitled*: World aeronautical chart ICAO 1:1,000,000 : [Canadian base map].)

Newfoundland excluded from the series.

Also shows northern United States.

Series consists of 2 sheets covering southern Canada and most of the United States.

FIR: Flight information region.
(*For a map entitled*: FIR plotting chart)

Strip maps.

Controlled mosaic.

Relief shown by contours; some sheets also have hachures and spot heights.

Relief variously shown on most sheets by contours and/or spot heights.

Frequency

Explanation: For serials, make a note on the frequency if this is not apparent from the rest of the description. If there are changes in the frequency, note this as well.

Issued every 56 days.
Semiannual.

Language

Later sheets bilingual.

Text in English only, 1960-Jan. 1980, text in English and French, Mar. 1980- .

Maps and marginal information in English; equivalents in other languages for some marginal information. Canadian sheets have English and French marginal information.

Source of Title Proper

Title supplied by cataloguer. Also known as: Township plans of the West.

Title from index.

Panel title.

Variations in Title

Explanation: When there is more than one collective title, record the ones not chosen as the title proper in a formal formatted note.

Title varies: Geophysical series, map of apparent conductance of bedrock conductors; Geophysical series, maps of apparent conductivity of overburden; Geophysical series, map of conductors and apparent conductivity of overburden.

Title varies: National topographic series, aeronautical edition; National topographic system, aeronautical edition; aeronautical edition; Aeronautical sheet.

Earlier sheets have title:

Sheets x and y titled:

If a more detailed explanation is required then use a non-formatted note. Also include notes on "unofficial" titles which are in common usage if these would be useful to researchers.

Also known as Chief Geographer's series, 1:250 000 and Old geographic series, 1:250 000.

Parallel Title and Other Title Information

Explanation: When this information is not recorded in the title and statement of responsibility area, it may be recorded here.

Statement of Responsibility

Explanation: If the name of the body named in the statement of responsibility changes over time, these changes may be recorded here if they differ from the publisher. If the body named in the statement of responsibility and in the publication area are the same, record variations under the note relating to the publisher. The name(s) of editor(s) of serials may be recorded here if considered important in identifying the serial.

Some sheets prepared in cooperation with provincial government agencies under the Federal-Provincial Aeromagnetic Survey Program.

Most sheets compiled by CLI; some by provincial departments.

Alberta sheet: Cartography by the Soil Research Institute, Research Branch, Agriculture Canada, with the support of the Lands Directorate . . . compiled by the Lands Directorate, Environmental Management Service, Environment Canada . . .

Cartography on later sheets by Land Resources Research Institute.

At head of title: Department of Transport (to 1970), Ministry of Transport (1970-1976), Transport Canada (1977-).

Sheets variously produced by: The Service; U.S. Army Topographic Command; Defense Mapping Agency Hydrographic/Topographic Center (U.S.); Directorate of Military Survey, Ministry of Defence, United Kingdom; Institut géographique national, France; and others.

Edition and History

Various editions of some sheets.

"Maps produced as required, for military training purposes with sheets centred on settlements."–Interview with DND.

Earlier sheets such as Winnipeg, Dartmouth, Nova Scotia, Newfoundland, St. Lawrence and the Great Lakes are no longer produced. Sheets still being produced: Torbay ; Charlottetown–Prince Rupert ; Esquimalt–Queen Charlotte Island ; Vancouver Island.

Straight grid north line and isogrivs ed.

Originally 6 maps planned, only 2 published.

"Between 1976-1978 known as the Isogonal edition."–RCAF chart catalogue.

Overprinted on 1:500 000 topographic maps.

First sheet in series: Montreal (Dorval). 1959.

Most sheets ed. 1; for sheets marked ed. 2, collection lacks ed. 1.

Relationship to Other Serials

Explanation: AACR2R 12.7B7 lists a number of different relationships: translation, continuation, continued by, merger, split, absorption, edition/numerous editions, and supplements. Standard introductory wording should be used where applicable.

Translation of:
Continues:
Continued by:
Merged with:
Continues in part:
Split into:
Separated from:
Absorbed:
Absorbed by:
Supplement:

Continues: National topographic series 1:506 880.

Continued by: Canada 1:500 000.

Replaces "Low-level pilotage chart." Continues: [Canada, aeronautical chart, 1:500 000].

Mathematical and Other Cartographic Data

Most maps before 1974 at scale 1:63 360; after 1974 at 1:50 000; some maps at scales 1:20 000, 1:24 000, 1:31 680, 1:125 000 and 1:126 720.

Most maps at 1:250 000; sheets at 1:125 000 and 1:126 720 cover British Columbia only.

Sheets published according to the NTS 1 in. to 1 mile quadrangles.

Contour interval 10 metres.

Numbering and Chronological Designation

Explanation: Notes may be made on any irregularities or peculiarities in numbering, chronology, etc., not already recorded. Also, if the publisher ceased publication for a time and then resumed publication, record the dates of the suspension.

Publication, Distribution, etc.

Explanation: If the name of the body changes over time, the various forms of the name may be recorded here. For series, and serials produced co-operatively, the names of the various publishers are recorded here.

Publisher's name varies: Topographical Survey.
[Publisher named in imprint is Geological Survey of Canada]

Publisher's statement from the index.

Publisher varies: 1871-1873 Dominion Lands Branch; 1873-1936 Dept. of the Interior; 1936-19– provinces of Manitoba, Saskatchewan and Alberta took over responsibility.

Publication dates from publisher's 1974 catalogue.

Physical Description

Some maps photomaps.

Maps originally printed on separate sheets, later, pairs combined and printed back-to-back.

Accompanying Material

Accompanied by 2 experimental maps of Val d'Or, Quebec, scale 1:50 000, nos. 036-037, 1981, entitled: Experimental colour compilations (VLF electromagnetic total field) / Geological Survey of Canada.

Accompanied by: Perret, N.G. Land classification for wildlife. 1970. (Report / Lands Directorate ; no. 7).

Series

Explanation: Any information relating to a series which may not be recorded in the series area may be given here.

Other Formats

Explanation: If the series/serial is available in another format (e.g., digital computer file), this may be noted here.

Indexes

Explanation: If a graphic index either accompanies the series or is available separately, note this here if desired. This includes indexes prepared by the institution.

Index map, radio facilities and aerodrome data on verso.

Contents

Explanation: For small series or serial, the content may be used to list the sheets in the series/serial.

Complete in 34 sheets. Various editions of some sheets.

Sheet IG has Loran C information shown in mirror print on verso of some eds.

Contents: Atlantic Provinces–Quebec south part–Ontario south part–Manitoba south part–Saskatchewan south part–British Columbia south part.

On verso are index map, radio facilities and aerodrome data.

Insets: Index to 1:1,000,000 aeronautical charts–Index to 8 mile aeronautical chart.

Contents: Chart 1. Vancouver-Edmonton. AIR 1631–Chart 2. Winnipeg. AIR 1632–Chart 3. Toronto-Montreal. AIR 1633–Chart 4. Moncton-Gander-Goose. AIR 1634.

Numbers

Publication numbers assigned to series: 1G-6999G; 8200G- .

Earlier sheets assigned AIR numbers 1606-1609, later sheets with AIR numbers 1621-1623.

Copy Described

Collection holds only Canadian coverage which is complete in 19 sheets.

"Issued With" Notes

Issued with: Road atlas of the United States, Canada and Mexico. (Note for *Commercial atlas & marketing guide*)

Item Described (i.e., Not the First Issue)

Explanation: If the first sheet/volume issued is not available, then the one used as the basis for the description is identified here.

Description based on:

Standard Numbers

Explanation: In the entry for the series/serial as a unit, record only the standard number which applies to the series/serial itself and not to the parts. The standard number for a part may be recorded in a separate description for the part (see the section on Analysis below).

ANALYSIS

For works consisting of multiple parts, be they sheets or volumes, there are a number of methods available for describing and thus gaining some intellectual control over the parts.[16] The method chosen should be appropriate for the needs of the researchers, balanced with resource limitations. In other words, provide the best access to the parts that is possible.

The simplest method is the contents note. Here, any pertinent bibliographic information for all or some of the parts may be included (e.g., titles, authors, edition, mathematical data, date(s) of publication) as required. Access to the individual parts in this instance could be effected, if deemed necessary, through title, or author/title added entries. However, if the parts are voluminous, this is not a practical option.

For some serials such as annuals, where it has been decided to catalogue each edition separately, the title of the whole work (the "comprehensive item" in AACR2R terms) is recorded as the series title in the series area.

For larger works, especially large series, multilevel cataloguing is a more effective technique if your system will support it. A parent record for the series as a whole is prepared. Then, a separate record for each sheet/volume/ title in the series is prepared and linked through a linking field to the parent record. Normally, common information is given in the parent record and not repeated in the child record. However, there may be a need to repeat some information in the child record if your system will not display the parent and child records together, or if they are not easily retrieved together.

For a fuller description, refer to *Cartographic materials* rule 13 or to AACR2 chapter 13.

ACCESS POINTS

Make access points for authors. Cartographic materials has a list of terms which indicate responsibility (1F1 Application) for which access points may be made. Not many series or serials have statements of responsibility; however, most publishers of cartographic serials and series are also the creator and are thus the main entry. When it is title main entry (see AACR2R 21.1C1), make an access point for the publisher(s) if they are known to be map producers. In the case of co-operative ventures, trace the body in the home country and any others deemed useful for retrieval in your institution.

Titles. Added entries should be made for variant titles and of course there should be an entry per title proper.

BI- AND MULTI-LINGUAL WORKS

For those working in a bi- or multi-lingual environment where access must be provided in two or more languages, bibliographic descriptions will be prepared according to the language of description policy of the institution. For those who may be unfamiliar with what such a policy may entail, the following is a sample of what is generally followed in bilingual institutions in Canada. Works in English are catalogued in English, works in French are catalogued in French. Works with substantial portions in both English and French are catalogued in English and again in French (i.e., two separate bibliographic entries). Bi- or multi-lingual works with English but not French as one of the languages are catalogued in English. If the work has French but no English, it is catalogued in French. I must add that, for the map collection of the National Archives of Canada, we follow this, but we always provide bilingual access points (name and subject) for works catalogued in one language only.

For those institutions cataloguing in more than one language, problems occur when a series or serial becomes bilingual later on in its life. The series/serial could become completely bilingual, or only parts of it could become bilingual. The following guidelines may be useful in knowing when to create separate records for each language and when a single catalog description with explanatory notes and additional added entries will suffice.

When a unilingual series/serial becomes bilingual, two separate records, one in each language, should be created to describe the bilingual period of its history. Theoretically this would mean that the old unilingual record could either be closed and two new records created (with notes to show the relationships), or it could be revised so that it might continue to function as the record in that language but have additional notes for the language change and on the relationship to the new other language record.

Added title: *[Title in other language]*

Added language note: In English only, 1950-1960; in English and French, 1961- .

This is the theory; what is done in practice depends on resources.

When a bilingual series/serial becomes unilingual, the two records will be revised to note the change. The record in the language which is no longer being produced could have the date closed and the extent completed.

MARC CODING

For MARC tags, see the MARC tag table at the beginning of the volume.

NOTES

1. Farrell 1984, 18.
2. *AACR2R* 1988, 620.
3. *Cartographic materials* 1982, 230.

4. Farrell 1984, 18.

5. *Cartographic materials* 1982, 177-180; Larsgaard 1987, 155-157; *Map cataloging manual* 1991, 7.1-7.5.

6. Farrell 1984, 18.

7. Based on Farrell 1984, 18.

8. Larsgaard 1987, 157.

9. *Cartographic materials* 1982, 7.

10. What is really meant here is "nontextual."

11. *AACR2R* 1988, 279.

12. *AACR2R* 1988, 315.

13. Institut géographique national (France) 1968, p. J7.

14. Institut géographique national (France) 1968, p. F4.

15. For more complete instructions see *Cartographic materials,* rules 5C.

16. See AACR2R chapter 13 and *Cartographic materials* rule 13.

REFERENCES

Anglo-American Cataloguing Rules. 2nd ed, 1988 rev. Ottawa: Canadian Library Association; London: Library Association Publishing Ltd.; Chicago: American Library Association, 1988.

Anglo-American Cataloguing Committee for Cartographic Materials. *Cartographic materials: a manual of interpretation for AACR2.* Chicago: American Library Association; Ottawa: Canadian Library Association; London: The Library Association, 1982.

Cataloguing Manual, Cataloguing Unit, Cartographic and Architectural Sector, Cartographic and Audio-Visual Archives Division, National Archives of Canada = Guide de catalogage, Unité de catalogage Secteur cartographique et architectural, Division des archives cartographiques et audio-visuelles, Archives nationales du Canada. Ottawa: National Archives of Canada, 1992.

Descriptive Cataloguing Manual = Guide de catalogage descriptif. Ottawa: Acquisition and Bibliographic Services Branch, National Library of Canada = Direction des acquisitions et des services bibliographiques, Bibliothèque du Canada, 1989- .

Farrell, Barbara and Aileen Desbarats. *Guide for a Small Map Collection.* 2nd ed. Ottawa: Association of Canadian Map Libraries, 1984.

Institut géographique national (France). *Catalogue des cartes de l'I.G.N.: cartes en service.* [Paris: I.G.N.], 1968.

Larsgaard, Mary Lynette. *Map Librarianship: an Introduction.* 2d ed. Littleton, Colo.: Libraries Unlimited, 1987.

Library of Congress. Geography and Map Division. *Map Cataloging Manual.* Washington, D.C.: Cataloging Distribution Service, Library of Congress, 1991.

APPENDIX

MARC CODING EXAMPLES

The following examples have been taken from the National Archives of Canada database (which is on the A-G Canada Sun workstations) and, for purposes of this article, have been modified to reflect the changes in coding since they were created.

CARTOGRAPHIC SERIALS

Example 1

Leader 5: n

Leader 6: e

Leader 7: s

Leader 17: [blank]

Leader 18: a

001	02744380
003	CaOEAGC
005	19971008135156.0
006	[sfr f 0 0]
007	a j | c |n | |
008	880720m19679999onc cc c f 0 eng d
034 0	$a a $d W1420000 $e W0450000 $f N0680000 $g N0400000
037	$a AIR 1631$bEMR
037	$a AIR 1632$bEMR
037	$a AIR 1633$bEMR
037	$a AIR 1634$bEMR
043	$a n-cn---
052	$a 3401
055 4	$a G3401.P6 $b s2500 .C36
110 1	$a Canada. $b Surveys and Mapping Branch.
245 10	$a FIR plotting chart : $b[Canada] / $c produced by Surveys and Mapping Branch, Department of Energy, Mines and Resources.

255 $a Scale [ca 1:2,500,000]. Approx. 35 nautical miles to 1 in. ; $b Lambert conformal proj., standard parallels 49°N and 77°N $c (W 142°--W 45°/N 68°--N 40°).

300 $a maps ; $c on sheets 92 x 122 cm. or smaller.

310 $a Semiannual.

500 $a FIR: flight information regions.

500 $a Sheets of various eds.

500 $a Base map of FIR plotting charts produced from: ATC plotting charts : [Canada].

500 $a Each sheet includes index map, and shows radio aids, airports, preferential routes, air defence identification zone boundaries, restricted areas, etc.

505 0 $a Chart 1. Vancouver-Edmonton. AIR 1631 -- Chart 2. Winnipeg. AIR 1632 -- Chart 3. Toronto-Montreal. AIR 1633 -- Chart 4. Moncton-Gander-Goose. AIR 1634.

690 $a Canada -- Aeronautical charts -- Plotting charts

690 $a Canada -- Cartes aéronautiques -- Cartes de tracé de navigation

710 1 $a Canada. $b Direction des levés et de la cartographie.

Example 2

Leader 5: n

Leader 6: e

Leader 7: s

Leader 17: [blank]

Leader 18: a

001 02744414?

003 CaOEAGC

005 199710081354656.0

006 [sar f 0 0]

007 a j | c | n | |

008 880804m19679999oncg ap c f 0 eng d

034 1 $a a $b3000000 $d W1400000 $e W0460000 $f N0780000 $g N0410000

037       $a AIR 1615 $b EMR

043       $a n-cn---

052       $a 3401

055 4 $a G3401.P6 $b s3000 .C35

110 1 $a Canada. $b Surveys and Mapping Branch.

245 10 $a Canada plotting chart : $b AIR 1615 / $c produced by Surveys and Mapping Branch,
          Department of Energy, Mines and Resources.

255       $a Scale 1:3,000,000 at 45°N ; $bPolar Stereographic proj. $c (W 140°--W 46°/N
          78°--N 41°).

310       $a Annual.

362 0 $a 1st ed. (1976)-   .

260       $a Ottawa : $b [The Branch], $c 1976-

300       $a   maps : $b col. ; $c 101 x 149 cm.

500       $a "To provide the Canadian Forces Air Navigation School in Winnipeg with a plotting
          chart for long range aircraft operations"--RCAF chart catalogue.

500       $a Relief shown by spot elevations.

500       $a "Between 1976-1978 known as Isogonal edition"--RCAF chart catalogue.

500       $a Includes radio aids, aerodromes, control zones etc.

690       $a Canada -- Aeronautical charts -- Plotting charts.

690       $a Canada -- Cartes aéronautiques -- Cartes de tracé de navigation.

710 1 $a Canada. $b Direction des levés et de la cartographie.

780 00 $a $t Winnipeg navigation plotting chart.

780 05 $a $t Canada plotting chart : AIR 1614, $c in 1979.

Cartographic Series

Example 3

Leader 5: n

Leader 6: e

Leader 7: m

Leader 17: [blank]

Leader 18: a

001   02746882

003   CaOEAGC

005   19940025131851.0

007   a j |c |n | |

008   940124r19651966oncgk bd b f 0   eng d

034 1 $a a $b 1000000 $d W0710000 $e W0502000 $f N0524000 $g N0405000

043    $a n-cna-- $a n-cn-qu $a n-cn-me

052    $a 3406

055 4 $a G3406.P6 $b s1000 .C37

110 1 $a Canada. $b Surveys and Mapping Branch.

245 10 $a R.C.N. Mercator plotting charts, east coast area / $c produced by the Surveys and
      Mapping Branch, Department of Mines and Technical Surveys.

255    $a Scale 1:1,000,000. At latitude 46°N $c (W 71°00'--W 50°20'/N 52°40'--N 40°50').

260    $a Ottawa : $b The Branch, $c [1965-1966]

300    $a 10 maps : $b col. ; $c on sheets 44 x 44 cm.

500    $a Relief shown by spot heights and bathymetric contours.

500    $a Title from index.

500    $a Lower left corner: Corrections.

500    $a "Base, 2nd ed., 1966 [and] 1965"--All maps.

500     $a Some base maps dated 1965 with air information for 1965 and for 1966. Some base maps dated 1965 with air information for 1966.

500     $a Complete in 7 maps.

500     $a Transferred from Aeronautical Chart Information Service, Canada Centre for Mapping, Surveys, Mapping and Remote Sensing Sector, Natural Resources Canada (RG 88M).

500     $a On verso are indexes to west and east coast charts and topographic and aeronautical symbols.

505 0 $a Cape.Cod-Yarmouth chart 100 -- Halifax chart 102 -- Sable Island chart 104 -- Cape Race chart 105 -- Newfoundland central chart 106 -- Gulf of St. Lawrence chart 107 -- Newfoundland north chart 108.

690     $a Canada -- Eastern Canada -- Aeronautical charts -- Plotting charts.

690     $a Canada -- Est du Canada -- Cartes aéronautiques -- Cartes de tracé de navigation

710 1 $a Canada. $b Division des levés et de la cartographie.

780 00 $a $t R.C.N. air plotting charts, northwest Atlantic.

Example 4

Leader 5: n

Leader 6: c

Leader 7: m

Leader 17: [blank]

Leader 18: a

001   027444267

003   CaOEAGC

005   19931020150249.0

007   a j |c |n | |

008   880319m19651983onc   b f 0   eng d

034 0 $a a $d W1300000 $e W0520000 $f N0590000 $g N0410000

041 0 $a engfre

043     $a n-cn---

052     $a 3401

055 4 $a G3401.J3 $b svar .C35

110 2 $a Canada Land Inventory.

245 10 $a Inventaire des terres du Canada, possibilités agricoles des sols / $c cartographie

        réalisée par l'Institut de recherche sur les sols, Direction de la recherche, ministère de

        l'Agriculture du Canada... = Canada land inventory, soil capability for agriculture /

        cartography by the Soils Research Institute, Research Branch, Canada Department of

        Agriculture...

246 31 $a Canada land inventory, soil capability for agriculture

255     $a Scales differ $c (W 130°--W 52°/N 59°--N 41°).

260     $a Ottawa : $b [Direction générale des terres = Lands Directorate, 1965-1983?]

300     $a   maps : $b col. ; $c 56 x 83 cm or smaller, on sheets 73 x 112 cm or smaller.

500     $a Most maps at 1:250 000; sheets at 1:125 000 and 1:126 720 cover British Columbia

        only.

500     $a Publisher's statement from index.

500     $a Includes legend, and key map.

500     $a General description of sheet area on verso.

500     $a Accompanied by: Canada Land Inventory. Soil capability classification for

        agriculture. 1969. (Report ; no. 2).

500     $a Cartography on later sheets by Land Resources Research Institute.

690     $a Canada -- Soil capability for agriculture -- Maps.

710 2 $a Land Resource Research Institute (Canada)

710 2 $a Soil Research Institute (Canada)

740 01 $a Soil capability for agriculture.

The Cataloging of Globes

Scott R. McEathron

SUMMARY. Globes are a unique cartographic format with special characteristics that need to be considered when providing bibliographic description and access. This article describes the challenges and characteristics of cataloging globes. It provides practical assistance in applying specific MARC 21 fields for globes. Special consideration is given to those fields that are unique to cataloging globes including: physical description areas, title, and subject access. *[Article copies available for a fee from The Haworth Document Delivery Service: 1-800-342-9678. E-mail address: getinfo@haworthpressinc.com <Website: http://www.haworthpressinc.com>]*

KEYWORDS. Globes, cataloging, subject analysis

INTRODUCTION

You may ask yourself: "why bother cataloging globes?" Globes need to be cataloged for the same reasons that other physical and electronic media in library and museum collections are cataloged. With good bibliographic description and access, catalogers provide the knowledge users need to make informed choices and improve the likelihood they will satisfy their search queries. Globes have had many functions in history: from symbols of knowledge, wealth and power to cartographic tools for instruction. The knowledge

Scott R. McEathron, MLIS, MA, is Map Catalog Librarian and Liaison for Natural Resources at the University of Connecticut Libraries, 369 Fairfield Road, Storrs, CT 06269.

[Haworth co-indexing entry note]: "The Cataloging of Globes." McEathron, Scott R. Co-published simultaneously in *Cataloging & Classification Quarterly* (The Haworth Information Press, an imprint of The Haworth Press, Inc.) Vol. 27, No. 1/2, 1999, pp. 103-112; and: *Maps and Related Cartographic Materials: Cataloging, Classification, and Bibliographic Control* (ed: Paige G. Andrew, and Mary Lynette Larsgaard) The Haworth Information Press, an imprint of The Haworth Press, Inc., 1999, pp. 103-112. Single or multiple copies of this article are available for a fee from The Haworth Document Delivery Service [1-800-342-9678, 9:00 a.m. - 5:00 p.m. (EST). E-mail address: getinfo@haworthpressinc.com].

103

that can be gained from a globe is fundamental to our understanding of many patterns or relationships on Earth and other celestial bodies, including such basic phenomena as time and weather. Globes document the growth of human geographic knowledge of the Earth and understanding of the cosmos.

Globes have rarely been collected solely for their content, but because they more closely portray the three dimensions of terrestrial and celestial bodies. Often, it is this unique three dimensional characteristic of globes that proves challenging to catalogers providing bibliographic description and access. The aim of this article is to provide both an overview of the process and lend practical assistance to catalogers unfamiliar with globes. Both early and modern globes will be discussed. The article begins with a brief overview of the tools used for cataloging globes and the challenges often faced when trying to catalog them. Primary focus will then be on discussing the unique AACR2R[1] rule applications encountered when cataloging globes and the application of specific USMARC fields for this form of cartographic material. The discussion will then be followed by actual examples of USMARC records for globes.

TOOLS FOR CATALOGING GLOBES

One of the primary challenges in cataloging globes is that seemingly little direction and the few examples are contained in common cataloging references such as AACR2R and Cartographic Materials: A Manual for Interpretation for AACR2. However, in this age of national bibliographic utilities, many good examples of globe records already exist.[2] The rules used to describe cartographic materials in chapter 3, AACR2R, are in most cases sufficient for providing good descriptions for globes. The exception being early globes, for which there are many good cartobibliographies that may be consulted.[3] These may be useful for assisting in the identification and description of globes.

DESCRIPTION USING AACR2R IN USMARC FORMAT

General Rules

The first rule to be concerned with when cataloging globes pertains to sources of information. As with other cartographic materials, the chief source of information is "a) the cartographic item itself . . ." but also "b) . . . the cradle and stand of a globe, etc. If information is not available from the chief source, take it from any accompanying materials (e.g., pamphlets, brochures)."[4] This is especially pertinent for many of the globes made in the twentieth century, as they often have little bibliographic information on the sphere itself. These modern globes often have title and statement of responsibility information on the stand or in an accompanying user manual. Globes are almost always described individually yet there are occasions when description could be made as a collection or pair of globes. Early globes, including those made well into the nineteenth century, were usually produced as matched pairs: terrestrial and celestial. Other instances of several globes being published or stored as a collection exist.[5] The needs of the cataloging agency determine if such globes are described as a collection or individually.

FIXED FIELDS

Unique fixed fields for the maps format in MARC 21 include Proj (projection), *Relf* (relief), and *Prme* (prime meridian). The *Proj* field may be left blank as map projections are not used for globes. When cataloging a globe, the *Relf* field is not treated any differently than any other map. The *Prme* field code should be entered if it is specified on, or can be determined from the globe. If the globe has more than one prime meridian enter the code for the first-mentioned or most prominent and describe them in a note field (500). This is especially relevant to early globes where sometimes more than one prime meridian was shown.

PHYSICAL DESCRIPTION FIXED FIELD (007)

Codes used in the unique physical description fixed field (007) for globes tend to vary more than those used in standard map cataloging as it is more common for globes to be made of different materials. Though most globes are now made of paperboard those made of metal, plastic, or plaster are still common in libraries and museums. Though one can usually tell by simply tapping lightly on the globe sphere it can be difficult to distinguish between the different physical mediums at times. (See Table 1.)

TITLE AND PUBLICATION FIELDS (2XX)

As previously noted, the title should be taken from the sphere itself, then the cradle or stand. If information is not available from the chief source take the title information from any accompanying materials if they exist. If need be, the title may also be determined from other sources, such as a cartobibliography reference book or a map dealer's catalog. If title information is taken

TABLE 1. Physical Description Fixed Field for Globes (007)

‡a Category of material	‡b Specific material designation	‡d Color	‡e Physical medium	‡f Type of reproduction
d *Globe*	a *Celestial globe* b *Planetary or lunar globe* c *Terrestrial globe* z *Other*	a *One color* c *Multicolor*	a *Paper* b *Wood*. Use also for particle board. c *Stone* d *Metal* e *Synthetics*. Plastic, vinyl, film, etc. f *Skins*. Parchment, vellum, etc. g *Textile*. Man-made fibers, silk, linen, nylon, etc. p *Plaster*. Use also for mixtures of ground solids and plaster. u *Unknown* z *Other*	f *Facsimile* n *Not applicable* u *Unknown* z *Other*

from a source other than the chief source of information it should be placed in square brackets. When the cataloger must supply the title, be sure to include the area covered, i.e., [Lunar globe].[6] Modern globes will often have numbers within the title such as: *"Replogle 16 inch library globe."* It is good practice to also spell out the numbers as an alternate form of title in the varying form of title field (246), thus: *"Replogle sixteen inch library globe."*

When transcribing statements of responsibility from globes it is important to note that the mapmaker is at times different than the manufacturer of the globe. For instance, the globe may be "compiled and drawn in the Cartographic Division of the National Geographic Society" and "manufactured by Replogle Globes, Inc., printed by A. Hoen and Co., Lithographers." In this example, only the portion "compiled and drawn in the Cartographic Division of the National Geographic Society" appears prominently on the item and therefore is the statement of responsibility. The remaining statements may be given in note fields. Input publication and distribution field (260) information as you would for any map.

MATHEMATICAL DATA (034, 255)

Give or determine the scale of a globe the same as you would any other cartographic item. Scale can always be reduced from a comparison between the diameter of the globe and the actual circumference of the Earth or represented body. Celestial globes do not have a linear scale. Coordinates may also be given as you would for any world map.

PHYSICAL DESCRIPTION (300)

More than any other field in the bibliographic record, the physical description area for a globe record varies from those of other formats. Since most globes are of the same subject (the Earth) and many do not have much bibliographic data, the physical description area can be very important for identification and description. The first element (subfield "a" of 300) in the physical description area is the extent of the item and its material designation as a globe. It is important to keep in mind that *globe gores not mounted are not designated as globes,* but rather as maps. The second element (subfield "b") is often lengthy as compared to other formats. Although there is no agreement or authority in terminology the following list includes the common terms used in describing the physical characteristics of globes. Care should be given to describe each of the components beginning with the color, the material of the gores and ball, then the mounting (see Figures 1 and 2).[7] As the final element (subfield "c"), measure the diameter of the globe at the equator or at its center and add a description of any container.

VOCABULARY OF COMPONENTS

base, cradle or stand–the primary support structure for the globe

horizon circle or horizon ring–circular ring representing the solar plane, generally supported by the stand, often graduated in various units of measure such as degrees, distance or time

hour circle and pointer–more often on early globes and generally attached to the meridian ring or at the top of the sphere

meridian ring–the vertical circular ring often graduated in degrees beginning with 0 degrees at the equator to 90 degrees at each pole, may be half (180 degrees or semi-circular) or full (360 degrees)

sphere or ball–the actual sphere or globe itself, on which the map gores are attached or relief is shown

NOTE FIELDS (5XX)

Notes should be added as they would be for any other cartographic item as described in *AACR2R*. For early globes, a citation/references note (510) or a publications about described materials note (581) can be very useful.

FIGURE 1. Sixteen Inch Physical Globe

SUBJECT FIELDS (6XX)

Assigning subject headings to globes can be simple but frustrating. Simple because few subject headings currently exist for globes. Frustrating because adequate subject access may be difficult to achieve. Assignment of subject headings for globes is an excellent case study of the content versus carrier discussion currently underway with regards to maps and other formats. In *Library of Congress Subject Headings,* 21st ed., the word "Globes" is a topical subject term. However, it is being applied by catalogers either as a topical, a geographic or a genre subject term. It is obvious that most globes are not about globes, but are given this subject heading because of their format. Furthermore, globes represent places, i.e., Earth and Moon, but be-

FIGURE 2. The Husun Star Globe

cause they are topical subject terms, free-floating subject subdivisions for places may not be added when they may be needed. The three primary subject terms currently used for globes are: *Moon–Globes* for lunar globes, *Celestial globes,* and *Globes* used for terrestrial globes. As libraries continue to collect more information in different formats in the future we can expect improved genre and subject access.

NOTES

1. Michael Gorman and Paul W. Winkler, ed., *Anglo-American Cataloguing Rules: Second Edition 1988 Revision,* (Chicago: American Library Association, 1988).
2. A subject keyword search (fin su globes/map) in *OCLC's WorldCat* retrieves 451 records in the maps format as of December 8, 1998.

3. The journal *Der Globusfreund: Journal for the Study of Globes and Related Instruments,* published by the International Coronelli Society, often contains detailed bibliographies of globes.

4. *AACR2R,* Rule 3.0B2, p. 94.

5. For example, a set of four thematic globes published by E. M. Walters of Hove, England circa 1920 in Scott R. McEathron and Sharon Hill, *The Sphere of the Cartographer,* American Geographical Society Collection Special Publication No. 5, (Milwaukee, Wis.: AGS Collection, 1996), p. 20.

6. *AACR2R,* Rule 1.1B7, p. 19.

7. *AACR2R,* Rule 3.5C1, p. 110.

BIBLIOGRAPHY

The following is a bibliography of useful references for identifying and describing early globes:

Dekker, Elly and Peter van der Krogt. *Globes from the Western World.* London: Zwemmer, 1993.

Fauser, Alois. *Altere Erd-und Himmelsgloben in Bayern.* Stuttgart: Schuler Verlagsgesellschaft, 1964.

Krogt, Peter van der. *Old Globes in the Netherlands: A Catalogue of Terrestrial and Celestial Globes Made Prior to 1850 and Preserved in Dutch Collections.* Utrecht: Hes Uitgevers, 1984.

Krogt, Peter van der. *Globi Neerlandici: The Production of Globes in the Low Countries.* Utrecht: Hes Uitgevers, 1993.

McEathron, Scott R. and Sharon Hill. *The Sphere of the Cartographer.* American Geographical Society Collection Special Publication No. 5. Milwaukee, Wis.: AGS Collection, 1996.

Muris, Oswald and Gert Saarmann. *Der Globus im Wandel der Zeiten: Ein Geschichte der Globen.* Berlin: Columbus Verlag, 1961.

Savage-Smith, Emilie. *Islamicate Celestial Globes: Their History, Construction, Use.* Smithsonian Studies in History and Technology No. 46. Washington, D.C.: Smithsonian Institution Press, 1985.

Schmidt, Rudolf. *Projections of Earth and Space.* Wien: International Coronelli Society, 1989.

Sider, Sandra. *Maps, Charts, Globes: Five Centuries of Exploration.* New York: The Hispanic Society of America, 1992.

Stevenson, Edward Luther. *Terrestrial and Celestial Globes.* New Haven, Conn.: Yale University Press, 1921.

Warner, Deborah Jean. "The Geography of Heaven and Earth." *Rittenhouse Journal of the American Scientific Instrument Enterprise.* No. 2 (1987): 14-32, 52-64, 88-104, 147-152.

Yonge, Ena L. *A Catalogue of Early Globes: Made Prior to 1850 and Conserved in the United States.* Library Series No. 6. New York: American Geographical Society, 1968.

APPENDIX

EXAMPLE 1

Type: c	ELvl: I	Srce: d	Relf: cd	Ctrl:	Lang: eng
BLvl: m	SpFm:	GPub:	Prme:	MRec:	Ctry: ilu
CrTp: d	Indx: 0	Proj:	DtSt: s	Dates: 1930,	
Desc: a					

040   GZN ‡c GZN

007   d ‡b c ‡d c ‡e p ‡f n

034 1   a ‡b 31680000 ‡d W1800000 ‡e E1800000 ‡f N0900000 ‡g S0900000

052   3170

090   G3170 1930 |b .G6

049   GZNA

100 1   Goode, J. Paul, ‡d 1862-1932.

245 10   Sixteen inch physical globe / ‡c by J. Paul Goode.

246 1   16 inch physical globe

255   Scale [1:31,680,000]. 1 in. to 500 miles ‡c (W 180^0--E 180^0/N 90^0--S 90^0).

260   Chicago : ‡b Rand McNally, ‡c [ca. 1930]

300   1 globe : ‡b col., paper gores over plaster, mounted on wooden floor stand with three legs supporting a wooden horizon and metal meridian ring ; ‡c 41 cm. in diam.

650 0   Globes.

710 2   Rand McNally and Company.

EXAMPLE 2

Type: e	ELvl: I	Srce: d	Relf:	Ctrl:	Lang: eng
BLvl: m	SpFm:	GPub:	Prme:	MRec:	Ctry: enk
CrTp: d	Indx: 0	Proj:	DtSt: s	Dates: 1920,	
Desc: a					

040   GZN ‡c GZN

007   d ‡b a ‡d c ‡e p ‡f n

034 0   a

052   3160

090   G3160 1920 |b .H8

049   GZNA

245 14 The Husun star globe.

246 1   ‡i Additional title on horizon circle: ‡a Paget star globe

255   Scale indeterminable.

260   London : ‡b H. Hughes & Son, ‡c 1920.

300   1 globe : ‡b col., paper gores over plaster, in wooden box with metal half meridian ring and horizon ; ‡c 19 cm. in diam. in box 28 x 28 x 24 cm.

500   Yellow background with black stars.

500   "No. 1007."

650 0 Celestial globes.

Cataloging Geologic Sections

Christopher J.J. Thiry

SUMMARY. In some existing cataloging records, there is evidence of considerable confusion in cataloging graphic representations of geologic measurements. The cataloging of geologic sections differs from the cataloging of maps in six areas: leader fields in a USMARC-formatted record; 0xx fields in a USMARC-formatted record; Scale; Physical description; Notes; and Subject headings. This paper will explain the use and importance of geologic sections, clarify why they should be cataloged in the USMARC Map Format, explain why they are called "sections," define what is meant by "geologic section," prescribe the rules for cataloging, and demonstrate the proper procedure for cataloging a geologic section. *[Article copies available for a fee from The Haworth Document Delivery Service: 1-800-342-9678. E-mail address: getinfo@haworth pressinc.com <Website: http://www.haworthpressinc.com>]*

KEYWORDS. Geologic sections, cataloging, geologic, maps, stratigraphic columns, profiles

Mr. Thiry has a Bachelor of Arts degree in History and a Masters degree in Information and Library Studies from the University of Michigan. He served as a map librarian at the New York Public Library for over 2 years. He has been the Map Librarian at the Colorado School of Mines' Arthur Lakes Library for over 3 years. He is a member of the Western Association of Map Libraries (WAML) and incoming chair of the Cartographic Users Advisory Council (CUAC). Mr. Thiry is a native of Dearborn, Michigan.

Address correspondence to: Christopher J.J. Thiry, Map Librarian, Colorado School of Mines, Arthur Lakes Library, 1400 Illinois, P.O. Box 4029, Golden, CO 80401-0029 (email: cthiry@mines.edu or visit website at http://www.mines.edu/library/maproom).

[Haworth co-indexing entry note]: "Cataloging Geologic Sections." Thiry, Christopher J.J. Co-published simultaneously in *Cataloging & Classification Quarterly* (The Haworth Information Press, an imprint of The Haworth Press, Inc.) Vol. 27, No. 1/2, 1999, pp. 113-145; and: *Maps and Related Cartographic Materials: Cataloging, Classification, and Bibliographic Control* (ed: Paige G. Andrew, and Mary Lynette Larsgaard) The Haworth Information Press, an imprint of The Haworth Press, Inc., 1999, pp. 113-145. Single or multiple copies of this article are available for a fee from The Haworth Document Delivery Service [1-800-342-9678, 9:00 a.m. - 5:00 p.m. (EST). E-mail address: getinfo@haworthpressinc.com].

CATALOGING GEOLOGIC SECTIONS

Considerable confusion exists regarding the cataloging of graphic representations of geologic measurements, that is, of vertical slices of a planet, most often the Earth. Librarians have not come to agreement on what to call these items; terms range from "maps" to "geologic sections" to "charts." The rules of AACR2R (*Anglo-American Cataloguing Rules,* second edition, revised) and even *Cartographic Materials: A Manual of Interpretation for AACR2* are vague on the subject of geologic sections. The purpose of this paper is to elucidate the use and importance of geologic sections, clarify why they should be cataloged in the USMARC Map Format, explain why they may only be termed "sections" in the physical description, define what they are, prescribe the rules for cataloging, and demonstrate the proper procedure for cataloging a geologic section.

USE AND IMPORTANCE OF GEOLOGIC SECTIONS

Geologic maps are common, useful, and relatively straightforward to catalog, but few know how to address all those "other" diagrams accompanying the map. These diagrams are different ways to display and interpret measurements of the layers of rocks (strata) below the surface of the Earth.

In the past few centuries, geologists have developed sophisticated ways to determine the composition of the Earth. In doing so, they have classified, cataloged and put in chronological order their items of study–the layers of the rocks beneath the Earth's surface. The length and order of time periods as well as terminology have been standardized; for example, the Mesozoic era occurred at and lasted the same amount of time in Kansas as it did in China.

These standardizations allow scientists to measure strata and report the findings in ways that can be understood by all geologists. One of the most common methods of measuring strata is to bore a hole deep in the Earth, preserving a core or section of the rock as it is drilled. The core is removed, analyzed, measured, and classified. Another method is to insert sensors in these holes and measure various physical characteristics, including gravity anomalies, gamma rays, and permeability. Yet another way is to set off charges on the Earth's surface and measure the effect of those charges using seismic monitors. Rarely do geologists take only one measurement in an area; usually they make multiple measurements. By comparing the readings from various locations, geologists are able to map the strata. Knowing how certain strata lie, geologists are able to predict with some accuracy where certain mineral resources will occur.

WHY SHOULD WE CATALOG SECTIONS USING THE MAP FORMAT?

Since all the types of measurements mentioned in this paper cover a specific geographic area, they must be considered "maps" for cataloging purposes. AACR2R describes a map as, "a representation, normally to scale and on a flat medium, of a selection of material or abstract features on, or in relation to, the surface of the Earth or of another celestial body" (Gorman and Winkler 1988, 619). Even though the cataloging is done in the Map Format, geologic sections are not "maps" as most would define them. While often showing surface features, all these items show some aspect of the subsurface of that specific area. A normal "map" will only show the horizontal surface of the Earth; a geologic section will also show the strata in a vertical slice or section.

"SECTION": THE BEST TERM

Many catalogers have misidentified geologic sections in the physical description field. OCLC contains many examples of these mistakes–"geologic cross sections," "diagrams," "well logs," "charts," and so forth. It is easy to understand the confusion. First, despite geologists' standardized terms, authors often give the items a variety of labels. Second, AACR2 permitted catalogers to assign their own terms if none of the specific material designations (SMDs) matched what the cataloger had in hand. AACR2R does not permit this, but rather directs the cataloger to forage in other chapters for suitable SMDs–which is no help in the case of geologic sections.

While geologic sections are cataloged in the MARC Map Format, sections are not given the SMD of "maps." The Map Format allows for the use of nine specific material designations: atlas; diagram; globe; map; model; profile; remote-sensing image; section; and view.

Of these, the term that best describes the graphic representation of geologic measurements is "section" because the definition of a "section (cartography)" is, "a scale representation of a vertical surface (commonly a plane) displaying both the profile where it intersects the surface of the ground, or some conceptual model, and the underlying structures along the plane intersection (e.g., a geological section)" (Gorman and Winkler 1988, 622). Other material designations appear to describe the measurements, but fall short of being an encompassing definition. A "map," as defined earlier, represents the "surface of the Earth." "Profile" is often misused by catalogers, but a "profile (cartography)" is defined as, "a scale representation of the intersection of a vertical surface (which may or may not be a plane) with the surface of the ground, or the intersection of such a vertical surface with that of a conceptual three-dimensional model representing phenomena having a continuous distribution" (Gorman and Winkler 1988, 621). A "map" shows the surface of the Earth, a "profile" the surface and above, while a "section" shows the surface of the Earth and below, with the emphasis on what is beneath that surface.

A "diagram," "model," or "view" may sound appropriate for these geologic items, but the definitions exclude their use. A "diagram" is described by the International Cartographic Association as, "a graphic representation of numerical data, or of the course or results of an action or process. The term is sometimes applied also to maps characterized by much simplified, or schematic, representation" (Stibbe, Cartmell, and Parker 1982, 226). An example of a "diagram" would be an item that shows the water cycle of the Earth, a block diagram of a portion of the Earth (with geologic formations possibly depicted on the vertical sides of the block, but with the emphasis on the Earth's surface), or a schematic of a petroleum processing plant. A "model" is defined by AACR2R as, "a three-dimensional representation of a real thing" (Gorman and Winkler 1988, 620). An example of this would be a raised-relief map. A "view" is described as, "a perspective representation of the landscape in which detail is shown as if projected on an oblique plane (e.g., a bird's-eye-view, panorama, panoramic drawing, worm's-eye view)" (Gorman and Winkler 1988, 624). A "view," by definition, may portray only the surface of the Earth.

Even though some items, such as well logs, are remotely sensed, they do not fall under the traditional definition of "remote-sensing images," which are considered to encompass aerial photographs and satellite images. "Globe" and "atlas" can be rejected out of hand as specific material designations.

As previously noted, while AACR2R states that if an item cannot be described by one of the cartographic terms, it is allowable to use terms from other chapters (Gorman and Winkler 1988, 108), when there are no other appropriate terms in AACR2R. "Chart" might be considered, but a chart, in cartographic terms, is, "a map designed primarily for navigation" (Stibbe, Cartmell, and Parker 1982, 225). Other terms such as "geologic sections," "cross sections," or "columns" do not appear in any part of AACR2R and therefore may not be used.

"Section" is the only term to use when describing the graphic representation of a geologic measurement. AACR2R's definition of the other eight SMDs preclude their use.

DEFINITIONS OF TERMS

Many items being cataloged will have terms on the item itself. These terms are not standardized; while one item may have a "geologic column," another will have a "stratigraphic column." These inconsistencies can lead to varied cataloging. The following list of terms has been included to provide clarification of the items being described in the process of cataloging. These definitions are based on the *Dictionary of Geological Terms,* third edition, by the American Geological Institute (Bates and Jackson 1984, 207, 118, 406, 207, 495, 494, 113, 494, 562, 405, 466, 181).

- Geologic time table

   A chronologic arrangement of geologic events. Almost all items that are used as examples have a time table. Most geologic maps, including ones in the U.S. Geological Survey *Geologic Quadrangle Series (GQ-),* have a geologic time table (Figures 1, 2).

- Cross section

   A diagram showing the intersection of a vertical plane and the surface of the Earth. The result shows the predicted strata in a generalized form. Cross sections often accompany geologic maps (Figures 3, 4).

- Profile section

   A diagram showing the intersection of a vertical plane and the surface of the Earth. Unlike a "cross section," the result is a line showing only the slope of the surface, paired with information on geologic strata below the profile (Figure 5).

- Geologic column–Stratigraphic column–Geologic section

   Diagrams that show a sequence of stratigraphic units below a specific point on the surface. The units are portrayed in a column. Most items in the U.S. Geological Survey's *Oil and Gas Investigations Chart Series (OC-)* are made up of geologic columns (Figure 6).

- Stratigraphic unit

   Stratum or strata used collectively for mapping, correlation, or description. Units may utilize the following determinants: fossils; radiation levels; lithology; age; and mineral content. Stratigraphic units compose a geologic column (Figure 7).

FIGURE 1. Geologic Time Table; from Wilmarth 1932

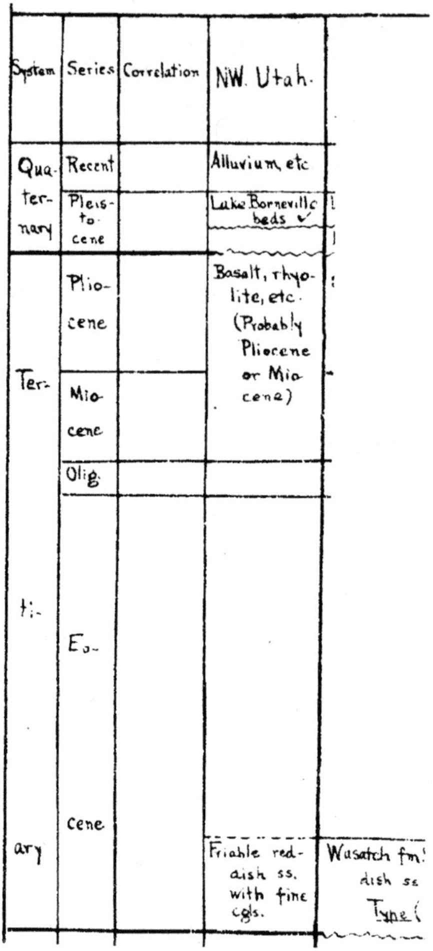

- Stratigraphic map

  A map that shows stratigraphic units. This is a geologist's general term for what is being discussed in this paper. The Library of Congress uses the term "Charts, diagrams, etc." as part of the subject string to describe this medium.

FIGURE 2. Geologic Time Table; from Swadley and Hoover 1989.a

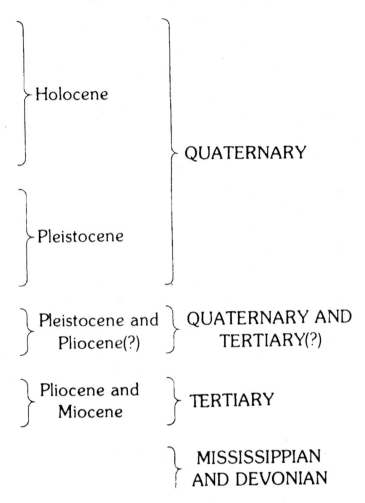

- Correlation lines

An item with two or more geologic sections usually has correlation lines connecting the sections. The lines show the relationship between stratigraphic units and how they compare in size and location (Figure 8).

FIGURE 3. Cross Section; from Geological Survey 1992

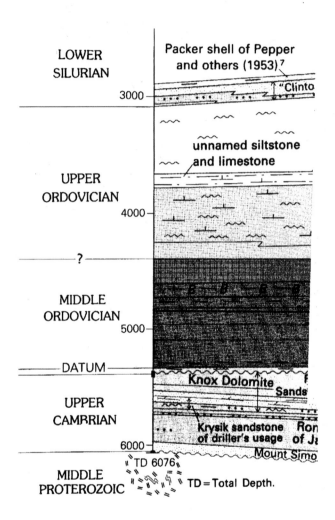

- Stratigraphic correlation–Correlation diagram

  When correlation lines are used, the item can be said to show stratigraphic correlation. Often generalized correlation diagrams are made of large areas, such as states (Figure 9).

FIGURE 4. Cross Section; from Krauskopf 1971

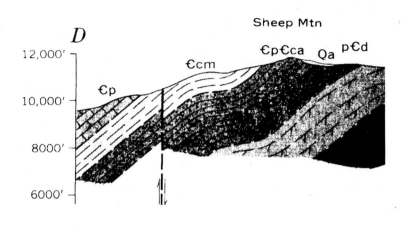

FIGURE 5. Profile Section; from Englund 1981

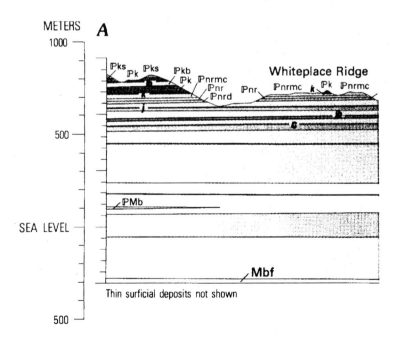

FIGURE 6. Geologic Column; from Nieschmidt 1954

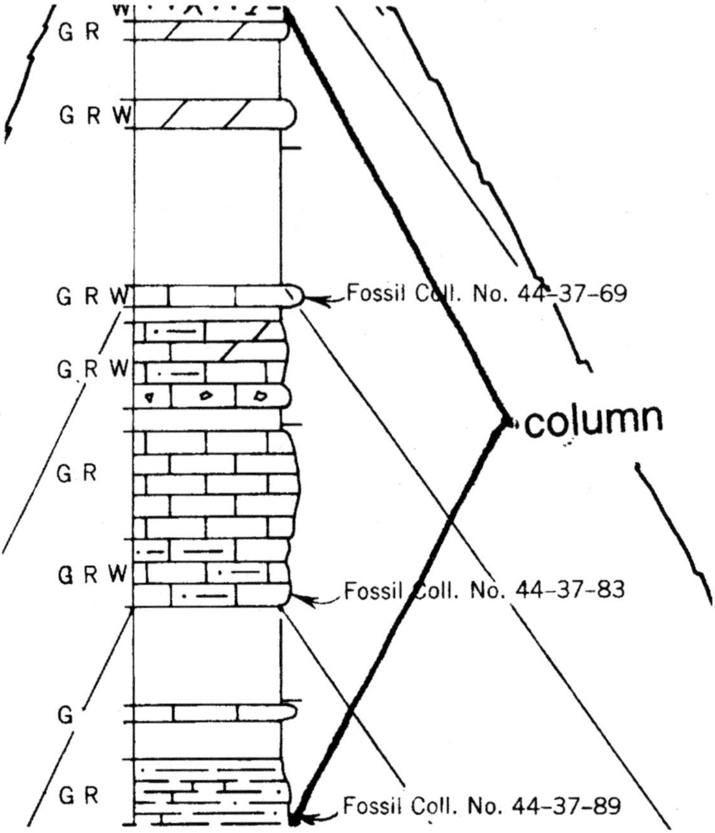

- Well logs–Geophysical logs

    Logs are created by lowering sensors into a deep hole that has been drilled. The sensors take continuous readings as they are raised. The resulting graph of the measurements is called a log. A log can show many geophysical properties including gravity anomalies, gamma rays, and radiation, and is commonly used in the U.S. Geological Survey's *Oil and Gas Investigations Chart Series (OC-)*. Well logs are made and used for both geology and mineral exploration, allowing geologists to get at information about petroleum and radioactive materials from the otherwise inaccessible rocks below the Earth's surface (Figures 10, 11, 12).

FIGURE 7. Stratigraphic Units; from Sloss and Laird 1945

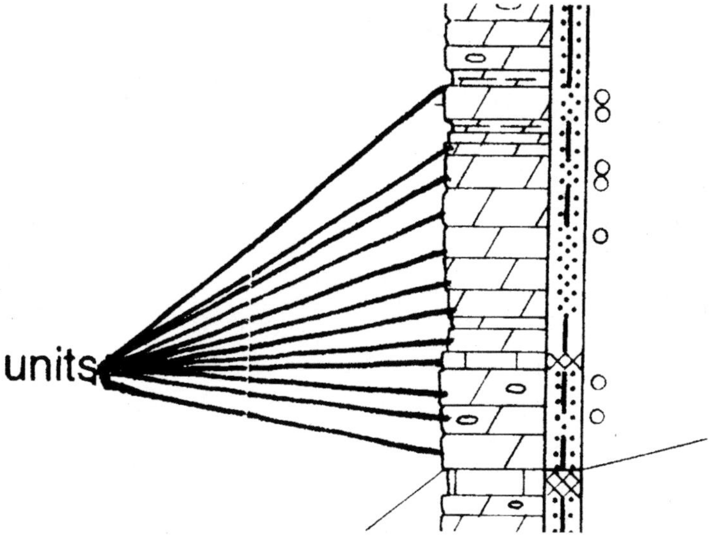

- Seismic profile, Profile

    When seismic energy is created by explosives, a sensor that previously has been inserted into a hole records measurements. A "profile" is created by compiling the results of many sensors over a wide area (Figure 13).

- Shot points

    The point at which a charge is exploded for the generation of seismic energy (Figure 13).

- Fence diagram

    A three-dimensional drawing of the subsurface, composed of three or more geologic sections. The intersection of the cross sections and correlation lines create the perspective. Fence diagrams are uncommon (Figure 14).

RULES FOR CATALOGING

For the purposes of simplicity, the term, "geologic section," will be used throughout the rest of this paper in place of all the terms above.

FIGURE 8. Correlation Lines; from Wanek 1954

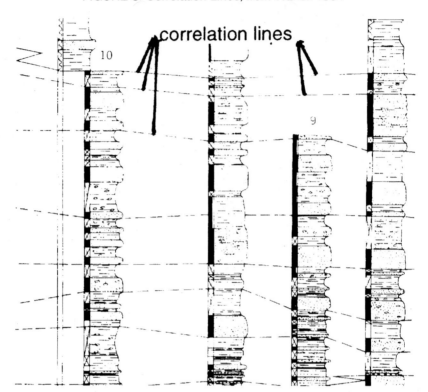

In general, sections should be treated as maps for cataloging purposes. There are no special rules for determining the title and statement of responsibility, publication, distribution, series, etc. However, there are six areas where the cataloging of geologic sections differs from the cataloging of maps: leader fields in a USMARC-formatted record; 0xx fields in a USMARC-formatted record; Scale; Physical description; Notes; and Subject headings.

LEADER FIELDS IN A USMARC-FORMATTED RECORD

In the Leader Fields, one variation occurs. It is possible for a geologic section to show relief, just as a map can. It does this by means of "profile"–showing a cross section of the ground above the sections (Capitol Reef 1980). Since "profile" is not one of the designates for "relief," "z" ("other") must be used. On some occasions, the relief may be identified with the

FIGURE 9. Correlation Diagram; from MacNeil 1947

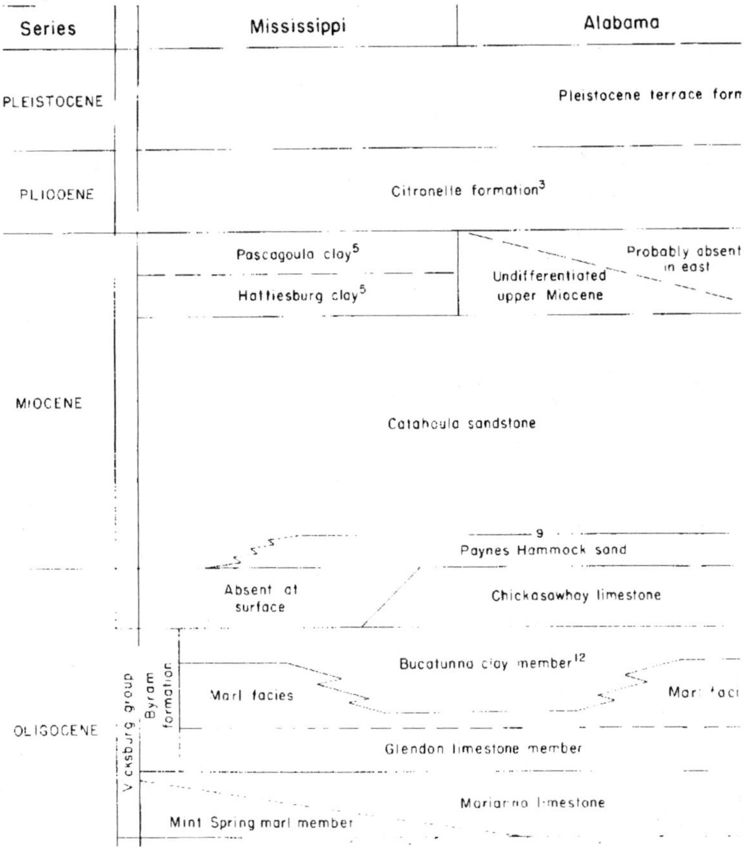

precise elevation figures, called "spot heights" (Capitol Reef 1980). Rarely, a section may show relief drawings of the land behind the profile area (Capitol Reef 1980); this is called relief that is depicted "pictorially." A section may accompany or be placed on top of a topographic map (Figure 15); if the map is an integral part of the item, report the relief as shown.

0XX FIELDS IN A USMARC-FORMATTED RECORD

The 0xx fields show two differences when comparing the cataloging of sections with that of maps. In the Physical Description Fixed Field (007), "s" ("section") must be chosen in the "b" subfield. As defined earlier by

FIGURE 10. Well Logs; from Markochick et al. 1982

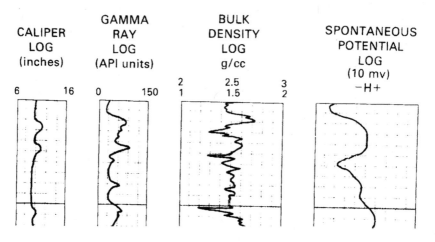

AACR2R terms, geologic sections can only be "sections." The second variant is scale and the 034 field, which will be explained later. The 043–Geographic Area Code–and especially the Geographic Classification Code (052) are recommended for use in records cataloging on the Map Format; use of the former may be dictated by local library practice. These fields are, "an aid to area specialists taking a subject approach to the item" (Online Computer Library Center, Inc., 1996, 0:108).

SCALE

As opposed to most maps, the horizontal scale of geologic sections is rarely included and often impossible to determine. A publication may have many sections that are placed evenly on the page, but while the interval between the sections on the item remains constant, the distance between them on the Earth's surface varies (Figure 16). Because of this, it is impossible to calculate the horizontal scale, even when the surface distance between the columns is known. Therefore the cataloger must use, "Scale not given," or, "Not drawn to scale." In a few cases, the horizontal scale is determinable when the scale takes the form of a bar (Figure 17), or is stated on the map.

Because sections, by definition, show some aspect of the subsurface, a vertical scale is frequently determinable. In some cases, a bar scale is placed on the item, away from the sections (see Figure 18); more commonly, the bar scale is superimposed on the sections themselves (Figure 19). Vertical scale is usually exaggerated by being considerably larger than the horizontal scale; it is not uncommon to find a vertical scale of 1:2,000 or larger. This vertical

FIGURE 11. Seismic Profiles; from Ryder 1981

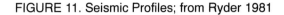

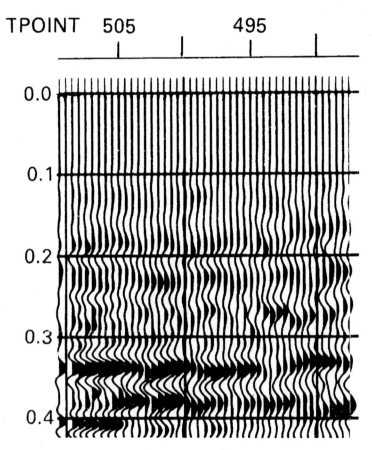

scale exaggeration is done to emphasize the information being presented in the sub-surface portion of the section.

In the 034 field of a USMARC-Formatted map record, the first indicator can only be "1" if a horizontal scale is determinable; otherwise, it must be a "0." The subfield "b" is used to indicate the horizontal scale, while subfield "c" is used to indicate the vertical scale. When cataloging a section, it is likely that subfield "b" will be blank, and subfield "c" will have a value.

In the 255 field of a USMARC-Formatted map record, some statement about the horizontal scale must be made, even if only the vertical scale can be calculated. An example could be: "Scale indeterminable. Vertical scale [ca. 1:1,000]." Vertical scale must be indicated by the word "vertical" so it is not confused with the horizontal scale.

FIGURE 12. Well Logs Matched with Geologic Units; from Huffman and Condon 1994

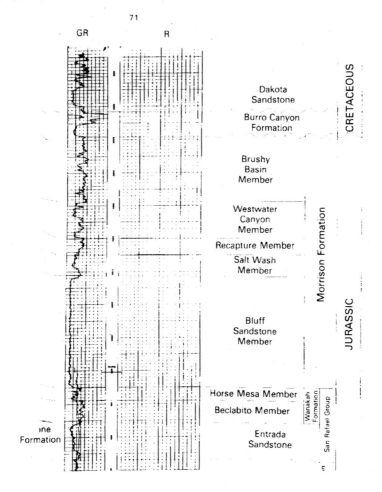

PHYSICAL DESCRIPTION

As when working with maps, describe the sections as being a number of sections on a number of sheets (e.g., "3 sections on 1 sheet"), unless there is one section on one sheet, in which case the correct phrase is, "1 section." Terms such as "geologic sections," "cross sections," or "columns" are not legitimate AACR2R descriptors.

FIGURE 13. Shot Points at Top, Seismic Profile Below Bar Scale; from Ryder 1981

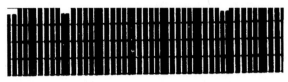

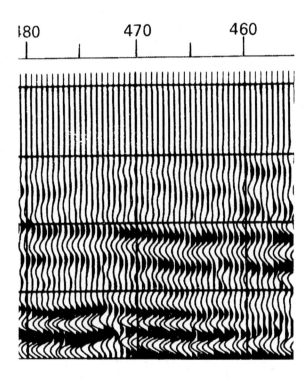

NOTES

Most note fields for a geologic section are determined and formatted just as are those for a map. Few unique incidences are likely to arise. In the notes fields, it may be necessary to elucidate what type of geologic section is being cataloged. Just as the phrase "bird's-eye view" is added, so may such phrases as "fence diagram," or "well log." These clarifications will assist patrons who are attempting to determine if a particular "section" fits their needs.

FIGURE 14. Fence Diagram; from Hardie and Van Gosen 1986

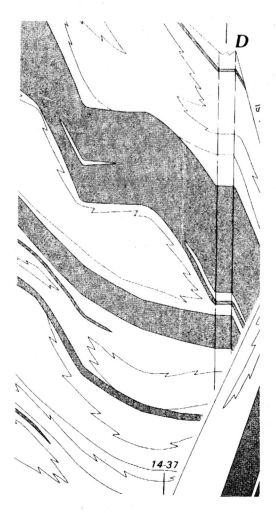

While relief should be mentioned as one would in cataloging a map, it may be necessary to use the phrase "Relief shown by profile" (Figure 5).

SUBJECT HEADINGS

As always, assign to items subject headings that are most appropriate for that particular item. When describing the format of sections, the term, "Charts, diagrams, etc.," is far more appropriate than "Maps."

FIGURE 15. Section; from Tabor and Ellen 1976

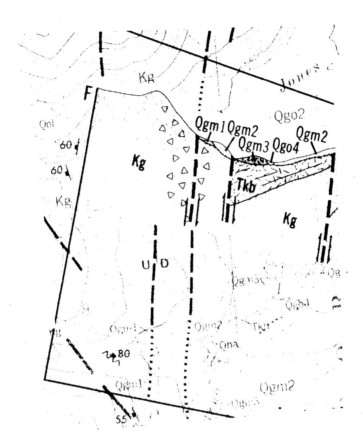

The Library of Congress' *Subject Cataloging Manual* prescribes: "Assign to the work being cataloged one or more subject headings that best summarize the overall contents of the work and provide access to its most important topics" (Library of Congress, Cataloging Policy and Support Office 1996, H180:1). Rarely will the general term "geology" be sufficient to describe a section that you are cataloging. As the Manual states, "Assign a heading that is broader or more general than the topic that it is intended to cover only when it is not possible to establish a precise heading. . . ." (Library of Congress, Cataloging Policy and Support Office 1996, H180:3).

As with maps, it is necessary to assign subject headings for the geographic location as well as the content of the item. A term that may be geographically subdivided and is appropriate for describing the content of geologic sections

FIGURE 16. "3 Miles" is Distance from Section Shown to Next Section to Right; from Love et al. 1945

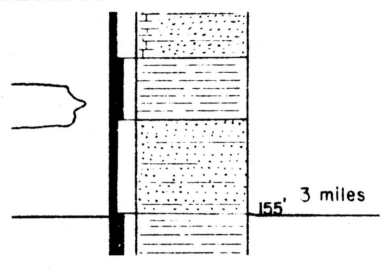

is "Stratigraphic correlation" (e.g., "Stratigraphic correlation $z Colorado $v Charts, diagrams, etc."). Sections are often specific to a geologic era, but geologic eras may not be expressed using "Stratigraphic correlation." Since "correlation" shows how several geologic sections line up with each other, "correlation" cannot be used if an item only has one section. "Geology, Stratigraphic" can be used if an item has one or many sections. In order to bring out the specific geologic era(s), "Geology, Stratigraphic" must be used, for it can mention geologic eras (e.g., " Geology, Stratigraphic $y Jurassic $v Charts, diagrams, etc."). However, "Geology, Stratigraphic" may not be geographically subdivided; for example, "Geology, Stratigraphic $z Colorado $v Charts, diagrams, etc." is an unacceptable subject heading. Therefore, a separate subject heading for area is needed, e.g., "Colorado $v Charts, diagrams, etc.," if the record has no subject headings with a geographic area.

Another subject heading that may be used and may be geographically subdivided is "Geophysics." When determining subject headings for an item that includes a well log, use the term "Geophysics." Do not use the terms "Oil well logging" or "Geophysical well logging," because they describe the methods; the items being cataloged show the results.

FIGURE 17. Bar Scales Showing Horizontal and Vertical Scales; from Wood, Johnson, and Dixon 1956

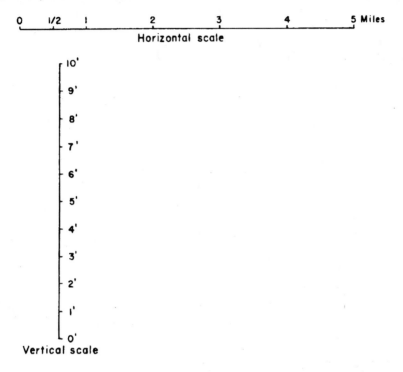

Horizontal scale

Vertical scale

EXAMPLE

Stratigraphy and Paleontology of Paleozoic Rocks, Hartville Area, Eastern Wyoming by J. D. Love, L. G. Henbest, and N. M. Denson will be used as an example. This 1953 section is number OC-44 in the U.S. Geological Survey's *Oil and Gas Investigation Chart Series*.

As stated earlier, sections should be treated as maps for cataloging purposes. There are no special rules for determining the title and statement of responsibility, publication, distribution, series, etc. The following shows the variant fields in a MARC record when compared to a map; these fields are emphasized with underlining. There are no changes in the Leader Fields between this and a typical map record. Note that the various figure numbers (e.g., "Figure 20") are inserted in this record to refer the reader to the correct illustration, and of course are not to be placed in a "real" catalog record.

007   a $b s $d a $e a $f n $g z $h n

034 0   a $c 480 (fig. 20)

043   n-us-wy

52   4261

255   Scale indeterminable. Vertical scale [ca. 1 :480]. (fig. 20)

300   16 sections on 2 sheets : $c sheets 138 x 104 cm.

500   Includes geologic time table (fig. 21), correlation lines (fig. 21), geophysical logs (fig. 21),

text (fig. 22), chart (fig. 23), location map (fig. 25), index map (fig. 24), 3 ancillary maps, and ill.

(fig. 26)

504   Includes bibliographical references (fig. 22).

650 0 Stratigraphic correlation $z Wyoming $x Charts, diagrams, etc.

650 0 Geology, Stratigraphic $y Paleozoic $x Charts, diagrams, etc.

650 0 Paleontology $y Paleozoic $x Charts, diagrams, etc.

CONCLUSION

There have been many irregularities in cataloging geologic sections and similar items. This paper has explained the use and importance of geologic sections, clarified why they should be cataloged on the USMARC Map Format, explained why they can only be called a "section" as an SMD, defined what they are, interpreted AACR2R and other manuals, and demonstrated the proper procedure for cataloging a geologic section. The intent is that catalogers will be able to use this paper as a reference as they wrestle with the complexities of geologic sections.

FIGURE 18. Vertical-Scale Bar; from Wanek 1952

Vertical scale

FIGURE 19. Vertical Scale Superimposed on Section; from Ryder 1992

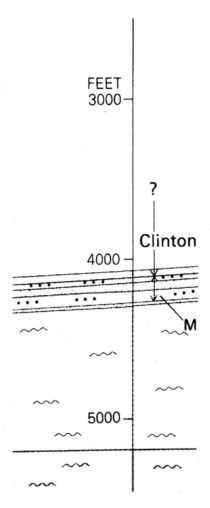

FIGURE 20. Vertical-Scale Bar; from Love, Henbest, and Denson 1953

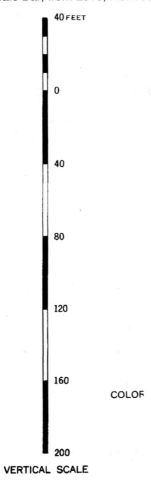

VERTICAL SCALE

FIGURE 21. Correlation Lines and Time Table; from Love, Henbest, and Denson 1953

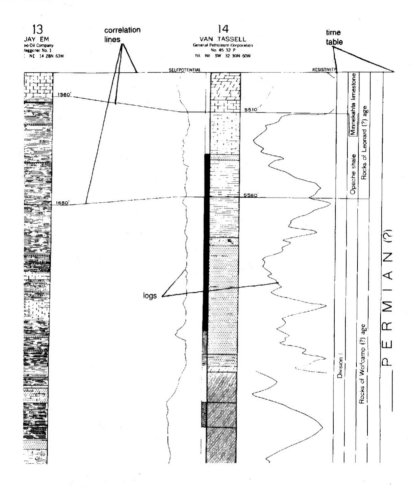

FIGURE 22. Text and Bibliography; from Love, Henbest, and Denson 1953

and Henbest,
s one of the
Nevertheless
zeringella by
Missouri age
and Keroher,
upper parts
ringella from
proceed with
?dekindellina
inid present.
re or absent.
the fauna of

and Keroher
· Pennsylva-
zata are gen-
y be distin-
gular coiling

he McLeans-
ibuted in the
zone near or
des *Fusulina*
i D. and H.,
a, *Bradyina,*
'obivalvulina

sion II because the age
so far obtained, the c
part of Division II must

Cloud, P. E., 1950, Wri
Condra, G. E., and Re
  of Wyoming: Nebr. G
Condra, G. E., Reed,
  Laramie Range, Har
  vey Bull. 13, 52 pp.,
Denson, N. M., and B
  U. S. Geol. Survey (
Dunbar, C. O., Henbe
  Illinois: Illinois St
Duncan, H., 1948, Ora
Elias, M. K., 1937,
  Kansas: Geol. Soc.
Hass, W. H., 1948, Wr
Henbest, L. G., 1946
  in adjacent parts
  Bull., vol. 30, pp.
Hunt, E. H., 1938, I
  leum Geologists Bu
Knight, J. B., 1948, V
Love, J. D., Denson,
  U. S. Geol. Survey
McCann, F. T., Ran
  Pennsylvanian, at
  Survey [Misc. Pub.
McCanne, R. W., ●94
  Niobrara County, W

FIGURE 23. Chart; from Love, Henbest, and Denson 1953

U.S.G.S. COLLECTIONS OF FORAMINIFERA (Geographic and stratigraphic position are indicated by location with reference to the columnar sections)		Ff. 6000	Ff. 6002	Ff. 6003	Ff. 6004	Ff. 6006	Ff. 6007	Ff. 6008	Ff. 6009	Ff. 6010
1	*Tolypammina, Calcitornella,* & allies					C				
2	*Trepeilopsis grandis* Cushman & Waters (2)					×				
3	Textulariidae (undifferentiated)									
4	*Spiroplectammina* sp.									
5	*Climacammina* sp.		×	2	×	×	?	2		
6	*Climacammina magna* Roth & Skinner (10)									
7	*Endothyra* sp.	?	C	A	×	C			×	
8	*Bradyina* sp.			×	R	R	×	×	R	
9	*Tetrataxis* sp.		×			R				
10	*Tetrataxis millsapensis* Cushman & Waters (2)						×			
11	*Polytaxis* sp.									
12	*Millerella* sp.									
13	*Millerella marblensis* Thompson) (15,16)	A	A	A						
14	*Millerella pinguis* Thompson (16)	C	C	R						

FIGURE 24. Index Map Showing Location of Geologic Mapping; from Love, Henbest, and Denson 1953

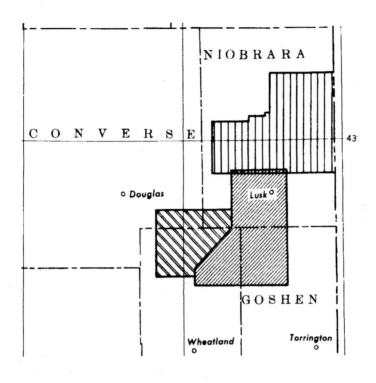

FIGURE 25. Map Showing Location of Section Lines; from Love, Henbest, and Denson 1953

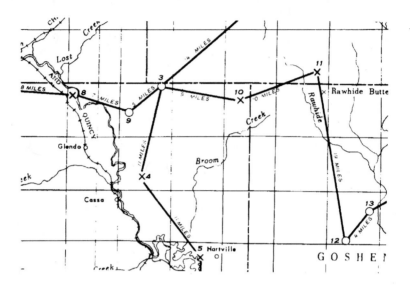

FIGURE 26. Illustrations; from Love, Henbest, and Denson 1953

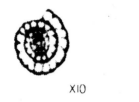

X10

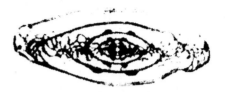

Triticites milleri

X10

X10

Triticites sp.

LIST OF REFERENCES

Bates, Robert L., and Julia A. Jackson, ed. *Dictionary of Geological Terms.* 3rd ed. Prepared under the direction of The American Geological Institute. Garden City, NY: Anchor Press/Doubleday, 1984.

Capitol Reef Natural History Association. *Geologic Cross Section from Cedar Breaks National Monument through Bryce Canyon National Park to Escalante, Capitol Reef National Park and Canyonlands National Park, Utah.* Torrey, Utah: Capitol Reef Natural History Association, 1980.

Englund, K. J. *Geologic Map of the Jewell Ridge Quadrangle, Buchanan and Tazewell Counties, Virginia.* U.S. Geological Survey *Geologic Quadrangle GQ-1550.* Reston, VA: U.S. Geological Survey, 1981. (Used in Figure 5.)

Geological Survey (U.S.). *Stratigraphic Framework of Cambrian and Ordovician rocks in the central Appalachian Basin from Lake County, Ohio, to Juniata County, Pennsylvania.* U.S. Geological Survey *Miscellaneous Investigations Series I-2200.* Reston, VA: U.S. Geological Survey, 1992. (Used in Figure 3.)

Gorman, Michael, and Paul W. Winkler, ed. *Anglo-American Cataloguing Rules.* 2d ed. rev. Ottawa: Canadian Library Association, 1988.

Hardie, J. K. and B. S. Van Gosen. *Fence Diagram showing Coal Bed Correlations within the Upper Part of the Fort Union Formation, in and adjacent to the Eastern Part of the Kaycee 30' x 60' Quadrangle, Johnson and Campbell Counties, Wyoming.* U.S. Geological Survey *Coal Investigations Map C-107.* Reston, VA: U.S. Geological Survey, 1986. (Used in Figure 14.)

Huffman, A. C.; and Condon, Steven M. *Southwest-northeast-oriented stratigraphic cross sections of Jurassic through Paleozoic rocks, San Juan Basin and vicinity, Colorado, Arizona, and New Mexico.* U.S. Geological Survey *Oil and Gas Investigations Chart OC-141.* Reston, VA: U.S. Geological Survey, 1994. (Used in Figure 12.)

Krauskopf, K. B. *Geologic Map of the Mt. Barcroft Quadrangle, California-Nevada.* U.S. Geological Survey *Geologic Quadrangle GQ-960.* Washington, D.C.: U.S. Geological Survey, 1971. (Used in Figure 4.)

Library of Congress. Cataloging Policy and Support Office. *Subject Cataloging Manual: Subject Headings.* 5th ed. Washington, D.C.: Library of Congress, Cataloging Distribution Service, 1996.

Library of Congress. Geography and Map Division. *Map Cataloging Manual.* Washington, D.C.: Library of Congress, Cataloging Distribution Service, 1991.

Love, J. D., L. G. Henbest, and N. M. Denson. *Stratigraphy and Paleontology of Paleozoic Rocks, Hartville Area, Eastern Wyoming.* U.S. Geological Survey *Oil and Gas Investigation Chart OC-44.* Washington, D.C.: U.S. Geological Survey, 1953. (Used in Figures 20-26.)

Love, J. D., R. M. Thompson, C. O. Johnson, and others. *Stratigraphic Sections and Thickness Maps of the Jurassic Rocks in Central Wyoming.* U.S. Geological Survey *Oil and Gas Investigation Chart OC-13.* Washington, D.C.: U.S. Geological Survey, 1945. (Used in Figure 16.)

MacNeil, F. S. *Correlation Chart for the Outcropping Tertiary Formations of the Eastern Gulf Coastal Plain.* U.S. Geological Survey Oil and Gas Investigation

Chart OC-29. Washington, D.C.: U.S. Geological Survey, 1947. (Used in Figure 9.)

Markochick, Dennis J., Robert E. Lanham, Hildie G. Bucurel, and Ben E. Law. *Summary Chart of Geologic Data from Amoco Tierney Unit 1 Well, SW 1/4 SE 1/4 Sec. 15, T. 20 N., R. 94 W., Sweetwater County Wyoming.* U.S. Geological Survey *Oil and Gas Investigation Chart OC-116.* Reston, VA: U.S. Geological Survey, 1982. (Used in Figure 10.)

Nieschmidt, C. L. *Subsurface stratigraphy of the Heath Shale and Amsden Formation in Central Montana.* U.S. Geological Survey *Oil and Gas Investigation Chart OC-50.* Washington, D.C.: U.S. Geological Survey, 1954. (Used in Figure 6.)

Online Computer Library Center, Inc. *Bibliographic Formats and Standards.* 2d ed. Dublin, Ohio: Online Computer Library Center, Inc., 1996.

Ryder, Robert T. *Processed and Interpreted U.S. Geological Survey Seismic Reflection Profile and Vertical Seismic Profile, Carter County, Montana, and Crook County, Wyoming.* U.S. Geological Survey *Oil and Gas Investigation Chart OC-115.* Reston, VA: U.S. Geological Survey, 1981. (Used in Figures 11, 13.)

——————. *Stratigraphic Framework of Cambrian and Ordovician Rocks in the Central Appalachian Basin from Lake County, Ohio, to Juniata County, Pennsylvania.* U.S. Geological Survey *Miscellaneous Investigations Series I-2200.* Reston, VA: U.S. Geological Survey, 1992. (Used in Figure 19.)

Sloss, L. L., and Laird, W. M. *Mississippian and Devonian Stratigraphy of Northwestern Montana.* U.S. Geological Survey *Oil and Gas Investigation Chart OC-15.* Washington, D.C.: U.S. Geological Survey, 1945. (Used in Figure 7.)

Stibbe, Hugo L. P., Vivien Cartmell, and Velma Parker, ed. *Cartographic Materials: A Manual of Interpretation for AACR2.* Chicago: American Library Association, 1982.

Swadley, W. C., and D. L. Hoover. *Geologic Map of the Surficial Deposits of the Topopah Spring Quadrangle, Nye County, Nevada.* U.S. Geological Survey *Miscellaneous Investigations Series I-2018.* Reston, VA: U.S. Geological Survey, 1989. (Used in Figure 2.)

Tabor, R. W. and S. Ellen. *Washoe City Folio: Geologic Cross Sections.* Reno, Nev.: Nevada Bureau of Mines, 1976. (Used in Figure 15.)

Wanek, A. A. *Geologic Map of Mesa Verde Area, Montezuma County, Colorado.* U.S. Geological Survey *Oil and Gas Investigation Map OM-152.* Washington, D.C.: U.S. Geological Survey, 1954. (Used in Figures 8, 18.)

Wilmarth, M. Grace, comp. *Tentative Correlation of the Named Geologic Units of Utah.* Washington, D.C.: U.S. Geological Survey, 1932. (Used in Figure 1.)

Wood, G. H. Jr., R. B. Johnson, and G. H. Dixon. *Geology and Coal Resources of the Gulnare, Cuchara Pass, and Stonewall Area, Huerfano and Las Animas Counties, Colorado.* U.S. Geological Survey *Coal Investigation Maps C-26.* Washington, D.C.: U.S. Geological Survey, 1956. (Used in Figure 17.)

Cataloging the Contemporary Printed Atlas

Paige G. Andrew

SUMMARY. Creating a bibliographic description for a contemporary atlas is perhaps more easily understood if one first gives consideration to what an atlas is. It is important to know that this article deals with cartographic atlases as opposed to atlases that focus on, e.g., the anatomy of human or other bodies, or of other types such as those about minerals. Of primary consideration is that cartographic atlases are first and foremost a means of displaying graphic information about the Earth's or other celestial body's surface and/or subsurface, with the physical nature of the item following in relevance when describing the item in hand. Following an overview of what defines an atlas this paper will serve to give the cataloger who has little or no experience with this format of cartographic information guidelines towards which fields are critical to its proper description and, therefore, its accuracy of retrieval. *[Article copies available for a fee from The Haworth Document Delivery Service: 1-800-342-9678. E-mail address: getinfo@haworth pressinc.com <Website: http://www.haworthpressinc.com>]*

Mr. Paige G. Andrew is the faculty Maps Cataloger and the Selector for the discipline of Geography at the Pennsylvania State University Libraries. He earned his bachelors degree in 1983 at Western Washington University, Bellingham, Washington, majoring in Geography. He earned his MLS in 1986 at the University of Washington, Seattle, Washington and is a native Seattleite. He can be reached via email at: pga@psulias.psu.edu

The author would like to thank the following individuals for editorial assistance and expertise in the creation of this article: Barbara Story, Cataloging Team Leader, Geography and Map Division, Library of Congress; Mary L. Larsgaard, Geospatial Data Librarian, Map and Imagery Laboratory and Alexandria Digital Library, Davidson Library, the University of California at Santa Barbara; and James W. Markham, Science & Atlas Cataloger, Davidson Library, University of California at Santa Barbara.

[Haworth co-indexing entry note]: "Cataloging the Contemporary Printed Atlas." Andrew, Paige G. Co-published simultaneously in *Cataloging & Classification Quarterly* (The Haworth Information Press, an imprint of The Haworth Press, Inc.) Vol. 27, No. 1/2, 1999, pp. 147-164; and: *Maps and Related Cartographic Materials: Cataloging, Classification, and Bibliographic Control* (ed: Paige G. Andrew, and Mary Lynette Larsgaard) The Haworth Information Press, an imprint of The Haworth Press, Inc., 1999, pp. 147-164. Single or multiple copies of this article are available for a fee from The Haworth Document Delivery Service [1-800-342-9678, 9:00 a.m. - 5:00 p.m. (EST). E-mail address: getinfo@haworthpressinc.com].

KEYWORDS. Bibliographic description, contemporary atlas, atlas, cartographic atlas, cataloging

INTRODUCTION

One of the most important outcomes of format integration in the cataloging world has been the growing acceptance that information content is more critical than its physical carrier or characteristics. Nowhere has this change been felt more prominently for map librarians than with atlases, those "book-like" items when bound, which are cartographic materials. Hence, our bibliographic descriptions must focus first on the unique aspects related to what an atlas contains, its cartographic components, followed then by the compilation of the remainder of its description. First and foremost, bibliographic records for atlases are now created on a workform for cartographic materials, that is, the code for "Type" in field 008 is now "e" as opposed to "a." In OCLC access to a blank workform of this type is achieved by typing "wfme" and <send> or <F11>. (Note that this paper will use OCLC field and tag labels and abbreviations throughout as this is the author's "home" bibliographic utility.) Therefore, the author aims to help move the cataloger's thought process away from an atlas' "book-ness" and towards its cartographic significance, and then to apply this knowledge to the task of describing the item in hand.

The article then will focus on which OCLC fields and subfields used in the bibliographic record have gained new levels of significance and how to correctly apply information about the atlas to them. This is not a radical leap away from the fields that a monograph cataloger is used to dealing with, in fact fewer than a dozen fixed and variable fields will be new. The end result will be twofold; the novice atlas cataloger will understand how to apply cartographic information found in the item to the bibliographic record while gaining respect for the primacy of the cartographic information displayed in an atlas.

A cartographic atlas, like its sister the sheet map, contains at its heart information related to an area of the Earth's, or other celestial body's, surface and/or subsurface. It has the same inherent problem as that of the single sheet map in that it attempts to display a three-dimensional surface in real life on a two-dimensional sheet of paper. Because of this transformation two items of information become critical in the bibliographic record, scale and projection. In one form or another scale must be stated while projection can be given if it is explicitly stated in the atlas. Geographic coordinates are also important to the atlas user but not as critical, at least as far as the printed atlas is concerned, as scale and projection.

Also similar to the sheet map or map set, the author of the atlas typically is attempting to create a work that reflects general information about a particular geographic area or chooses to display one or more specific topics about a given geographic area. General atlases most usually focus on a political area somewhere on the Earth, be it a country, a state or province, or a city or town. Common topical atlases include those created to show the geology, soils, climate, or other physical nature of an area or the area's history (natural or human) of either a single location or comparing/contrasting two or more locations. No matter which kind of atlas we are working with, general versus topical, it is imperative the cataloger understands that this is a cartographic creation.

ATLAS DEFINED

Anglo-American Cataloguing Rules, Second Edition, 1988 Revision (AACR2R) defines an atlas as "A volume of maps, plates, engravings, tables, etc., with or without descriptive text. It may be an independent publication or it may have been issued as accompanying material."[1] Additionally, the definition for atlas found in *Cartographic Materials: A Manual of Interpretation for AACR2* (Cartographic Materials) adds that it is "A collection of maps designed to be kept (bound or loose) in a volume."[2] For many years atlases were treated bibliographically as monographs in the cataloging community, the thinking being "if it looks and reads like a book it must be a book!" Contrast this thought process with two key components of the definitions above, that an atlas is a volume of maps and that this volume may or may not be bound, and one is able to see the significance of the atlas as a cartographic form. Format integration, achieved in 1996, brought the geospatial information found in an atlas to the fore and reinforced its standing as cartographic-material content first, followed in importance by its often unusual physical size and make-up.

The cataloger must now pay attention to information in the atlas that spells out its cartographic aspects, primarily scale, projection, and coordinates, as much as the information more commonly focused on in the past; title, author, publication information, and physical description. One other aspect that the cataloger must be aware of, but not overly concerned about, is the issue of monographic versus serial publication. This is an area that parallels the publisher's intent in the monographic world and thus must be given and correctly displayed in the bibliographic record. Fortunately, as with sheet maps, this is a fairly unusual circumstance. For an excellent review of seriality and series treatment related to cartographic materials see the title *Cataloging Map Series and Serials* found elsewhere in this volume.

The remainder of this article will describe the differences and/or additions, at the field level, used for describing an atlas, starting with the fixed fields, moving to the coded variable fields and the textual variable fields, and concluding with a brief look at subject access. Classification will be covered during the discussion of the 052 (Geographic Classification Code) field.

FIXED FIELDS OF SIGNIFICANCE
IN THE BIBLIOGRAPHIC RECORD

Once format integration was put in place the type of bibliographic record created for atlases changed from the monographs format to the cartographic format. This change means that the cataloger must become familiar with some new variable and fixed fields and the latter's appropriate codes. We will deal with the variable fields later on. While one new to atlas cataloging may perceive this to be a major hurdle, be assured that the number of new fixed fields to be learned is relatively small, and includes: Type of Cartographic Material (CrTp:), Relief (Relf:), Prime Meridian (Prme:), and Projection (Proj:).

Additionally, the following fixed fields will now require a different set of applicable codes, or a reinterpretation of existing codes, for atlases: Type of Record (Type:), Bibliographic Level (BLvl:), and Index (Indx). Note also that very often many of these fixed fields have a sibling variable field, meaning that the information provided in coded form in the fixed field is brought out descriptively in the variable field and information in both must match to be correct (e.g., the Index fixed field and 500 note in line 24 of Example 2).

A. New Fixed Fields

Type of Cartographic Material (CrTp)

This fixed field is used to determine what kind of cartographic item is being cataloged. The code to be used for printed or digital atlases in this field is "e." For manuscript atlases the correct code is "f." (For guidance in cataloging digital atlases see *Cataloging Digital Cartographic Materials* by Grace Welch and Frank Williams and *Cataloging Cartographic Materials on CD-ROMs* by Mary Larsgaard later in this volume.)

Relief (Relf)

This fixed field is to be coded for the appropriate methodology used to show either elevation above the Earth's surface or depths under a body of water. There are several ways in which cartographers graphically depict elevation or depth, nine of which we identify in the bibliographic record. The most common method is by contour lines and their opposite, bathymetric lines; by spot heights and their opposite, soundings; and by gradient tints (the use of different colors between either contour lines or bathymetric lines). For a complete list of codes see page FF:63 in OCLC's *Bibliographic Format and Standards*.[3] To see what these relief forms visually look like see the plate between pages 150 and 151 of *Cartographic Materials*. (Also see Example 1.)

Since depictions of relief may vary, the Relief fixed field is coded for one or more types of relief portrayed consistently throughout the main maps in an atlas. The field allows for up to four codes to be used for relief and/or depth; one can bring out more than this in a note. Showing an area's physical nature, or physiography, is very important in maps, so the first note given in the record very often will be the "relief note" or a note detailing how relief and/or depth is depicted. Remote-sensing images, however, are actual recordings of the Earth's surface and thus have no relief code or relief note recorded because it is not a visual representation of reality.

Projection (Proj)

As mentioned in the Introduction above, atlases (as well as sheet maps) must somehow overcome the problem of displaying a three-dimensional

EXAMPLE 1

```
OCLC:    37154051        Rec stat:    n
Entered:    19970619        Replaced:    19970619      Used:      19970619
Type:   e      ELvl:   I    Srce:   d    Relf:   dg   Ctrl:        Lang:  eng
BLvl:   m      SpFm:        GPub:        Prme:        MRec:        CLry:  wau
CrTp:   e      Indx:   1    Proj:        DtSt:   s    Dates: 1997,
Desc:   a
  1   040      WAU |c WAU
  2   007      a |b d |d a |e a |f n |g z |h r
  3   034 1    a |b 31680
  4   043      n-us-wa
  5   052      4283 |b J4
  6   090      G1488.J4 |b M42 1997
  7   090      |b
  8   049      UPMM
  9   100 2    Metsker Maps.
 10   245 10   Metsker's atlas of Jefferson County, Washington.
 11   246 30   Atlas of Jefferson County, Washington
 12   246 1    |i Title on generic title page: |a Jefferson
 13   246 1    |i Title on generic title page: |a Metsker's county atlas
 14   250      Brand new ed.
 15   255      Scale [1:31,680]. 2 in. = 1 mile.
 16   260      [Tacoma, Wash. : |b Metsker Maps, |c 1997]
 17   300      1 atlas (41 [i.e. 46] leaves) : |b maps ; |c 38 x 47 cm.
 18   500      Cadastral maps.
 19   500      Relief shown by hachures and spot heights.
 20   500      Cover title.
 21   500      Blue line print.
 22   500      "Copyright by Thomas C. Metsker."
 23   500      Includes index map and indexes to places, plats, roads, and
donation land claims.
 24   650  0   Real property |z Washington (State) |z Jefferson County |v
Maps.
 25   651  0   Jefferson County (Wash.) |v Maps.
 26   650  0   Landowners |z Washington (State) |z Jefferson County |v Maps.
 27   700 1    Metsker, Thomas Charles.
```

EXAMPLE 2

```
OCLC:   39671295        Rec stat:    n
Entered:    19980528    Replaced:    19980811   Used:    19981021
Type:   e   ELvl:       Srce:       Relf:       Ctrl:       Lang:   eng
BLvl:   m   SpFm:       GPub:       Prme:       MRec:       Ctry:   enk
CrTp:   e   Indx:   1   Proj:       DtSt:   s   Dates: 1998,
Desc:   a
   1   010       98-675210/MAPS
   2   040       DLC |c DLC
   3   007       a |b d |d c |e a |f n |g z |h n
   4   020       3828302726
   5   034 1     a |b 1000000
   6   041 0     engfreger
   7   050 00    G1797.21.P2 |b W54 1998
   8   052       5701
   9   072   7   P2 |2 lcg
  10   082 00    912.4 ß2 21
  11   090       |b
  12   049       UPMM
  13   110 2     Hallwag AG.
  14   245 10    Europa : |b 1:1000000 = Europe : 1:1000000.
  15   246 1     |i Title on p. [2] of cover: |a Collins road atlas Europe 1998
  16   250       Rev. ed. 1998.
  17   255       Scale 1:1,000,000.
  18   260       London : |b Collins, |c c1998.
  19   300       1 atlas (145 p.) : |b col. maps ; |c 39 cm.
  20   546       Legend in English, French, and German.
  21   500       Cover title.
  22   500       "Collins road atlas Europe 1998"--Verso cover.
  23   500       "Copyright HarperCollins Publishers 1998, Copyright Hallwag AG,
Bern, road maps, urban area maps & city maps"--Verso cover.
  24   500       Includes index, index map, distance table, and city maps in
various scales.
  25   650 0     Roads |z Europe |v Maps.
  26   650 0     Cities and towns |z Europe |v Maps.
  27   710 2     Collins (Firm : London, England)
```

surface on a two-dimensional sheet(s) of paper. One or more of four properties become distorted in this process, including true distance, true area, true direction, and true shape.[4] Therefore, numerous "projections," or mathematical means of correcting up to three of these properties in the transfer from three- to two-dimensions, have been established. Perhaps the most familiar one to the reader would be "Mercator," named after its inventor Gerhard Mercator. For an excellent bibliography of texts and other materials related to cartographic projections (and scale) see the following Web site, http://www.konbib.nl/kb/skd/skd/mathemat.htm (and for a beginner's look at what map projections are, their use, etc., go to http://everest.hunter.cuny.edu/mp/mpbasics.html).

In order to code this field all of the main maps must be of one type of projection, or a majority of the maps be one type, and that type must be specifically stated in the atlas, either in the text or on the maps themselves. For a list of specific codes that match the type of projection used see the list under "008–Maps" at the MARC 21 Website,[5] which is current, (or page

FF:60 of *Bibliographic Formats and Standards* or see http://www.oclc.org/oclc/bib/ffflist.htm#proj which is still useful). It may be necessary to code this field as "zz" if the stated projection found in the atlas is not one found on this list.

Prime Meridian (Prme)

In order to find one's place at a particular point on the Earth's surface some type of grid is usually used to create an "x/y" axis on a map. The most commonly used grid system, but certainly not the only, is called "latitude and longitude."[6] In this system degrees of latitude start at the Equator (given the value of zero) and move either north or south 90 degrees. Conversely, degrees of longitude are counted going either east or west from a starting line. (Since the Earth is a spherical object it is measured in degrees, minutes, and seconds, also known as a sexagesimal system whereby a circle is divided into 360 degrees.) This "starting line" is known as the Prime Meridian. In 1884 this line was drawn from the Earth's North pole to its South pole and passing through a place called Greenwich, England. Hence, modern maps use Greenwich as the prime meridian, and in this fixed field can be coded "e," but ONLY if Greenwich is explicitly stated. However, since Greenwich is the accepted standard prime meridian it is also acceptable to not code the field at all, which then means that it is understood to be the Greenwich prime meridian.

Historically however, prime meridians were established by the country creating maps for its own use, typically done by a national military agency. Other prime meridians gained prominence because of their frequent use such as Paris, London, and Washington, D.C. It is possible, in the case of a historical atlas of a given country, that all or most of the main maps have a prime meridian different from Greenwich. In such a case see the list of prime meridians and their accompanying codes to be used on page FF:59 of *Bibliographic Formats and Standards*.

B. Reinterpreting Codes for Familiar Fixed Fields

Type of Record (Type)

This fixed field is now coded "e."

Bibliographic Level (BLvl)

Appropriate codes for this field are "m" for monograph and "s" for serial. (Additionally, depending on the circumstance of publication and sometimes local need, the following codes can be used: "a" for monographic component part; "b" for serial component part; "c" for collection; and "d" for subunit of a collection.)

Index (Indx)

In the case of a monograph this field would be coded "0" if an index was not included that one could use to find where a topic was located in the work and "1" if such an index were included. In the case of atlases the two codes, "0" and "1" are the same, but their interpretation is different. *MARC 21 Concise Format for Bibliographic Data,* under the section on "008–Maps" tells us that the Index fixed field contains "A one-character code that indicates whether the item or accompanying material includes a location index or gazetteer." *Bibliographic Formats and Standards,* page FF:51, says that the Index fixed field "Indicates whether the work has an index, location index, or gazetteer to its own contents."

Use code "1" only if a gazetteer or textual place name index exists as part of the atlas (e.g., see all four examples). A gazetteer is a listing of places found on a map and usually includes their geographic, or latitude/longitude, location. The places listed can be exclusively populated places such as cities, towns, and villages; but commonly include rivers, lakes, mountains and mountain ranges, seas, and oceans. If a gazetteer or textual index is not present in the work code this field "0." Location maps or diagrams and index maps that show geographic location of a particular state, country, or region in reference to its surrounding localities do not qualify as an "index." Indexes are meant to have a one-to-one relationship between a place on a map (or word or topic in a book) and the listing of the place so that place can be easily referenced by the user.

CODED VARIABLE FIELDS OF NOTE

The following coded variable fields must be addressed while creating a bibliographic record for atlases: 006 (Fixed-Length Data Elements-Additional Material Characteristics), 007 for maps (Physical Description Fixed Field), 034 (Coded Cartographic Mathematical Data), and 052 (Geographic Classification Code). Optionally, one may add the 043 (Geographic Area Code) as a potentially indexable piece of coded information that could be used for retrieval of groups of items representing a particular place (e.g., see Examples 1 and 3).

006 Field

With format integration and the move towards placing content of information into prominence over carrier a new field was established so that physical

EXAMPLE 3

```
OCLC:  31161111          Rec stat:    c
Entered:  19940923       Replaced:   19970929      Used:    19981222
Type:  e      ELvl:  I    Srce:  d    Relf:     Ctrl:       Lang:   eng
BLvl:  m      SpFm:       GPub:       Prme:     MRec:       Ctry:   abc
CrTp:  e      Indx:  1    Proj:       DtSt:  s  Dates: 1994,
Desc:  a
  1  040    GIS |c GIS |d UIU |d IXA |d IQU
  2  006    [aab    b    s001 0 ]
  3  020    0920230539
  4  034 0  a |d W1260000 |e W0960000 |f N0620000 |g N0490000
  5  043    n-cnp--
  6  052    3465
  7  090    G1151.C57 |b M67 1994
  8  092 0  557.120223 |b G292 ß2 20
  9  090    |b
 10  049    UPMM
 11  100 1  Mossop, Grant D.
 12  245 10 Geological atlas of the Western Canada Sedimentary Basin / |c
compiled by Grant Mossop and Irina Shetsen ; [senior scientific editors,
Grant Mossop, Jim Dixon ; associate editors, John Kramers, Mika Madunicky].
 13  246 30 Western Canada Sedimentary Basin
 14  255    Scales differ |c (W 126 --W 96 /N 62 --N 49 ).
 15  260    Calgary, Alta. : |b Canadian Society of Petroleum Geologists ;
|a Edmonton, Alta. : |b Alberta Research Council, Alberta Geological Survey,
|c c1994.
 16  300    1 atlas (x, 510 p.) : |b ill. (chiefly col.), maps (chiefly
col.) ; |c 45 x 59 cm. + |e 1 plastic overlay (1 sheet ; 36 x 54 cm.)
 17  500    "The headquarters of the Atlas project was located at the
Alberta Geological Survey (AGS) in Edmonton; [The] co-compilers ... were
responsible for the scientific work associated with the building of the Atlas
database and the production of the contour and lithofacies maps; All aspects
of project coordination ... were directed and discharged by Mika Madunicky"--
p. iv.
 18  500    "Virtually all of the authors and contributors are members of
the Canadian Society of Petroleum Geologists"--p. v.
 19  500    "Published jointly by the Canadian Society of Petroleum
Geologists and the Alberta Research Council, in sponsorship association with
the Alberta Department of Energy and the Geological Survey of Canada."
 20  500    Issued in 6-post binder.
 21  500    Errata sheet inserted.
 22  504    Includes bibliographical references and index.
 23  651 0  Western Canada Sedimentary Basin |v Maps.
 24  650 0  Sedimentary basins |z Prairie Provinces |v Maps.
 25  650 0  Geology |z Prairie Provinces |v Maps.
 26  650 0  Paleogeography |z Prairie Provinces |v Maps.
 27  650 0  Petroleum |x Geology |z Prairie Provinces |v Maps.
 28  650 0  Geology |z Canada, Western |v Maps.
 29  650 0  Paleogeography |z Canada, Western |v Maps.
 30  650 0  Petroleum |x Geology |z Canada, Western |v Maps.
 31  650 0  Geology, Stratigraphic.
 32  700 1  Shetsen, I.
 33  700 1  Madunicky, Mika.
 34  710 2  Canadian Society of Petroleum Geologists.
 35  710 2  Alberta Geological Survey.
```

description of the carrier could be brought forth in a coded means when it differs significantly from content. In the case of atlases, since coding of the fixed fields is aimed at bringing out cartographic content, the 006 field serves to define specifics of its physical features, i.e., its "book-like" features such as illustrations or level of audience. This is a required field in OCLC for member institutions (although the Library of Congress does not use it for its bibliographic records of atlases). See pages 0:2-0:5 in *Bibliographic Formats and Standards* for the procedures in setting up this field. (See Example 3.)

007 Field

"Use field 007 to code for the physical characteristics of an item."[7] There are nine sets of codes for the type of material one is cataloging, e.g., one for maps, one for globes, etc. This is a required field for full-level cartographic cataloging. Use the 007 for Maps, found on pages 0:39-0:41 in *Bibliographic Formats and Standards*. In cataloging sheet maps the most common set of 007 codes would be for a multi-colored map on paper or a single-colored map on paper and would look like:

007 a $b j $d c $e a $f n $g z $h n (for a multiple-color map)

007 a $b j $d a $e a $f n $g z $h n (for a single-color map; usually, but not always, black)

The only significant change to these most-common occurrences for physical characteristics is to change subfield "b" to code "d" (instead of "j"), which designates it as an atlas. (See Examples 1 and 2.)

034 Field

This field's content is inherently linked in a one-to-one relationship with the 255 (Cartographic Mathematical Data) field (usually called the "scale" field by catalogers). The information portrayed in the 034 field is simply an abbreviated form of the information given in the 255 field (discussed later), with the exception of type of projection. (See all four examples.) Up to two 034 fields can be given (paralleling up to two 255 fields) in the case where all of the main maps in an atlas are of two primary scales for a given single geographic area or when two primary geographic areas are displayed throughout the atlas.

Starting with 1st indicator values, which can be "0," "1," or "3," they mean the following: "0" means no scale is given, a scale cannot be easily determined, or a statement such as "Scale not given," "Scales differ," or

"Not drawn to scale" is given in the 255 field (see Examples 3 and 4); "1" means that a single scale is present and its value is then given in numeric form in subfield "b" (see Examples 1 and 2); and "3" means a range of scales is given, each numeric value is then displayed using two subfield "b"s. [Note that "a range of scales" means that scale changes in value going from the center of a map to its outer content *on a single map,* therefore, one does not have to worry about indicator value "3" for atlases unless the entire atlas were made of these kinds of maps!]

034, subfield "a" indicates "category of scale" and one of three codes can be applied. In the case of atlases this will always be code "a," meaning that it contains a linear scale.

034, subfield "b" contains the denominator value of the representative fraction (RF) given in field 255, subfield "a." An example might be:

255 Scale 1:250,000.

034 1 a |b 250000

Note that in 034 |b the comma is *not* transferred from the scale information displayed in the 255 field.

EXAMPLE 4

```
OCLC:   36665279          Rec stat:    n
Entered:    19970402      Replaced:    19970402    Used:     19980904
Type:   e      ELvl:  I   Srce:  d     Relf:       Ctrl:         Lang:   eng
BLvl:   m      SpFm:      GPub:        Prme:       MRec:         Ctry:   nyu
CrTp:   e      Indx:  1   Proj:        DtSt:   s   Dates: 1997,
Desc:   a
  1  040     CKM |c CKM
  2  006     [aab           001 0 ]
  3  020     0723004927
  4  034 0   a
  5  052     3200
  6  090     G1021 |b .T55 1997
  7  090     |b
  8  049     UPMM
  9  245 04  The Times atlas of the world.
 10  246 30  Atlas of the world
 11  250     9th comprehensive ed., reprinted with revisions 1997.
 12  255     Scales differ.
 13  260     New York : |b Times Books, |c 1997.
 14  300     1 atlas (xlvii, 123 [i.e. 245], 218 p.) : |b col. ill., col.
aps ; |c 46 cm.
 15  500     "Maps prepared in Great Britain by Bartholomew,
arperCollinsPublishers, Glasgow"--T.p. verso.
 16  500     Duplicate legend laid in.
 17  500     Maps on lining papers.
 18  500     Includes index.
 19  650  0  Atlases.
 20  710 2   Times Books (Firm)
 21  710 2   Bartholomew (Firm)
```

034 |d, |e, |f, and |g are coded for coordinate values, if given, and contain the abbreviations for "west," "east," "north," and "south" (W, E, N, S), and the numerical values given for each coordinate in field 255 subfield "c." (See Example 3.) In the case of atlases, if all of the main maps show one or two geographic areas and the bounding coordinates for that area are given or can be easily ascertained then the coordinates for one or both areas are given in the record in separate 034 fields. For instructions on how to correctly order and display coordinates see pages 57-61 in *Cartographic Materials.* Assuming the atlas being cataloged is for one geographic location, such as the United States, an example of a complete 034 field would look like this:

> 034 1   a |b 7000000 |d W1250000 |e W0650000 |f N0500000 |g N0250000

Note that in subfields "d" through "g" there are always seven numerical place-holders, whether those places contain numbers greater than zero or not. In the example above the zeroes following the number of degrees, i.e., 125, 65, 50, and 25, are holding places for minutes and seconds and the zero before 65, 50, and 25 degrees holds the place for if the degrees were 100 or higher. The matching 255 field for the example above would look like this:

> 255   Scale 1:7,000,000 |c (W 125°--W 65°/N 50°--N 25°).

[This 255 field could also contain the name of a projection and additional related information in a subfield "b," but remember that projection information is not part of the 034 field, instead it is given in coded form in the (Proj) fixed field.]

052 Field

This coded field is used to display the particular classification number(s) and, if necessary, geographic area cutter portion of a class number derived from the Library of Congress Classification, Class G schedule (G schedule).[8] There are two parts to the field, subfield "a" is the place for the four-digit class number that represents the geographic area being classified and subfield "b" is the place for the geographic cutter, representing a subarea such as a city (see Example 1).

The G schedule is arranged with a beginning section on the discipline of geography, geographers, voyages and travels, etc., then followed by a large section used to classify atlases, a small section devoted to classifying globes, and finally the largest section for sheet maps. As much more broadly explained in the article *Navigating the G Schedule,* found later in this volume, the range of classification numbers delegated to atlases is G1000-G3122 (followed by G3160-G3182 for globes and G3190-G9980 for maps). In all cases at least one classification number must be given when cataloging an atlas, it should be derived from the list of class numbers for maps, and it must match the subject heading(s) given in the 65X field(s). If the atlas is about more than one specific place not only should there be 65X fields designated for subject access, but there should be more than one 052 field. This field is repeatable, as is the subfield "b" within a given 052 field.

The G Schedule is a superb system that is flexible enough in its use, combined with a method of both geographic and subject cuttering, for us to be able to uniquely identify places on the Earth, starting with the Earth itself. The G Schedule also allows us to classify other planets and the Earth's Moon. In the case of places on Earth, the system takes the largest area, the Earth, and methodically subdivides it into hemispheres, continents, geographic regions, countries, and states or provinces. Areas smaller than states or provinces, such as counties in the case of the United States, geographic regions, features, etc. (i.e., national parks, rivers, lakes, mountains), and individual cities require the use of cutters combined with a method of changing the fourth digit in a particular classification number, to further refine this continuous subdividing of the Earth. There *are* a few problems with the G Schedule; these are also noted in *Navigating the G Schedule*.

Currently, the only published list of geographic cutters that has been accepted as the national standard, is a microfiche volume created by the Library of Congress Geography and Map Division titled *Geographic Cutters*.[9] Unfortunately, the list is exclusively for places in the United States. In order to obtain cutters for places outside the U.S. one can go to the Library of Congress' online catalog via the WWW at: http://www.lcweb.loc.gov/catalog, search for a record of the same place and borrow the cutter from the atlas or map record found; search for a bibliographic record for the same place created by the LC G&M Division in a bibliographic utility such as OCLC and borrow from there (its institution symbol is "DLC" as in the 040 field in Example 2); or contact the Cataloging Team Leader using her email address found on the G&M Division's homepage at http://lcweb.loc.gov/rr/geogmap/catteam.html and place a request. There are plans in the near future for an electronic version of the G Schedule, to be mounted as part of the Library of Congress' *Cataloger's Desktop* software, and it will include, or link to, a database of geographic cutters that include places worldwide.

VARIABLE FIELDS OF NOTE RELATING TO THE DESCRIPTION OF ATLASES

There is one variable field in particular that may be new to the novice atlas cataloger. This is the 255, or Cartographic Mathematical Data field, touched on briefly during the discussion of the 034 Coded Cartographic Mathematical Data field above. Below you will find a discussion on its use and application. In addition, variable fields one may be familiar with using in the realm of monographs, but which are used differently in the context of cartographic materials, will be elaborated on below beginning with the 1XX Main Entry field.

1XX (Main Entry) Field

Contemporary atlases, and cartographic material in general, are most usually created, or caused to be created, by a corporate body as opposed to a single individual. Due to this particular circumstance a new category was added to Rule 21.1B2 in the 1988 revision of the second edition of Anglo-American Cataloguing Rules, though corporate body main entry was valid in some cases before this rule change. Category "f" allows for corporate body main entry in the case of cartographic materials, "cartographic materials emanating from a corporate body other than a body that is merely responsible for their publication or distribution."[10] In a good many cases a corporate body is both responsible for the creation of the atlas and its publication. (See Example 1.) However, if the atlas' creation is responsible by an individual then personal main entry is correct. (See Example 3.) This was more the case for atlases created before the twentieth century; for more specific information on this situation see the article *Cataloging Early Atlases: A Reference Source* later in this volume. Note that, if in the case of personal main entry the atlas was published by a recognized cartographic corporate body, governmental or private, it is wise to also make an added entry for such a body (e.g., the U.S. Geological Survey, Rand McNally & Company, the National Geographic Society, etc.).

255 (Cartographic Mathematical Data) Field

This field has been largely covered in the area above relating to the 034 field thus I will not go into great detail here. It is important to reiterate the fact that the information given in subfields "a" and "c" must match, except in form (i.e., punctuation and spacing), that given in the 034 field (as in all four examples).

This is a repeatable field and two 255 fields (along with two 034s) can be created in the case where all of the main maps in an atlas are of two scales. In the case where all of the main maps are of three or more scales a single 255 field is created and one uses the phrase "Scales differ" to indicate this (see Examples 3 and 4). And finally, in the case where scale is not given at all a 255 is created using the phrase "Scale not given." For more detailed information and applications see pages 51-55 in *Cartographic Materials*.

260 (Publication, Distribution, Etc. Area) Field

Note once again that the publisher is often the primary author of an atlas, but in cases where this is not true it is valuable to our patrons to create a 7XX tracing for prominent or well-known publishers (e.g., see Examples 2 and 4, also note the added entries for the three government agencies in Example 3).

300 (Physical Description) Field

Unfortunately, this field remains more closely aligned with describing a book than it does with describing a volume of maps that are bound into one physical item. First, in AACR2R, rule 3.5B1 we are instructed to "Give the extent of a cartographic item. In the case of atlases . . . give the number of physical units." Then rule 3.5B3 says, "Add, to the statement of extent for an atlas, the pagination or number of volumes as instructed in 2.5B" before considering the maps involved. The number of maps, if counted, is relegated to subfield "b," along with such matters as color or mounting. Counting maps in an atlas can be problematic at best as often a map covers two or more pages, (i.e., do we count these as one map or more and how many are really involved?) and besides, which maps does one count in a particular atlas? For example, general world atlases often have brief sections (e.g., 50 pages) of thematic material that include ancillary maps. Does one count these "maps," even though they are ancillary in nature, in addition to the main maps? Turning to rule 2.5C3 and rule 2.5C4 one of the examples given in each indicates that maps are merely "illustrations." But, rule 2.5C4 also says "Give the number of illustrations *if their number can be ascertained readily . . .*" [emphasis added by author], allowing us to do what is more appropriate and that is to not count the maps in an atlas unless they are very few in number. Where the line is drawn as far as when one will or will not count the number of maps is, of course, up to local needs and practice; at the Pennsylvania State University the practice is that 20 is the cutoff point.

Also, "ill." or "col. ill." comes before "maps" or "col. maps" in subfield "b," which contradicts the importance of the cartographic content . . . but I digress. To get to the crux of the matter see pages 94-102 and 104-106 in *Cartographic Materials.*

Recording the extent of an atlas is more straightforward. Rule 3.5D2 in AACR2R refers one to rule 2.5D where in most cases one records the height of the volume in centimeters. Rule 5D2 in *Cartographic Materials,* pages 115-116, gives clear guidance on recording physical dimensions of an atlas.

Of particular concern though are those cases in which one of the physical dimensions is more than twice the size of the other (regarding height vs. width). In this case one is referred to rule 2.5D2 in AACR2R, "If the width of the volume is either less than half of the height or greater than the height, give the width following the height preceded by a multiplication sign." (See Examples 1 and 3.) In summary, the pattern is as per the following examples:

300   1 atlas (   ) : |b col. maps ; |c 29 cm.

300   1 atlas (   ) : |b col. ill., col. maps ; |c 32 x 14 cm.

5XX (Notes) Fields

Applying notes to the bibliographic record for contemporary atlases does not vary much from doing the same for monographs, and by definition follows the pattern for sheet maps. In particular, the order of notes as outlined in Rule 3.7 in AACR2R should be applied, beginning with "Nature and scope of item." However, here is the place to bring out further specific information regarding the cartographic aspects of the maps included in the atlas. For instance, a note regarding the specific type of relief symbolization (and depth) should be given at or near the top of the list as the only other appearance of this kind of information is given in coded form in the Relief fixed field (e.g., "Relief shown by spot heights. Depths shown by soundings," also see Example 1). Notes that bring out more specific mathematical information such as any prime meridian used other than Greenwich (in the form "Prime meridian: Paris.") or variations regarding scale can be very important to the user.

Notes regarding variations in title or other title locations, publication/distribution-specific information, edition(s), series, accompanying material, "includes" notes, contents, language and the like all follow from what the cataloger is used to while working with monographs.

6XX (Subject Heading) Field

Assigning subject headings to a work that contains primarily maps lends itself to being, thankfully, fairly straightforward. The rule of thumb here is to first determine the kind of atlas you are working on, general vs. topical, and then proceed. If one has a general atlas of France then your primary subject heading will be France--Maps (see below for those situations when one can use the subject *subdivision* "Atlases," unfortunately, not with cartographic materials). In the case of the *subject heading* "Atlases," the scope note in LC Subject Headings indicates a very specific use, "Here are entered geographical atlases of world coverage . . ."[11] (see Example 4). The cataloger may wish to bring out other topical aspects following this if it is warranted, e.g., if the atlas contains a notable selection of ancillary maps for individual cities in France the subject heading Cities and towns--France--Maps may also be appropriately added. (See Example 2 for a similar situation.)

In the case of topical atlases the world of subject headings broadens rather quickly. In most cases these atlases are about the physical geography of a place (e.g., geology or hydrology), or specific topics such as its political boundaries and entities (Political atlas of . . .), economic conditions (Economic atlas of . . .), looking at the area's historical development (Historical atlas of . . .), or based on the location of attributes such as population. Using the example of a geological atlas of Chile, the primary subject heading would be Geology--Chile--Maps, potentially followed by related types of geological forms subject headings if necessary. (See Example 3 for a similar situation.)

The *subject subdivision* "Atlases" can only be used for non-geographic works containing plates, charts, diagrams, etc., of a scientific nature, usually atlases of the anatomy of an animal or human beings. See H1935, no. 2f in volume 4 of the Library of Congress' *Subject Cataloging Manual: Subject Headings* for a complete explanation of this use of the term "Atlases."[12]

7XX (Personal or Corporate Body Added Entries) Field

As noted above in the discussion on the 1XX and 260 fields, oftentimes it is appropriate to add a tracing for a prominent or well-known cartographic publisher as it increases access to the atlas (as is also true with sheet maps). Additionally, in circumstances where the 1XX is for corporate body main entry and there is also a person named as editor or there is more than one corporate body involved one may add a 7XX(s) as appropriate to provide access to the work from these persons or bodies.

CONCLUSION

In 1996 format integration was fully implemented in the cataloging world, following a change of philosophy concerning which is more important, the information content of an item or its physical container. Atlases, which until this time were considered monographs primarily because of their "booklike" physical features, finally were moved into the arena and under the purview of cartographic materials cataloging. This welcome, and long argued for, change suddenly awakened a need for monograph catalogers to learn new fields in the bibliographic record or recognize new meanings for existing fields. The above article should prove useful to the cataloger who has never dealt with describing contemporary atlases, the cataloger who has been used to describing atlases as monographs, and even to those map catalogers with little or no experience working with atlases.

NOTES

1. *Anglo-American Cataloguing Rules, Second Edition, 1988 Revision.* Edited by Michael Gorman and Paul W. Winkler. (Ottawa: Canadian Library Association; Chicago: American Library Association, c1988).

2. *Cartographic Materials: A Manual of Interpretation for AACR2.* Hugo L.P. Stibbe, general editor. (Chicago: American Library Association, 1982).

3. *Bibliographic Formats and Standards.* 2nd ed. (Dublin, Ohio: OCLC, c1996).

4. Roman Drazniowsky, *Map Librarianship: Readings.* (Metuchen, N.J.: Scarecrow Press, 1975), 88.

5. *USMARC Concise Format for Bibliographic Data.* 1994 ed. with updates. See the Website: http://lcweb.loc.gov/marc/bibliographic/ecbd008s.html

6. Peter H. Dana, *Geodetic Datum Overview, Global Coordinate Systems.* See the Website: http://www.utexas.edu/depts/grg/gcraft/notes/datum/datum.html

7. *Bibliographic Formats and Standards,* 6.

8. *Library of Congress Classification, Class G: Geography, Maps, Anthropology, Recreation.* (Washington, D.C.: Library of Congress, 1976).

9. Geography and Map Division, Library of Congress, *Geographic Cutters.* 2nd ed. (Washington, D.C.: Library of Congress, Cataloging Distribution Service, 1989).

10. *Anglo-American Cataloguing Rules, 2nd Ed., 1988 rev.,* 314.

11. *Library of Congress Subject Headings,* 20th Ed. Prepared by the Cataloging Policy and Support Office, Library Services, (Washington, D.C.: Cataloging Distribution Service, 1997), 371. (Additionally, headings in LCSH are available through the LC Name Authority file in OCLC's WorldCat and an electronic version of LCSH is available as a part of *Cataloger's Desktop* software.)

12. *Subject Cataloging Manual: Subject Headings.* 5th ed. Prepared by the Cataloging Policy and Support Office, Library of Congress. (Washington, D.C.: Cataloging Distribution Service, 1996).

Cataloging Aerial Photographs
and Other Remote-Sensing Materials

HelenJane Armstrong
Jimmie Lundgren

SUMMARY. Remote-sensing images are valuable library resources, which provide highly useful information to a variety of library patrons. They are graphic representations of spatial relationships recorded by a device that was not in physical contact with the geographic entity being studied. Effective access and description in the library catalog is necessary so that these images may be found and used. This article discusses characteristics of remote-sensing images and maps: how to identify

---

HelenJane Armstrong, University Librarian, is Head of the Map & Imagery Library in the George A. Smathers Libraries and Adjunct Professor, Geography Department, University of Florida. In addition to her MLS she holds a PhD in Geography. She founded the Map Library in 1973 and was responsible for map cataloging until December, 1997. She is one of the United States Representatives to the Anglo-American Cataloguing Committee for Cartographic Materials. Address correspondence to: HelenJane Armstrong, P.O. Box 117011, Gainesville, FL 32611-7011 (e-mail: hjarms@mail.uflib.ufl.edu).

Jimmie Lundgren, Assistant Librarian, is Head of the Science Cataloging Unit, Resource Services Department, and is responsible for cataloging atlases, maps and imagery in the George A. Smathers Libraries, University of Florida. In addition to her Master's of Library Science, she holds Master's and Specialist Degrees in Counselor Education. Address correspondence to: Jimmie Lundgren, P.O. Box 117007, Gainesville, FL 32611-7007 (e-mail: jimlund@mail.uflib.ufl.edu).

The authors wish to acknowledge the support of our library managers and the assistance of: Jorge Gonzalez, Mil Willis, Scot Smith and the members of the Revision Committee of the AACCCM.

[Haworth co-indexing entry note]: "Cataloging Aerial Photographs and Other Remote-Sensing Materials." Armstrong, HelenJane, and Jimmie Lundgren. Co-published simultaneously in *Cataloging & Classification Quarterly* (The Haworth Information Press, an imprint of The Haworth Press, Inc.) Vol. 27, No. 1/2, 1999, pp. 165-227; and: *Maps and Related Cartographic Materials: Cataloging, Classification, and Bibliographic Control* (ed: Paige G. Andrew, and Mary Lynette Larsgaard) The Haworth Information Press, an imprint of The Haworth Press, Inc., 1999, pp. 165-227. Single or multiple copies of this article are available for a fee from The Haworth Document Delivery Service [1-800-342-9678, 9:00 a.m. - 5:00 p.m. (EST). E-mail address: getinfo@haworthpressinc.com].

165

them; and how to catalog, classify and provide subject access for them. A variety of remote-sensing items and their catalog records are reproduced and discussed, including application of the new Remote-sensing 007 field. *[Article copies available for a fee from The Haworth Document Delivery Service: 1-800-342-9678. E-mail address: getinfo@haworthpressinc.com <Website: http://www.haworthpressinc.com>]*

KEYWORDS. Aerial photographs, cataloging, cartographic format, satellite imagery, remote-sensing images, photo map, Landsat, SPOT, spatial data

I. INTRODUCTION

How humans see and represent the world around them has drastically changed since the first caveman sketched animal hunting grounds on cave walls by firelight. Over the years their world grew to be the universe. The crude drawings became maps, which then drastically changed during the twentieth century with the arrival of computer technology and digital images. During the last century and a half, there has been a parallel development of the remote-sensing images from cameras in hot air balloons to satellite digital sensors. The capability for studying the Earth and the Universe has expanded tremendously. A number of these images and maps have found a home in libraries, and thus they have entered the realm of the cataloger.

Differentiating a remote-sensing image from a map can occasionally be a challenge for the cataloger. Usually the basic differences are clearly discernible but a blurring of the lines may occur in transition zones. Maps and remote-sensing images both show spatial relationships of geographic entities in a graphic representation. The map shows selected features, while a remote-sensing image reproduces a picture of a place as it appears at one given moment. The appearance of the final product is what determines the physical description and the characteristics to be described by the cataloger.

The purpose of this article is to assist the novice cataloger in making these determinations. It has been divided into four topics. The first provides insights into how to determine if the item in hand is a map, a remote-sensing map, an aerial photograph or a remote-sensing image, and also discusses how the level of cataloging reflects the anticipated use of the item in the library. The second topic centers on the elements of remote-sensing image description and cataloging; the appropriate MARC 21 fields and notes will be used to structure these comments. The third subject concerns classifying the images and possible organizational schemes. The fourth focuses on accessibility through subject headings. To illustrate these principles several case studies are provided with reproductions of images, and records for aerial photographs and satellite images that were cataloged from the University of Florida Map & Imagery Library are used as examples. As an illustration, a scheme that was designed to provide an expeditious method for reducing the uncataloged collection of almost 300,000 images held by the University of Florida Map & Imagery Library is discussed; the case-studies encompass both successes and unsatisfactory procedures.

In writing the article, certain assumptions have been made. It is assumed that the reader has basic familiarity with: (1) map terminology, e.g., scale, map set, multi-sheet map; (2) Library of Congress terminology and classification schemes; and (3) the MARC 21 format for maps. For those unfamiliar with the terminology it would be useful to refer to *Cartographic Materials: A Manual of Interpretation for AACR2* (Chicago, etc.: American Library Association, etc., 1982), hereafter referred to by the acronym CM. A list of selected references has been provided which include works concerned with remote-sensing imagery. In the Appendices there is a glossary of selected remote-sensing terms. Many of the definitions have been taken from the second edition of the previously mentioned CM (to be published).

II. CHARACTERISTICS AND SCOPE OF MATERIALS FOR CATALOGING REMOTE-SENSING IMAGERY

What is it anyway? That is probably the first question that springs to mind for all but the most experienced cataloger of remote-sensing imagery. Even this rare-breed of catalogers would have to admit that they still occasionally ask the same question. One begins debating as to whether the item is a map or an image, and may even find oneself one day asking if it is a thermal infrared or Infometric SAR (Synthetic Aperture Radar) image. Do not worry if you have no idea what the last two terms mean, for even some remote-sensing specialists might have difficulties explaining the difference. As with any new subject, one starts with the basics and builds on them. With a solid foundation and several good remote-sensing reference books, a cataloger can surmount most challenges presented by the latest images acquired by the library.

Of these remote-sensing texts that do not assume one is an expert in the field but are inclusive enough for those who have passed the introductory stage, the two mentioned in this article are used on the University of Florida campus for introductory remote sensing classes. These are the third edition of *Remote Sensing and Image Interpretation* (New York: Wiley, 1994) by Thomas Lillesand and Ralph Kiefer, and the second edition of *Introduction to Remote Sensing* (New York: Guilford Press, 1996) by James Campbell. A third book, helpful for those dealing in depth with aerial photographs, is the fifth edition of *Fundamentals of Remote Sensing and Airphoto Interpretation* (New York: Macmillan, 1992) by Thomas Avery and Graydon Berlin.

The old saying of a picture being worth a thousand words is at the core of remote sensing and is relevant to why imagery is found in libraries and should be cataloged. The cataloger must learn to note image characteristics that will help the researcher select the appropriate "picture" for study. Lillesand and Kiefer in their *Introduction* give one of the better explanations of remote sensing:

> Remote sensing is the science and art of obtaining information about an object, area or phenomenon through the analysis of data acquired by a device that is not in contact with the object, area, or phenomenon under investigation. As you read these words you are employing remote sensing. Your eyes are acting as sensors that respond to the light reflected from this page. The 'data' your eyes acquire are impulses corresponding to the amount of light reflected from the dark and light areas on the page. These data are analyzed, or interpreted, in your mental computer to enable you to explain the dark areas on the page as a collection of letters forming words. Beyond this, you recognize the words form sentences, and interpret the information that the sentences convey.
>
> In many respects, remote sensing can be thought of as a reading process. Using various sensors we remotely collect data that may be analyzed to obtain *information* about the objects, areas, or phenomena being investigated. (p. 1)

In the broadest sense, remote sensing is the measurement or acquisition of information of some property of an object or phenomenon, by a recording device that is not in physical or intimate contact with the object or phenomenon under study. The technique employs such devices as the camera, lasers, radio frequency receivers, radar systems, sonar, seismographs, gravimeters, magnetometers, and scintillation counters; for definitions, see the American Society of Photogrammetry's *Multilingual Dictionary of Remote Sensing and Photogrammetry* (Falls Church VA: The Society, 1984). The development of multi-country space programs and satellites has markedly affected the number and variety of sensors and platforms, which record the information in digital and photographic formats.

When cataloging remote-sensing imagery using the cartographic format, there are many similarities to cataloging maps. The principal differences occur in recording the source of the image and how it was acquired. Remote-sensing images are acquired either in photographic or digital format. The equipment or device which records (that is, acquires) the image is called a sensor. Most images found in libraries come from airborne and space platforms, which have electromagnetic-energy sensors. These sensors acquire data on how different features and objects on the Earth's surface emit and reflect electromagnetic energy. With few exceptions, all objects on the Earth's surface emit radiation or reflect radiation emitted by another object:

> By recording this emitted or reflected radiation and applying a knowledge of its behavior as it passes through the earth's atmosphere and interacts with objects, remote sensing analysts develop a knowledge of the character of features. . . . (Campbell, p. 22)

Electromagnetic energy can be expressed either by frequency or wavelength. In remote sensing the whole range of solar energy is usually defined by wavelength areas, and the spectrum of energy is divided into regions on this basis. The electromagnetic spectrum (EMR) regions (see Table 1) are given names, such as ultraviolet and microwave, for convenience:

> Most common sensing systems operate in one or several of the visible, IR, or microwave portions of the spectrum. (Lillesand and Kiefer, p. 4)

These regions or bands appear in the coding for the 007 fixed fields. When

TABLE 1. Electromagnetic Spectrum

cataloging these images it is wise to have a clear and uncomplicated reproduction of the electromagnetic spectrum in front of you, which gives the wavelengths by numbers and names. The majority of digital images will have the bands recorded numerically, as well as giving the wavelength in the summary file. When referring to satellite images, the regions are called "bands" or sometimes "channels." The number assigned to the band differs from one sensoring system to another, so it is useful to obtain a chart displaying the characteristics for each spectral band with its wavelength; for example, Landsat TM Band 2, 0.52-0.60–Green is in the visible spectrum. See Appendix I for the spectral-band information for NASA/USGS Landsat products (MSS and TM).

While catalogers will be asked to note in the record various codes which relate to the electromagnetic spectrum, it is not necessary to understand fully the physics involved. Once the terminology is conquered and one is familiar with the available guidelines, there is no need to ponder the secrets of the Universe. Both the remote-sensing image and remote-sensing map will have the same foundation images; it is the processing done by the producer to the image after it is acquired which differentiates one from another. The remote-sensing map will have information added to it by the publisher or user, but the image itself will not have been altered from the way it was acquired by the sensor. Alteration does not apply to a format change such as digital to photograph, but rather to the integrity of the image data–that is, have there been any changes, additions, or deletions.

The oldest of remote-sensing images is also the most likely to be found in libraries. Aerial photographs, or as they are often called, air photographs, have been around since 1858. It was the development of standard aerial photographic programs and their use by state and federal government agencies that made photos readily available in many libraries. In the past they sometimes have been termed aerial remote-sensing images, which simply means an image produced by a remote-sensing device located in an aircraft. With changes in Library of Congress cataloging and subject headings this is obsolete, replaced by the term, "Aerial photographs." CM states that aerial photographs are any photographs taken from a platform in the air. When used in a cartographic context, this term normally refers to photographs of the surface of the Earth (or other celestial body) taken downwards, vertically, or at a predetermined angle from the vertical. The differences in altitude, film type and format are all characteristics that must be recorded in the physical-description and associated fields (most often in USMARC fields 007, 300, and 500).

The remote-sensing image that appears in the library cataloging unit may be in a variety of physical formats. The most common as previously mentioned is the aerial photograph. The first known aerial photograph was taken from a balloon over Bievre, France in 1858. Since that time cameras have been mounted on a number of airborne platforms including kites, birds and airplanes. The earliest existing aerial photograph was taken from a balloon over Boston in 1860. This view of Boston was described in the *Atlantic Monthly* by Oliver Wendell Holmes:

> Boston, as the eagle and the wild goose see it, is a very different object from the same place as the solid citizen looks up at its eaves and chimneys. (Lillesand and Kiefer, p. 38)

Modern aerial photography began as a reconnaissance tool for the military during World War I. Technical features and interpretation techniques were refined during World War II, which were translated into the civilian arena after the war. Especially in the past fifty years, there has been a significant increase in variety and quality of aerial photographs and other remote-sensing images produced by both military and non-military governmental agencies. To catalog aerial photographs, it is advantageous to recognize some of the terminology and platforms on which cameras are mounted, since this information will be recorded in both the note fields and the 007 subfield.

The most extensive aerial photographic program of the United States Government is from the U.S. Department of Agriculture's National Conservation Resources Service. This is the latest name for an umbrella agency that includes the Soil Conservation Service, the Agricultural Stabilization and Conservation Service and other agencies that began using aerial photographs to measure cropland acreage. The earlier program began around 1938 with photographs taken directly over a site from a low-flying aircraft. (See Figure A.) Today, aerial photographs have a multiplicity of uses ranging from soil maps to urban planning. Most counties in the United States have aerial photography produced on a fairly regular basis by Federal or state agencies or a combination of the two. These are the photographs which are most likely to be obtained by libraries.

Aerial photographs are classified according to the orientation of the camera in relation to the ground at the time of the exposure. Low-altitude vertical images are some of the more important of these images, and we will talk more about them later. Briefly, they show more detail than do most high-altitude images, although certainly there are exceptions to this, such as the Corona program photos; they are taken with the camera pointed straight down at the Earth, with the line of sight perpendicular to the Earth's surface. If the camera is oriented toward the side of the aircraft, it is called an oblique aerial photograph. A high-oblique photograph shows the horizon, while a low-oblique photo does not. When a cataloger views an image and sees both sky and earth with a horizon "line" then this a high-oblique image. If the image has no horizon but features appear to have a perspective view, it is a

FIGURE A. ASCS Aerial Photograph from Alachua County

low-oblique and not a vertical image. For example, if the side of a ship in its natural position can be seen in the photograph, it is an oblique image; but if the top of the ship is seen as though one is directly over it, the image is a vertical one.

A bit more on vertical photos. The majority of vertical aerial photography flights are arranged so that the multiple photographs of the area coverage overlap on a precise layout. By the use of special equipment these images can be seen as three-dimensional images. "Stereo pairs" are a desirable feature of an air-photo flight, and must be noted by the cataloger. Another feature of the vertical multiple image flights will be the possibility of a mosaic. A series of photographs of adjacent ground area is joined together to form a smaller scale image of the far larger photographs. These can be roughly aligned in correct sequence, which would be an uncontrolled mosaic, or–if the photographs are precisely arranged to preserve correct positional relationships–in a controlled mosaic. Frequently these mosaics are made as indexes for use with what is called a flight of aerial photographs flown for a given purpose. (See Figure B.)

In 1980 a high-altitude, aerial photography program was begun, with the purpose of consolidating the programs of various Federal agencies. Often a cataloger will see the letters "NHAP" on the photograph, which indicate that the photograph is part of that program, the National High Altitude Aerial

FIGURE B. Photo Mosaic Index for Alachua County Aerial Photographs

Photography Program. There are data sheets available from the NHAP manager, US Geological Survey (USGS), which give standard sizes and related scales for the NHAP photographs. In 1987, a replacement program was begun, which continues at this time, and is now in its third "cycle." The National Aerial Photography Program (NAPP) was designed to produce aerial photographs for the coterminous U.S. according to a systematic plan that ensures uniform standards. The USGS manages the program, which is funded by a number of agencies. Further information about the program is available at the USGS Internet site, http://edcwww.cr.usgs.gov/napp/napp_examples.html

There are a number of other sources that may produce aerial photographs. Many state and local agencies have their own programs. Also, photographs are available to the general public from different Space Shuttle missions, the SkyLab, space and celestial observatories, and weather-related programs. For further information on the source or technical elements of these programs, refer to the books mentioned at the beginning of this section. The scope and variety of remote-sensing imagery is tremendous. Only the surface has been scratched in this discussion. There are a number of special-purpose satellites such as GOES (Geostationary Operational Environmental Satellite), a U.S. National Oceanic and Atmospheric Administration (NOAA) meteorological satellite program whose images frequently appear in weather reports on television. There is another category of satellite, primarily concerned with space and not the terrestrial imaging of planets such as with Landsat or SPOT (the latter is a French satellite, whose acronym stands for Système Pour Observation de la Terre). Furthermore, there have been ocean-monitoring satellites such as Seasat, which provides synoptic views of the ocean over large areas and extended time periods. Finally there are images from sensors that record in the non-visible bands of the electromagnetic spectrum. Examples of this would be radar, which records and/or displays the signal as an image:

> Active microwave sensors are radars–instruments that transmit a microwave signal, then receive its reflection as the basis for forming images of the Earth's surface. (Campbell, p. 201)

Campbell's *Introduction to Remote Sensing* spends several chapters discussing these types of sensors and provides a number of excellent illustrations.

III. ELEMENTS OF REMOTE-SENSING IMAGERY SIGNIFICANT FOR CATALOGERS

In cataloging remote-sensing images and maps, there are twelve basic elements that should be recorded whenever possible on a cataloging record. This section will discuss these elements with reference to the appropriate USMARC fields and types of notes. Some of these are obvious and have their own field, such as title, while others are specific to imagery and are recorded in the 007 fixed field and/or a 500 note. The newly issued *Technical Bulletin 227: OCLC-MARC Bibliographic Update 1998* (Dublin OH: OCLC, Inc., 1998; URL http://www.oclc.org/oclc/tb/tb227/frames_man.htm) contains the new 007 Physical Description Fixed Field (Remote-Sensing Image). The data sheets provide badly needed information as to which elements should be recorded, including a guideline as to what types of notes catalogers should make for the reader as well as the fixed fields that may be used in some online catalogs as search elements.

While learning to apply these designators, have the corresponding pages from USMARC Bibliographic July, 1997 (Remote-sensing image 007). These provide a brief guideline for applying the codes (see Appendix II; URL is http://www.loc.gov/marc/bibliographic/ecbd007s.html#mrcb007r). Since the elements included in the 007 fixed field closely relate to each other, they will be discussed together below, after the other information needed to catalog at the "full" level a remote-sensing image. These are elements numbered 9 to 12.

1. Title

Determining the title for remote-sensing images is often a creative experience. Unlike the remote-sensing map–which has such map-type information as graticule, grids, and place names–the image by definition will not have any cultural overlays on it. If the photographic reproduction of the image were designed with a title and other bibliographic information outside of the image's border, the cataloger is having a lucky day. Usually those images acquired directly from a governmental agency do not have the title, or much other desirable information either. (See Figures A and C.) Such information is available more frequently on specialized images from commercial firms or

FIGURE C. Mt. St. Helens Color Infrared Photograph (Reproduced in Black and White)

those designed for a distinctive purpose, e.g., following the progress of Hurricane Andrew or of oil fires burning in Kuwait during a battle.

For cataloging purposes, every object must have a name, and therefore a main title of some kind must be entered in the 245 |a field, with any subtitle in 245 |b and statement of responsibility or remainder of title-page information in 245 |c fields. A bracketed general-material-designation (GMD) term from the prescribed list may be entered in the 246 |h field, following the title proper if the item is a computer file, microform, or transparency; local practice as to whether to use GMDs or not differs. There is some discussion in the map-cataloging world that the GMD for all cartographic material should be "cartographic material," followed by a physical-carrier GMD in parentheses. Alternate titles appearing anywhere on the item may be entered as title-added entries in the 246 field.

If there is no title presented on the chief source, one must be generated by the cataloger. Such "made-up titles" are enclosed in brackets in the 245 field, and a 500 note should give an explanation such as, "Title supplied by cataloger." Sometimes the remote-sensing image lacks all the usual chief sources for cataloging such as panel title, legend, container, etc. When the necessary data is missing from the image, the accompanying information, such as indexes or order forms, becomes an important source of information. Most imagery flown as part of a standardized program will have flight and/or directional coordinates, which can be used to determine the geographic area. As with maps, its geographic area is the prime element in determining a title and classification number. From this information and using any unique numbers, the cataloger can supply the title.

Most aerial photographs flown as part of a governmental agency's program will be similar to the U.S. Department of Agriculture Natural Resources Conservation Service materials. (See Figure A.) These will have:

1. a code number for the administrative or geographic area, e.g., county name;
2. an individual sequential alphabetic and/or numeric code for each photograph; and
3. the year, month and day, or some portion thereof.

Some aerial photos are flown for special purposes such as soils or mineral surveys. These occasionally will have accompanying text or indexes. For example, in alternate years the Florida Citrus Industry commissions a Fruit and Crop Inventory flight, which is flown only for areas of the state that have citrus groves. There are no subdivisions, so the title must be assigned to the whole state from accompanying indexes and agency correspondence. A cataloger may have to seek documentation before an aerial photograph(s) can be cataloged. For example, when the aerial photograph is for an urban area with no qualifying information available, the cataloger will need to make a decision as to the amount of time which should be spent researching the photograph. Perhaps it may be as simple as contacting the aerial surveying corporation or as difficult as searching photographic files seeking the code number.

Current satellite images will have latitude and longitude plus some form of X-Y coordinates, e.g., path and row. Using these with the image's unique number, a cataloger can determine the geographic area and supply a title. Earlier satellite images may be lacking one of the coordinates but still have sufficient information to figure out the geographic area. (See Figure E on page 185.) When satellite images are obtained on CD-ROM or magnetic tape, the coordinates and unique numbers will usually be included in the header or beginning, or in a histogram. A title supplied by the cataloger includes the unique number somewhere. (See Figures H and I on page 213.)

2. Area Coverage

As noted above, remote-sensing images are chiefly defined by the area they cover. The area may be a geographic area such as a river valley, an administrative area such as a county, or an area described by limits of latitude and longitude. Frequently Landsat and SPOT images on CD-ROMs or magnetic tapes will arrive unlabeled except for the identification number or scene ID. Access to an extensive map collection is an invaluable tool when cataloging satellite imagery, but even a detailed atlas can be very helpful in verifying the geographic area of an image.

As emphasized in the previous section, establishing the complete area covered by the image and calling it by a comprehensive name is vital to supplying the correct title and subject headings. Even if there is an original title it may be partial, or indicate only the generic subject, e.g., "Hurricane in the Gulf." The purpose for cataloging remote-sensing images is to provide the widest possible accessibility to the library user; therefore, the spatial extent of the image as well as the hurricane itself should be indicated. The title may be very general and thus not useful, e.g., it may state, "Air Photograph of the Atlantic Coast," while the area shown is much smaller and is actually the Cape Hatteras National Seashore. The same rules about area which affect the assigning of titles and subject headings to maps are applicable unless otherwise noted.

Most satellite imagery will provide the center-point latitude and longitude of an image as well as the coordinates for each of the four corners. These geographic coordinates may be found on a hard-copy image, on the header and histogram of a computer compatible tape, or as a separate file on a CD-ROM. Since each scan of the satellite differs slightly with the angle of the platform and day/time of the year, coordinates are an excellent way of defining coverage. Thus it is important whenever possible to include the latitude and longitude in the 255 and 034 fields.

The area captured in a remote-sensing image does not always correspond to a particular administrative area or to an area defined by a feature such as a mountain or valley. In these cases, to provide subject access it may be advantageous to choose an established geographic subject term that lies within the area covered and to add "region" on the end. Thus a rectangle area which might include X River but not be limited to the river or its valley or its watershed could be described by, "X River region." Or optionally, use the next larger geographic area.

3. Unique Identifiers

The majority of remote-sensing images have a unique identifier that relates to their production, origin and organization. When many images are recorded in the same flight or mission, it is very often one number that provides the unique designator for the image. The code numbers for aerial photography in a standardized program have been discussed under 1. *Title.* There are different types of code numbers, which have been assigned to those photos taken from Space Shuttles, the military and national high altitude flights (U-2) or special purpose flights, e.g., surveys of coastal erosion in California. But regrettably, the information on the face of these photos can vary from sketchy to nonexistent. Those with local numbering systems usually will have an inventory number and date, but they are often not printed on the image and are included only in accompanying materials–so the cataloger must be sure that these materials are with the images, and if not received that they are requested from the producer.

Most imagery recorded from the Space Shuttle missions is photographic and most of it is hand-held. Seldom does the word Space Shuttle or any other text appear on the image. Instead, each Shuttle mission is assigned a unique number always beginning with "STS" for "Space Transportation System," and the code "STS" with number is combined with a string of flight recording numbers. Usually the number string on the negative will be reproduced within the photograph's border. A typical Shuttle photograph number would be:

891126 094212 STS-33 72 071

Decoded, this means:

891126 = Date; November 26, 1989

094212 = GMT or Greenwich Mean Time of photographic acquisition

STS-33 = Mission # 33

72 = Rl or Number assigned to each roll of film

071 = FR for Frame

Each Space Shuttle mission has a catalog of all the photographs taken during the mission. The compilation is sorted by both row/frame and geographic name. With the catalog in hand the cataloger would find:

- the sample image, whose code is given above, is a view over India of the Himalayas;
- the center point of the image in latitude and longitude and the nadir or point directly beneath the spacecraft;
- the altitude of the shuttle;
- its inclination;
- azimuth and sun elevation; and
- the launch and landing dates.

Satellite images may have a variety of identification numbers depending on the source, type and age of the satellite. The most commonly found surface-observing satellite images in United States libraries are those from two earth observation systems. These are the U.S. Landsat program, begun in 1972 with Landsat 1 (originally called ERTS-1), and the French SPOT, begun in 1986 with SPOT 1. (See Figures H and I on page 213.) Each has their own coding for identification numbers and for locating images using the X-Y coordinates. These identification codes contain information needed to catalog the image. There are translations available from both organizations, on-line and in print, that the cataloger may obtain.

A. Landsat

When working with a Landsat image it is important to know that there have been images produced from five different Landsat satellites. Landsat 1, 2 and 3–active between 1972 and 1983–had two types of recording devices whose images can be found in libraries. The Multispectral Scanner Subsystem (MSS) was the primary sensor. The second device was the Return Beam Vidicon (RBV), a camera system which generated high-resolution television-like images; it started to malfunction early in the program. Most images will have these codes–"MSS" or "RBV"–in a prominent location, which will assist the cataloger in providing needed information for the user as well as for encoding the 007 Physical Description. Landsat 4 and 5 also had two sensors, which were the MSS and the Thematic Mapper (TM). The TM is an advanced multispectral scanner, which has a number of improvements over the MSS.

For the following discussion of unique identifiers, it will be useful to refer to the image accompanying Example #4 with its cataloging record. (See Figure E on page 185.) It is a hard copy of a Landsat MSS image centered on the Florida Everglades. At the image bottom is a wedge-shaped brightness scale that is common to all Landsat imagery. Since it is an expected feature of Landsat images, it does not have to be noted. The alphanumeric codes located above it are called the "annotation block." Using these codes and reading from left to right, one can translate the code and identify those, which should be recorded. Items followed by an * in the list below are optional even when doing full level cataloging.

Information from Figure E. Landsat Image of Everglades Region Band 5

Date: 22MAR73 = March 22, 1973

Lat/Long in degrees at the image point: C N25-59/W080-55 = 25° 59min North Latitude by 80° 55min. West Longitude.

Nadir (Ground point directly beneath the satellite): N25-59/W080-48

Sensor: MSS

Band: 5

Reception mode*: D = direct instead of recorded

Sun Elevation: SUN EL 50

Azimuth: AZ124

Other orbital/processing parameters*: 189-3373-N-1-N-2L

Satellite ID: NASA ERTS = National Aeronautics and Space Administration Landsat 1 before the retroactive name change.

Unique scene ID number: E-1242-15240-5 = E124215240 was the unique number of that scanned image which could be viewed in one of four bands or a composite of bands to simulate color infrared.

Around the margins of the image are a series of numbers and tick marks, which represent an approximated latitude and longitude grid. The brightness scale and chain of numbers that appear above the image are from the adjoining image. This information should not be used in cataloging, and care must be taken that it is not recorded in place of the annotation block at image bottom. The shape of the image is not a square but a parallelogram because of the rotation of the Earth and the time it takes to scan the image.

When one is cataloging images from Landsat 4 and 5, the MSS and TM hard-copy images will appear very similar. The annotation block has been changed several times but shows essentially the same information. There are two major differences. The first is the addition of index numbers of Path and Row from the Worldwide Reference System (WRS). The Path and Row numbers are either printed in large numbers in front of the annotation block or precede the date as the first code number. The WRS is a concise designation of nominal center points of Landsat scenes used as a location index. There are 233 Paths that represent the north-south track of the satellite while 119 Rows represent the image's centerline of latitude. The WRS index system is frequently used to organize and access Landsat imagery in collections of satellite images. The combination of the Path and Row creates a unique numerical identifier. The number frequently appears without the words Path and Row, e.g., "017-039."

The second major difference is due to the Thematic Mapper (TM) added to Landsat 4 and 5. It was designed to be an upgrade of the MSS sensor and incorporates a number of spectral, radiometric, and geometric improvements. These advances are recorded in several codes, which should be noted by catalogers. The first is the "T" which appears in the image annotation block in place of the "MSS." It is important to the users to know quickly if they are working with MSS instead of TM images. This information normally is recorded in the title using the word Thematic Mapper. (See Figure H on page 213.)

There are several basic differences between the MSS and TM sensors on Landsat 4 and 5, which affect information to be recorded by catalogers. Imagery users normally prefer the TM images because they have finer resolution (thirty instead of ca. 100 meters), giving better quality and more detailed information per pixel (picture element). This is achieved partly by the lower orbit elevation of the satellite, which caused changes in the expanded coverage cycle. With Landsat 1, 2 and 3 coverage of the Earth was completed in 14 days, but with the expanded coverage Landsat 4 and 5 complete a cycle in 16 days. This required a new WRS indexing system to adequately show paths and rows. There are now 233 paths and 248 rows; while the row numbers have remained the same, there has been a slight shift in path numbers. Thus a cataloger who wishes to create the lacking but important x-y coordinates or WRS index numbers for a Landsat image must ensure that the correct WRS system is in use. When in doubt remember that for all TM images the Landsat 4 and 5 system should be used. Any MSS image dated before Landsat 4 was launched on July 16, 1982 should use the Landsat 1, 2 and 3 index. Any images after September 7, 1983 should use the Landsat 3 and 4 index. The gap between the 1982 and 1983 dates is the period when sensors on Landsat 2 and 3 were still recording but Landsat 4 was the major source of desirable imagery. Librarians working with Landsat images hope they are lucky enough to receive only images from this period, which clearly identify the specific satellite.

The second major difference between MSS and TM is that the number of spectral bands has increased from four broadly defined MSS bands to seven TM bands tailored for specific scientific investigations. The band numbering is entirely different for each system. Spectral bands, or the selection of wavelengths, are at the heart of Landsat imagery. From the earlier discussion of the electromagnetic spectrum, one can see that understanding the significance of each band can express to the user how radiation reflected and/or emitted from different objects and features on the Earth's surface has been recorded. See Appendix I for a comparison of the band numbering and their description from the two sets of Landsat satellites.

If the physical carrier of the Landsat TM image being cataloged is a CD-ROM, the scene or entity identification number will be written in a slightly different manner. (See Figure H on page 213.) In Example #8 the Entity ID number that appears in the 245 field in subfield "n" is LT5001082008624310. The explanation of the number given below is based on information provided by Eros Data Center on their webpage and on paper order forms.

TM Entity ID: LT5001082008624310

> L = Landsat
>
> T = Thematic Mapper
>
> 5 = Satellite number.
>
> 001 = Path
>
> 082 = Row
>
> 00 = WRS row offset (set to 00)
>
> 86 = Last two digits–year of acquisition
>
> 243 = Julian date of acquisition
>
> 1 = Instrument mode
>
> 0 = Night Acquisition
>
>> 1 = TM Bands 1-7 (Landsat 4-5)
>>
>> 2 = TM Panchromatic plus Bands 4, 6, 7 (Landsat 7)
>>
>> 3 = TM Panchromatic plus Bands 4, 5, 6 (Landsat 7)
>>
>> 9 = MSS Data (EDC use only–Landsat 5)
>
>> 0 = Instrument multiplexor (MUX):
>>
>> 0 = Landsat 4-5
>>
>> 1 or 2 = Landsat 7

B. SPOT

The United States Land Satellite (Landsat) program was so successful that other countries began developing their own system or joining forces for a multi-national system. The best known is the one for which libraries would most likely have images in their collections. This system is SPOT, designed by a French organization in collaboration with other European nations. It was the first commercial satellite and was designed to have finer quality image coverage with a quicker turn-around time in data acquisition than Landsat. It has further capabilities that Landsat does not, such as stereo imagery and sensitivity in additional spectral bands. The general principles for recording digital data are the same as Landsat. SPOT has two identical sensors known as HRV (high-resolution visible) instruments. Currently there have been three SPOT satellites with the fourth planned for launch soon (as of this writing). SPOT 1 was launched February 11, 1986, SPOT 2 on January 21, 1990 and SPOT 3 on September 13, 1993. SPOT Image is also working with the Canadian Space Agency, to development applications for the world's first commercial radar spacecraft, Canada's Radarsat.

The company, SPOT Image, has its own indexing system and does not use Path and Row for identifying geographic locations. The SPOT Grid Reference System, or GRS, is made up of the intersection of columns (K) and rows (J). For a further explanation of the SPOT products and reference systems one should see the two-volume SPOT Handbook or visit their Internet Homepage. As of this writing it was: http://www.spotimage.fr/welcome.htm. When cataloging SPOT images, one will usually find an annotation block similar to Landsat. The SPOT Image Corporation name or the name of the satellite is commonly prominent so there should be little difficulty in establishing the authority for the 110 MARC field. (See Figure I on page 213.) When cataloging a SPOT image, the 007 fixed field will be dependent upon which mode of the HRV sensor has been recorded. It is possible to have the panchromatic (PN) mode or the multispectral (xS).

On the hard-copy version in addition to the annotation block, the cataloger customarily will find a statement such as, "SPOT 1 HRV 617-291 26 Jun 86." This code means:

SPOT 1 = Name and number of the Satellite

HRV 1 = Name and number of HRV Instrument

617-291 = Location of Scene on the SPOT Grid Reference System (GRS), i.e., KKK-JJJ

26 Jun 86 = Viewing date in Universal Time (DD-Day; MM-month; YY-year)

P = Panchromatic mode

C. Remote-Sensing Maps

When the item being cataloged is a remote-sensing map instead of an image, the identifying numbers frequently have been removed from the face of the image. Hopefully the information will be included on the map with the compilation or source text. It may be buried deep within strings of numbers and agency names, but the unique identifiers should resemble the form described for the images. Look for keywords like: Band, Landsat, Spot, AVVR, Orbit, Flight Number, Mission, etc.

In some instances, there is little information to indicate that the map contains a modified satellite image or when it was recorded. However, the word "Landsat" or "SPOT" may appear in the margin or text. With experience the cataloger should gain the ability to recognize it as a simulated color-infrared image. This information would then appear in a 500 note. Example #3 of McMurdo Station, Antarctica actually has "PHOTOMAP" in the title but other than dates there are no unique identifiers. (See Figure D.) A

FIGURE D. McMurdo Station, Antarctica

FIGURE E. Landsat Image of Everglades Region Band 5

500 note would show the source of the photograph and the dates of the individual photographs.

D. Researching and Recording Satellite Imagery Identification Numbers

Any unique identifiers associated with the resource should be recorded either in the title or in a note field, usually as a quoted note. Some examples are LANDSAT code, flight line/photo ID, and ASCS invitation or contract number. It is important to record these numbers, as they may prove crucial to identification of the content of the resource for users. In situations where there is insufficient information and an explanation of the code number is needed, several sources are available for researching these flights. Two major repositories of remote-sensing imagery have user service divisions that will give assistance. These agencies are the U.S. Geological Survey's EROS Data Center and the Cartographic and Architectural Branch of the National Archives and Records Administration. Every Regional Library of the Federal Depository Library Program has a CD-ROM from USGS, which lists aerial photography for each state and county flown by a governmental agency; there have been several different editions of this CD, which is called *APSRS*

(Aerial Photography Summary Record System). If a library has extensive unidentified aerial photography and did not select this item on depository, it may wish to purchase its own copy. In addition, many states and administrative sub-divisions have lists of aerial photography flown for their agencies.

Researching images can be a long and frustrating experience. The value of pursuing an explanation of the identification number for cataloging purposes must be judged against the patron needs and the level of cataloging to which the library is committed. In libraries with smaller collections of remote-sensing imagery, when the geographic area, date and perhaps source are known, it is acceptable to simply record the identification number in the note field.

4. Date and Time

For the user of remote-sensing imagery, the date of a specific image is extremely important. The researcher may want to compare an image of Miami immediately before Hurricane Andrew with one from immediately afterwards. Another person may want the latest image of an area in the Amazon Basin to locate boundaries of rain-forest destruction. Since the information is recorded exactly as it appeared at a specific date and time, the recording of this information can be crucial.

Many aerial photographs will show at least the flight year and some the month and date. Sometimes it is coded, e.g., "NHAP 83." The various satellites record the date and time of acquisition (recording) in their identification numbers. These appear on most contact prints as well as at the beginning of magnetic tapes and CD-ROMs. How to determine the time and date of satellite images was discussed previously in *Unique Identifiers.*

The date the image was recorded is considered the date of situation. This date would appear in the classification number. If no date of publication is available, the year of recording will be used as the year of publication in the 260 field as well as the dates field. It must be emphasized that the date of recording is all-important in the context of images, as the exact same conditions will not repeat themselves.

5. Size

Size is an element that is extremely important when cataloging maps, remote-sensing maps and some but not all images. When discussing hard-copy or photographs from cameras, the size in combination with the scale can immediately provide the educated user with an understanding of the level of detail present on the image. There are standardized sizes for photographs from the U.S. National Resources Conservation Service (NRCS–the old ASCS and SCS) and U.S. Geological Survey NAPP and NHAP flights. These can be purchased in specific customized sizes. It can be a challenge to catalog an enlargement when it is only part of a whole. For instance, the ASCS county aerial photographs in 10" x 10" size may have each quarter enlarged to a 36" square. However, the flight information is not repeated on each of the four enlargements. There are information sheets available for products from NRCS and USGS, and may be used to verify necessary information.

Photographic copies of most land satellites and many space satellites may be purchased and also have information sheets available which would be useful for the cataloger. However, satellite images in digital form have no size component in the traditional sense. The physical carrier is described, such as CD-ROM or computer file, but it is the resolution and density of the file that is more important to the user.

6. Scale

Scale along with size is an element that is extremely important when cataloging maps, remote-sensing maps and some images, but, again, not all images. The level of detail available in the image can quickly be determined when the scale is available. Unfortunately scale is one of the elements least likely to be included on hard-copy images and not at all in digital formats; in the latter, resolution effectively takes its place as far as many users are concerned. The data sheets mentioned in *Size* are useful in determining scale. If the cataloger is really lucky, there will be accompanying information which includes the missing scale. When scale is known, it should be recorded in the 255 and 034 fields. In the case of most digital remote-sensing images, scale becomes a slippery concept and may not be applied. The statement, "Scale indeterminable," or, "Scale not given," should be used in this case.

7. Accompanying Materials

It cannot be overly stressed how important indexes, finding aids and accompanying texts can be in helping patrons utilize remote-sensing images. When these are received with the materials, they should be described in notes. It is recommended that any locally created indexes or other aids be acknowledged. Recording them in local notes or holdings information will ensure that patrons and all staff may benefit from them.

8. Cultural Information/Added Features

The departure from remote-sensing image to remote-sensing map is the addition of information onto the face of the image. This can be seen by comparing the two cataloged examples of Miami and south Florida. The addition of place names makes Example #6 a remote-sensing map, as compared to Example #7 which is a remote-sensing image published on poster paper. An image may be published on a map sheet with an enormous amount of information in the margins or outside the borders as in Example #7. However, as long as there is no information added to the face of the image it is still considered to be a remote-sensing image. As always, it is important that this information be scrupulously described in the note fields.

Elements #9 to #12.

Elements #9 to #12 correspond to the subfields of the 007 Description Fixed Field (Remote Sensing Image). It would be useful to refer to Appendix II during the following discussion. Portions of the explanations are based on the USMARC Bibliographic 1997 guidelines distributed by the Library of Congress. All the information coded in the 007 Fixed Field should also be expressed in the part of the catalog record viewable to the public.

9. Orientation

Orientation in remote-sensing materials is more complex than with maps, since we are concerned not just with the cardinal directions of North, South, etc., but with the position of the sensor platform relative to the Earth or physical entity being remotely sensed. Imagery is acquired from three basic sources; these are coded in RSI 007 subfield d as:

(a) Surface–Indicates that the remote-sensing image was made from a device located on the surface of a physical entity or body. Usually this is a planet or moon. An example would be images recorded on Mars in 1998 from the Sojourner.

(b) Airborne–Indicates images made from a device above the surface of a body and generally in the atmosphere. The device or platform could be an aircraft, balloon, or some other airborne device. Included in this category would be pigeons with cameras and Space Shuttle photography taken within the earth's atmosphere.

(c) Spaceborne–Indicates the image was recorded from a device located in space outside the primary layers of an atmosphere or far beyond. Usually the platform for the sensor is in orbit. Satellites and the Space Shuttle missions while in orbit around the moon would be examples.

If the item being cataloged does not fit any of these categories, then one of the three remaining codes should be used. If the altitude of the sensor does not matter, then the code would be "n" for not applicable. If one really cannot determine where the sensor was located, apply "u" for unknown. Use "z" when the sensor's location is known but none of the other defined codes are appropriate.

The angle of the sensor recording the image is extremely important because of its potential use for scientific analysis. Images recorded at certain angles are preferable for specific areas of study, e.g., low angle to display building features. There are three basic categories for describing the general angle of the device from which the remote-sensing image is made; these are coded in RSI 007 subfield "e" as:

(a) Low oblique–Oblique is any position or direction that is slanted, inclined or not perpendicular to the surface being imaged. Low oblique indicates the angle of the remote-sensing device is closer to being parallel with the surface being imaged than to being perpendicular. The horizon is deliberately not shown in the image.

(b) High oblique–Indicates the angle of the remote-sensing device is closer to being perpendicular than not. The apparent horizon line appears in the image.

(c) Vertical–Indicates that the remote-sensing device is directly above or vertical to the surface being imaged.

(n) Not applicable–Generally applied when the attitude does not affect the image.

(u) Unknown–The ever-popular option for when no information is known to the cataloger about the angle of the device.

10. Cloud Cover

When a person wishes to use a remote-sensing image for scientific investigation or analysis, there is a variety of images that may be chosen. A significant factor in deciding which image to use is the amount of the surface being viewed that is unobstructed. The majority of sensors do not penetrate clouds to show what is located below them. On the other hand, there are other scientists who study clouds and are seeking images with clouds. Furthermore there are some sensors such as radar and sonar that are designed to penetrate clouds, tree cover, and ocean depths. Cloud cover is not germane for images produced by these sensors.

When a cataloger is coding the amount of cloud cover for aerial photographs and satellite images, a decision must be made as to the percent of cloud coverage visible on the image. This information is usually available with a satellite image either from the container or ordering information but can sometimes occur on the image. Aerial photography seldom contains a statement concerning cloud cover. If there is no official information available, it is important for the cataloger to at least make an attempt to estimate the coverage if the image can be viewed. This information should appear in a 500 note with a statement estimated by the cataloger. The cataloger should not agonize over the difference between 59 percent and 61 percent, but concentrate on expressing whether clouds cover a small part or most of the image. The researcher will appreciate the effort made to assist in choice of image. If the library has a number of similar images for the same area, a good additional option for the cataloger is to state in which quadrant(s) of the image the clouds occur. For example, "Cloud cover is estimated by the cataloger to be 35% located primarily in the north west quadrant" or ". . . located principally over the ocean."

There are eleven choices for recording the amount of cloud coverage. The last two are the two old favorites, "n" for not applicable–e.g., radar images– and "u" for unknown. The other nine categories are a range based on 9 percent for each category except the last. These are:

0–0-9%,

1–10-19%,

2–20-29%,

3–30-39%,

4–40-49%,

5–50-59%,

6–60-69%,

7–70-79%,

8–80-89%,

9–90-100%,

n–Not applicable,

u–Unknown

11. Source

The source of the images may speak volumes about the purpose for which the images were recorded, as well as place them in context of well-defined standardized programs. The platform from which the recording took place is likewise a significant piece of information, such as a particular satellite. A platform is any structure that serves as a base upon which remote sensors are mounted. The structure does not have to be a flat surface. There is documented evidence that cameras mounted on birds and dolphins have produced good photographs. This source of remote-sensing images could account for a creative 500 note.

When encoding the 007 fixed field, there are three important categories concerning source of the image. These are Platform Construction Type recorded in the "g" subfield, Platform Use recorded in the "h" field, and Sensor Type recorded in the "i" field.

‡g Platform Construction Type. In addition to nine codes listed below, there are the usual "n," not applicable, and "u," unknown, plus "z" for other (remember the dolphin). The guidelines are based on information from the July, 1997 USMARC Bibliographic.

(a) Balloon–The base for the remote-sensing device was a balloon or similar lighter-than-air platform.

(b) Aircraft, low altitude–The base was a dynamic-lift aircraft designed for low-altitude flight (below 29,500 ft. (8,962m.)).

(c) Aircraft, medium altitude–The base was a dynamic-lift aircraft designed for high-altitude flight (between 29,500 ft. (8,962m.) and 49,000 ft. (14,810 m.)).

(d) Aircraft, high altitude–The base was a dynamic-lift aircraft designed for high-altitude flight (above 49,000 ft. (14,810 m.)).

(e) Manned spacecraft

(f) Unmanned spacecraft

(g) Land-based remote-sensing device

(h) Water surface-based remote-sensing device–The base was designed to stay on the surface of a body of water, e.g., a ship or floating platform.

(i) Submersible remote-sensing device–The base was designed to be submerged beneath the surface of a body of water, e.g., a submarine or floating platform.

‡h Platform Use category refers to the primary use intended for the base described in ‡g. A satellite or other platform may be designed for a variety of uses but the cataloger will not have to decide between categories for uses of apparently equal importance. In addition to a 500 note there is code to express this in the 007 field. The content designators are:

(a) Meteorological–The primary use for the platform is to make remote-sensing images of meteorological events and conditions. This can include images recorded in aircraft as well as weather satellites. While an image of a hurricane may show part of the earth's surface, it should be encoded as meteorological if the hurricane tracker plane and GOES Satellite imaged it as a weather event.

(b) Surface Observing–The primary use for the platform is to make remote-sensing images of the surface of a planet, moon, etc. This could encompass Landsat TM images as well as county aerial photographs.

(c) Space Observing–The primary use for the platform is to make remote-sensing images of space. The emphasis in this category is on devices mounted on a base, which is designed to make images of space. It is not for the platform in space whose principle purpose is to record features of Earth. Photographs of Halley's Comet, and digital images of the asteroid belt, would be recorded with this code.

(m) Mixed Uses–The platform for mounting the sensors was designed for a variety of uses (i.e., uses covered by two or more of the other codes). It is possible that the platform was designed to observe both space and the Earth. Also it is possible for a satellite to have different sensors for imaging the earth's surface (b) and beneath the ocean (z) for other. When in combination these would be recorded as (m)

(n) Not Applicable

(u) Unknown

(z) Other

‡i refers to the recording mode of the remote-sensing device, specifically whether the sensor is involved in the creation of the transmission it eventually measures. The content designators are:

(a) Active–Indicates a sensor that measures the strength of reflections of its transmissions sent to a remote target. It must be "actively" sending a transmission such as a sonar signal.

(b) Passive–Indicates a sensor that measures the strength of transmissions (e.g., radiation) emitted by a remote target without stimulation by the sensor.

(u) Unknown

(z) Other–Indicates a sensor recording mode for which none of the other defined codes are appropriate.

12. Data Type

A two-character code that indicates the spectral, acoustic, or magnetic characteristics of the data received by the device producing the remote-sensing image. It can be used to indicate both the wavelength of radiation measured and the type of sensor used to measure it.

Visible

   aa-Visible light,

Infrared

   da-Near infrared,

   db-Middle infrared,

    dc-Far infrared,

    dd-Thermal infrared,

    de-Shortwave infrared (SWIR),

    df-Reflective infrared,

    dv-Combinations,

    dz-Other infrared data,

Microwave (radar)

    ga-Sidelooking airborne radar (SLAR),

    gb-Synthetic aperture radar (SAR)–Single frequency,

    gc-SAR–multi-frequency (multichannel),

    gd-SAR–like polarization,

    ge-SAR–cross polarization,

    gf-Infometric SAR,

    gg-Polarmetric SAR,

    gu-Passive microwave mapping,

    gz-Other microwave data,

Ultraviolet

    ja-Far ultraviolet,

    jb-Middle ultraviolet,

    jc-Near ultraviolet,

    jv-Ultraviolet combinations,

    jz-Other ultraviolet data,

Data fusion (combinations)

    ma-Multi-spectral, multidata,

    mb-Multi-temporal,

    mm-Combination of various data types,

Acoustical (elastic waves)

    nn-Not applicable,

    pa-Sonar–water depth,

    pb-Sonar–bottom topography images, sidescan,

    pc-Sonar–bottom topography, near surface,

    pd-Sonar–bottom topography, near bottom,

    pe-Seismic surveys,

    pz-Other acoustical data,

Gravity

    ra-Gravity anomalies (general),

    rb-Free-air,

    rc-Bouger,

    rd-Isostatic,

    Magnetic field

    sa-Magnetic field,

Radiometric surveys (gamma rays)

    ta-radiometric surveys,

    uu-Unknown,

    zz-Other.

IV. CLASSIFICATION AND SUBJECT HEADINGS
FOR REMOTE-SENSING IMAGERY

To provide efficient and easy access to remote-sensing images is a fundamental goal of librarians responsible for these materials. As with maps, spatial data in remote-sensing imagery format are organized primarily by their geographic area and special topic. The organization can be by a geographic name or coordinates, which represents the area coverage. The images must be grouped in a standardized system, which allows for expansion of coverage, ease of retrieval and changes of name, topic or format. With the majority of libraries having some form of on-line catalog, it is important to have a system that can be adapted to the local catalog. There are several major systems in use in most libraries that house remote-sensing images. However, some libraries have developed local systems based on specific needs. Both the classification system and subject headings will be affected by the level of cataloging that the library decides to do for remote-sensing images.

A. Classification

Remote-sensing imagery collections cataloged using the cartographic materials format should be classed by the Library of Congress "G" Classification system whenever possible. The standardized system provides a formula for assigning a unique number to the photograph or group of photographs based on geographic area. The number can be assigned for a county flight with the subject class number indicating remote-sensing images (A4)* and a geographic cutter for the county. Aerial photographs and other remote-sensing images with no topical subject are given the subject Cutter A4. But maps are assigned Cutters according to the topic that they present; thus, a remote-sensing map with a subject theme such as weather should be given the appropriate subject cutter. This is an effective way of arranging materials covering the same place so that they are retrieved together. The date of situation is prominently displayed in the call number so all the relevant information is clearly evident.

In Example #1, the 1937 Alachua County, FL, flight would be classed according to administrative division, at G3933.A4A4 1937.M2. If the photograph were simply of Gainesville, it could be placed in the city category and be assigned the number of G3934.G2A4 1937.M2. A photographic flight for a geographic feature such as an island would be classed in G3932 for physical feature.

Even if collection-level cataloging were done for the entire state, a LC Class number could be assigned. The state-level number for this Florida flight would be G3931.A4 1937.M2. This option is not recommended for aerial photography unless one uses it as a temporary device, since user accessibility would be extremely limited. The state-level classification number would be more appropriate when cataloging satellite imagery, each of which images generally covers, relatively speaking, a much larger area (e.g., 100 miles by 100 miles) than does an air photo (e.g., 10 miles by 10 miles). Satellite images seldom conform to the smaller or well-defined administrative areas. Most states and countries cannot be subdivided by geographical subgroups such as northern or panhandle Florida. Thus the principal accessibility feature will be the subject heading.

There have been attempts in some libraries to organize remote-sensing imagery by the Superintendent of Documents item number. This has been an awkward system for images because of the emphasis on agency. There are hundreds of thousand of remote-sensing images issued by some agencies and the method for subdividing the materials by geographic area has been frus-

---

*A47 is now used by the Library of Congress' Geography and Map Division for remote-sensing images. However, many libraries have items classed with the former code. Therefore, the former code is used in this paper.

trating. The need to organize by agency can be handled within the LC classification scheme by the author/publisher fields.

In specialized libraries a local system can be devised which follows the principles of LC classification but is still adjustable for a specialized thesaurus. Some libraries will organize their photographs by geographic place names. These frequently reflect the names coded to the LC Classification "G" schedule. For example, if all the images in a collection were for that state they might be arranged by county and/or city.

If all the photographs are for only two counties in a planning district, then an elaborate system of numbers would not be necessary but rather a finding device based on local needs. This could be an in-house index or special subject classifications, e.g., weather satellite, coastal erosion, or FEMA (U.S. Federal Emergency Management Agency) aerial photographs before and after a disaster. It is difficult to maintain a home-made system if one is involved in any type of cooperative effort. Developing a system that will provide for adequate future expansion can be time-consuming. Using established land division systems–such as Township/Range/Section or Census Divisions–might be an alternative. There are several world reference systems for locating satellite images by X and Y coordinates. These are discussed in more detail in the Elements section. The two major satellites do not use the same referencing system so it would be difficult to inter-file these images. They could be arranged by their latitude and longitude. There are systems being developed for searching imagery collections online using a graphic interface. When this is accepted in imagery collections the access of the collections will expand tremendously. It will not preclude the use of the LC Classification "G" schedule. Decisions for how individual libraries will adapt their search techniques will need serious investigation. When deciding on any organizational scheme, serious thought should be given to the time and effort that will be necessary to maintain and explain the system to patrons and staff.

The Library of Congress classification-number construction allows for the inclusion of a special location or information code following the author cutter. This can be a word like Vault or initials such as RSI (Remote Sensing Image). If you have a large number of standard-size aerial photographs of your state or region, regular office file cabinets with hanging file folders, or standard archival boxes on standard stacks, may be used to establish a sublocation for these materials. Whenever possible, one should avoid filing them with the other maps in order to avoid damage to the other maps from the sharp edges of the photographic prints.

An advantage of cataloging images or groups of images separately as compared with collection-level cataloging or use of in house indexes is the ability to arrange by the reliable, geographic-based Library of Congress classification system and to furnish subject access for specific place names.

B. Subject Headings

1. Library of Congress

The Cataloging Policy and Support Office has made notable progress in the past few years by providing Library of Congress subject headings and subheadings needed to address this rapidly developing genre. Some pertinent subheadings available as subdivisions after places include: "Aerial photographs," "Aerial views," "Photographs from space," "Remote-sensing images," and "Remote-sensing maps." Recently these are being used with delimiter v instead of the traditional delimiter x to show that they are form/genre subheadings. They may also be used as form/genre headings or subheadings with the tag 655. Headings tagged 650 such as "Remote sensing" are reserved for works about remote sensing rather than for works which are themselves remote-sensing images or maps. Remote-sensing maps are distinguished from remote-sensing images by the presence of cultural information. Aerial photographs are a subset of remote-sensing images, distinguished by having been recorded from aircraft using photographic equipment. "Aerial views" is a free-floating subdivision for use, "under names of countries, cities, etc., or under individual educational institutions, to maps or atlases depicting those places from the air." "Photographs from space" is a free-floating form subdivision that may be used under names of countries, cities, etc., for collections of photographs, or reproductions of them, taken from outer space, but is not to be used with cartographic materials.

2. Local

In the local online catalog at the University of Florida, we add to these subject headings the specific reversed geographic headings based on the Library of Congress headings. These are tagged 690 or 691 and are retrieved by a "sg=" search in our NOTIS online-catalog system. In addition to direct and reversed order headings, we add subheadings for date of situation and scale.

V. CASE STUDIES

A. Aerial Photographs

The Maps and Imagery Library of the University of Florida offers a large number of aerial photographs of Florida, a resource which is very heavily used. For years we had been trying to get these cataloged, but it seemed an overwhelming task. The primary tool for controlling the collection had been a loose-leaf notebook kept up to date as additional prints were received. The decision to initiate a project for cataloging the collection came at a time when technical advances in efficiency of cataloging appeared to make it a feasible endeavor even with limited staff. The team approach, which had been used to great advantage in other cataloging projects in our library in recent years, seemed to hold promise for these materials. The Science Cataloging Unit, which is currently responsible for map cataloging, works closely as a team and meets on a monthly basis, frequently including a training session with the meeting. For example, details for scale and coordinates and map format fixed fields were reviewed at unit meetings. The unit is composed of two catalog librarians and two very capable and experienced paraprofessionals, with occasional support from student assistants. For this project one librarian, one paraprofessional, and a student assistant from the team worked closely together under the wing of the head of the Map and Imagery Library, to benefit from her wisdom and experience.

Example 1. ASCS County Level, Florida (See Figures A and B.)

ASCS County Level Aerial Photograph of Alachua Co., Fla.

OCLC: NEW　　Rec stat: n

Entered: 19980903　Replaced: 19980903　Used: 19980903

Type: e　ELvl: I　Srce: d　Relf:　Ctrl:　Lang: eng

BLvl: m　SpFm:　GPub: f　Prme:　MRec:　Ctry: utu

CrTp: a　Indx: 0　Proj:　DtSt: m　Dates: 1937,1938

Desc: a

040 FUG |c FUG

007 |a r |b u |c b |e c |f u |g b |h b |i b |j aa

034 1 a |b 31680 |d W0824000 |e W0820000 |f N0300000 |g N0292500

035 (FU)mapRS

052 3933 |b A4

090 G3933 .A4A4 1937 |b .M2

110 1 United States. |b Agricultural Adjustment Administration.

245 10 Alachua County, Florida (part) / |c U.S. Department of Agriculture, Agricultural Adjustment Administration ; Edgar Tobin Aerial Surveys.

246 3 Aerial photographs, Alachua County, Fla., contact prints.

255 Scale 1:31,680 ; |c (W 82 40'-W 82 00'/N 30 00'-N 29 25').

260 [Salt Lake City, Utah] : |b The Administration, |c 1938.

300 604 remote-sensing images ; |c 24 x 24 cm.

500 Complete in ca. 746 images.

500 Title and publication information from photomosaic index ; accompanied by hand-made paper index.

500 Flights flown Dec. 1937 through Jan. 1938.

500 Photos numbered ITT-1-1 through IT-10-72.

651 0 Alachua County (Fla.) |v Aerial photographs.

651 0 Alachua County (Fla.) |v Remote-sensing images.

655 7 Remote-sensing images |z Florida |z Alachua County |2 lcsh.

691 9 Alachua County (Fla.) |v Aerial photographs |y 1938.

691 9 Florida, Counties |z Alachua County |v Aerial photographs |y 1938.

710 2 Tobin (Edgar). |b Aerial Surveys, San Antonio.

For more that six decades, various companies have been contracted by federal agencies as part of a national program to fly over and photograph most counties of the United States. The Map & Imagery Library serves as a state archives for those photographs of Florida produced by the U.S. National Resources Conservation Service under its former names. The Library owns a large number of standard 9 x 9-inch photos, as well as various enlargements. Each of the 67 counties of Florida was flown an average of five times.

Initially, the most desirable approach to provide access for this collection appeared to be a separate bibliographic record for every flight for each of the counties. The first small step towards providing access in the catalog was to create a local collection-level record supplying basic subject and keyword access to the materials as a group. The project would then involve several stages. The first stage was to fully catalog one flight of one county as a test case and examine the fields in the record. Much of the record would remain constant across counties and years and could be duplicated to form the basis of each of the other records. Another group of fields could be added methodically based on the index and a chart of county coordinates and cutter numbers. The next stage would require examination of the prints and photo-mosaic indexes to complete the notes, corporate-body added entries, and extent-of-item fields (see Example #1). The final stage would be overall editing, authority control, and revision of each record, and submission to the bibliographic utility, in our case OCLC. Because of the large number of records involved, this was projected to take some months to complete. Enlargements would require additional records, but could be accomplished with relative ease at a later date based on the records for the 9 x 9-inch prints. Full cataloging at this level would have the advantage of letting patrons both on campus and those using the catalog over the Internet throughout the state know that the photographs exist and are available at the University of Florida Library.

Careful preparation of the model record required consideration of all the questions and fields described above. What did we have? Examination of the collection showed that we had positive photographs taken at a vertical angle from airplanes and printed on sheets of paper. There was no cultural information such as place names or latitude or longitude or any other markings added to the simple photographs other than a numeric code. These enabled use of the accompanying photo-mosaic index to locate target locations or order enlargements. This would place them in the form category, "Aerial photographs," a subset of the more general category, "Remote-sensing images."

The accompanying photo-mosaic indexes show north orientation and county name. (See Figure B.) Cataloging each print separately was not a feasible or even desirable option, but if we only had a few miscellaneous prints, we might have chosen differently. Our patrons would find the county and year analytic level quite useful. The cartographic format is the best choice for these materials (instead of the graphics format one might use for other photographs), since it better enables expression of geospatial content characteristics of the resource. Our library uses Library of Congress classification. For this collection each title will receive the class G3933, the cutter for the particular county (an administrative division), the subject cutter A4 and the year of the fly-over. Since we already have a separate section of the library where the Florida county aerial photographs are filed together alphabetically, it will not require rearrangement to place them in classification order.

The corporate-body main entry for these items had to be considered carefully. The federal agency, which contracted for the making of the photographs, wrote the specifications, rectified the images (adjusted them for distortion), and later distributed the photographs was the first possibility. The company employed to actually fly over and take the photographs would have been used as main entry in the absence of such an agency. The rule in AACR2R, 21.2B2, would dictate we enter under a corporate body, "cartographic materials emanating from a corporate body other than a body that is merely responsible for their publication or distribution. . . . In some cases of shared responsibility and mixed responsibility, enter such a work under the heading for the corporate body named first" (AACR2R p. 314). Typically the federal agency was presented first and prominently on these photo-mosaic indexes we relied upon as chief source. The company that took the photographs was given in small print near the bottom.

The source of information can sometimes provide a challenge in cataloging remote-sensing images. The *title* for most of these aerial photographs would be able to be derived from the legend of the photo-mosaic index that accompanied them. Unfortunately, these were unavailable for more recent years and would require assignment by the cataloger. Source of title would be indicated in a note, and for those assigned by the cataloger would require brackets in the 245 field. The *area of coverage* would obviously be the county. The geographic area would be given in the 651 field and the latitude and longitude coordinates in the 034 and 255 fields. Planned omission of parts of some counties would be determined from the photo-mosaic index and detailed in a note. The invitation number of the contract between the sponsoring agency and the company conducting the fly-over would be used as a *unique identifier* and recorded in a note. The *orientation* of the photography was vertical in this case. The *date* was usually given on each photograph and recorded on the photo-mosaic index. The photographs cataloged in the initial phase of the project varied between 9 x 9 inches and 10 x 10 inches, so they required measurement and recording in centimeters, as well as counting for size and extent of item information for the 300 field. The *scale* was usually given on the photo-mosaic index, which was the only *accompanying material.* Florida is sadly lacking in mountains, but does have hills and valleys. While these can sometimes be discerned from examining the photographs, the record must show that *relief* is not indicated. Relief means how that characteristic is represented by symbols, and therefore is not recorded for an image. Most of the small prints were in black and white. Some recent photographs are color infrared, which is recorded as color even through it is also a format.

For efficiency, and in order to utilize the team approach in the mass record-creation stage, we decided to use the "Thesis Button." This is an easy-to-use, fill-in-the-blank template macro devised by Gary Strawn as an extension of his CLARR Catalogers Toolkit. Initially this was used in creating bibliographic access to our theses without requiring use of MARC tagging by the in-putters. If the reader wishes to undertake a similar project but cannot use this macro, the OCLC constant data record could be used to advantage. The strategy was to use the information about the photographs from the in-house index notebook and a table of coordinates and Cutters for each county in Florida to allow a student using the Thesis Button to create incomplete records for all the counties for each year of flight. One powerful feature of this macro is that it eliminates the need for repeated keying of data that needs to be plugged into multiple fields, such as the year and the name of the county. These partial records would already include full subject access, dates, and classification and a default value for other elements. Creating the custom template for the Thesis Button was accomplished by a creative and experienced cataloging paraprofessional. This person would also be participating in the project by overseeing a student assistant in the basic inputting stage and by completing the detailed portions of the records in the next stage.

The next stage was to complete each of these partial records based on information from the photographs, the available photo-mosaic images, and from agency sales information for those years lacking photo-mosaic images. This was a complex task wherein the exact names of the responsible corporate bodies (as authoritative forms) would be added to the record, as well as appropriate notes describing the materials. Extensive consultation with the Map Librarian was required initially to accomplish this task. Also there had to be a high tolerance for ambiguity and frustration due to the inconsistency in availability, format and source for data elements. Counting and measuring the images occurred at this time also. Finally, each record would be reviewed and edited by the professional cataloger before being loaded to OCLC using the CLARR software.

The project was approached methodically beginning with Alachua County, and progressing finally to Washington County. Monitoring of the project was facilitated by an indexed 035 field to enable retrieval of all these records in the local online catalog. The photomosaics being used as indexes fit the category of "uncontrolled" and were not used for other purposes. Therefore the decision was made to catalog them as accompanying indexes and not as separate entities requiring individual records. There are reasonable arguments for cataloging them either way and local use is the deciding factor.

Example 2. High Altitude (U2) Mt. St. Helen's Color Infrared (CIR) (See Figure C.)

OCLC: NEW   Rec stat: n

Entered: 19980917   Replaced: 19980917   Used: 19980917

Type: e ELvl: I Srce: d Relf: z Ctrl: Lang: eng

BLvl: m SpFm: GPub: Prme: MRec: Ctry: xxu

CrTp: a Indx: 0 Proj: DtSt: s Dates: 1981,

Desc: a

040 FUG |c FUG

007 r |b u |d b |e u |f u |g d |h b |i b |j aa

034 0 a

043 n-us-wa

052 4282 |b S25

090 G4282.S25A4 1981 |b .A3

049 FUGG

245 00 [Mount Saint Helens after eruption, color image]

255 Scale not given.

260 [S.l. : |b s.n., |c 1981]

300 1 remote-sensing image : |b col. ; |c 50 x 50 cm.

500 High-altitude aerial photograph in color of the Mount Saint Helens region following the 1980 eruption.

500 "UAg II 153.46"--Edge.

500 "0038 02891"--Edge.

651 0 Saint Helens, Mount (Wash.) |v Aerial photographs.

650 0 Volcanoes |z Washington (State) |v Aerial photographs.

655 7 Remote-sensing images |z Washington (State) |z Saint Helens, Mount |2 lcsh

691 9 Saint Helens, Mount (Wash.) |v Aerial photographs.

691 9 Aerial photographs |z Washington |z Saint Helens, Mount.

This photograph of Mount Saint Helens taken from a high-altitude flight is useful to compare with similar photographs taken before the volcanic eruption and with those taken later. The edge notation, "UAg II 153.46," tells us that the photograph was taken from a U2 reconnaissance plane and must be recorded in a note. It is also reflected in the 007 for remote-sensing in subfield d as "b Airborne," in subfield g as "d Aircraft-high altitude" and subfield h as "b Surface observing." Subfield i for sensor type is coded "b passive," since the light energy received in the recording device is passively received rather than produced in the airplane (no flashbulbs were used).

Example 3. Remote-sensing Map (Photograph) of McMurdo Station, Antarctica. (See Figure D.)

OCLC: 34560258   Rec stat: c

Entered: 19960412   Replaced: 19960531   Used: 19980819

Type: e ELvl: I Srce: d Relf: Ctrl: Lang: eng

BLvl: m SpFm: GPub: f Prme: MRec: Ctry: vau

CrTp: a Index: 1 Proj: DtSt: s Dates: 1996,

Desc: a

040 GPO |c GPO |d UUM

007 r |b u |d u |e c |f u |g u |h b |I u |j aa

034 1 a |b 1800

052 9804 |b M24

074 0619-G-19

086 0 I 19.100/6:M 22/996

090 G9804.M24A4 1996 |b .G4

049 FUGG

110 2 Geological Survey (U.S.)

245 10 Antarctica photomap. |p McMurdo Station / |c produced by the United States Geological Survey.

246 30 McMurdo Station

246 1 |i Filing title: |a McMurdo Station, Antarctica, photomap

255 Scale ca. 1:1,800.

260 Reston, Va. : |b The Survey ; |a Denver, Colo. : |b For sale by the Survey, |c [1996]

300 1 map : |b col. ; |c 76 x 76 cm., on sheet 115 x 81 cm.

500 "Photograph obtained by the U.S. Geological Survey in cooperation with the National Science Foundation."

500 Photograph date: 23 Nov. 1993.

500 Includes 2 ancillary photomaps of Station from 15 Feb. 1960 and 5 Feb. 1975, each with scale of 1:2,400.

500 Includes feature identification index and location diagram.

651 0 McMurdo Station (Antarctica) |v Remote-sensing maps.

655 7 Remote-sensing maps |z Antarctica |z McMurdo Station |2 lcsh.

691 9 McMurdo Station (Antarctica) |v Remote-sensing maps |y 1996.

691 9 Antarctica |z McMurdo Station |v Remote-sensing maps |y 1996.

690 9 Remote-sensing maps |z Antarctica |z McMurdo Station |y 1996.

710 2 National Science Foundation (U.S.)

The McMurdo Station Photomap was received from U.S. Geological Survey through the GPO Federal Depository Program for Libraries. It was chosen for inclusion because it is a good example of the transition between a remote-sensing map and remote-sensing image. There are three images on the map sheet. The largest image has numbers superimposed on specific features and an index in the margin for the features. No other cultural features or text have been added to the photograph. The two smaller images are dated photographs to which nothing has been added. The title states the publisher's intent for the entire item to be considered a "photomap." The GPO catalogers have cataloged it as a "Remote-sensing map," undoubtedly after considering these points. It would not be surprising if the numbers were overlooked and someone cataloged it as 3 remote-sensing images, 76 x 76 cm or smaller on sheet 115 x 81 cm.

B. Satellite Imagery Examples

Landsat Images of the Everglades Region of Florida

Example 4. Landsat Image of Everglades Region Band 5. (See Figure E.)

OCLC: 30502954   Rec stat: c

Entered: 19940527   Replaced: 19980119   Used: 19940527

Type: e ELvl: I Srce: d Relf: Ctrl: Lang: eng

BLvl: m SpFm: GPub: f Prme: MRec: Ctry: sdu

CrTp: a Indx:.0 Proj: DtSt: s Dates: 1973,

Desc: a

040 FUG |c FUG |d OCL

007 r |b u |d c |e c |f u |g f |h b |i b |j aa

034 1 a |b 1000000 |d W0820000 |e W0800000 |f N0263000 |g N0253000

052 3932 |b E89

052 3934 |b M5 |b N2

090 G3932 .E89A4 1973 |b .E7

110 2 EROS Data Center.

245 10 [Everglades region, Florida, LANDSAT satellite image]. |n E-1242-15240-5.

255 Scale 1:1,000,000. |c (W 82   00' --W 80   00'/N 26   30' --N 25   30').

260 [Sioux Falls, S.D.] : |b NASA, |c 1973.

300 1 remote-sensing image ; |c 22 x 22 cm.

500 Shows southern portion of the Florida peninsula, Lake Okeechobee to Everglades to Florida Bay (N-S) and Miami to Naples (E-W).

500 "Path 16, row 42"--index.

500 "22 MAR 73."

500 Title devised by cataloger.

500 Oriented with north to the upper left.

500 "C N25-59/W080-55 N N25-59/W080-48 MSS 5 D SUN EL50 AZ124 189-3373-N-1-N-D-2L NASA ERTS."

500 Glossy monochrome photograph on Kodak paper.

651 0 Everglades Region (Fla.) |v Remote-sensing images.

651 0 Miami Region (Fla.) |v Remote-sensing images.

651 0 Naples Region (Fla.) |v Remote-sensing images.

655 7 Remote-sensing images |z Florida |z Everglades Region |2 lcsh.

691 9 Everglades Region (Fla.) |v Remote-sensing images |y 1973 |x 1,000,000.

691 9 Miami Region (Fla.) |v Remote-sensing images |y 1973 |x 1,000,000.

691 9 Naples Region (Fla.) |v Remote-sensing images |y 1973 |x 1,000,000.

691 9 Florida |z Everglades Region |v Remote-sensing images |y 1973 |x 1,000,000.

691 9 Florida |z Miami Region |v Remote-sensing images |y 1973 |x 1,000,000.

691 9 Florida |z Naples Region |v Remote-sensing images |y 1973 |x 1,000,000.

710 1 United States. |b National Aeronautics and Space Administration.

Example 5. Landsat Image of Everglades Region Color Infrared (CIR) (See Figure E for black and white reproduction.)

OCLC: 30502989   Rec stat: c

(Same as previous record)

007 r |b u |d c |e u |g f |h b |i b |j da

(Same as previous record)

110 2 EROS Data Center.

245 10 [Everglades region, Florida, LANDSAT satellite image]. |n
E-1242-15240-CIR.

(Same as previous record)

500 Glossy color infrared photograph on Kodak paper.

(Same as previous record)

There are many examples in libraries of products of the Landsat satellite
program. These two Examples, #4 and #5, are hard-copy photographs of two
separate expressions of the same scene but one is band 5 (in black and white),
and the other a false-color composite which simulates infrared. Their identification numbers vary slightly to indicate spectral bands used, and are recorded
in the annotation block. The date, coordinates, and other details such as the
elevation of the sun also appear on the edges. A note records this on the
record as it is given. "NASA ERTS [Earth Resources Technology Satellite]"
tells us that this is a product of the EROS Data Center and United States
National Aeronautics and Space Administration, and these are entered in
corporate-body entries, the first as main entry. The coordinates and date also
are reflected in the 034, 255, 260 and dates fields. Since no title was given,
the cataloger devised one and enclosed it in brackets in the 245 field, stating
in a note that the title was, "devised by cataloger." We have used several
geographic subject headings that are based on the Everglades and nearby
cities with the wild-card term "region" added to indicate that the region
around the feature or city itself is included. As form subheading, we have
used "Remote-sensing images." The inclusion of coordinates, gray scale,
etc., in the edges of the photographs does not place these items in the category of maps instead of images because they are not placed on the face of the
image.

Example 6. Remote-sensing Map of Miami (Poster) (See Figure F.)

OCLC: 21060411   Rec stat: c

Entered: 19910816   Replaced: 19980119   Used: 19960715

Type: e ELvl: Srce: Relf: Ctrl: Lang: eng

BLvl: m SpFm: GPub: Prme: MRec: Ctry: cau

CrTp: a Indx: 0 Proj: DtSt: s Dates: 1989,

Desc: a

010 91-685058/MAPS

040 DLC |c DLC |d OCL

007 r |b u |d c |f u |g f |h c |I b |j uu

034 0 a

045 0 |b d1988

050 00 G3934.M5A4 1988 |b .T4

052 3934 |b M5

110 2 Terra-Mar Resource Information Services.

245 10 Miami & south Florida / |c the image was digitally processed by Stephanie M. Tanaka of Terra-Mar Resource Information Services, Inc., Mountain View, CA.

255 Scale not given.

260 Corte Madera, Calif. : |b Portal Publications Ltd., |c c1989.

300 1 map : |b col. ; |c 61 x 50 cm., on sheet 92 x 61 cm.

500 False-color satellite image map of Miami region.

500 "The scene was recorded . . . by . . . the Thematic Mapper (TM) sensor onboard the LANDSAT 5 satellite on March 5, 1988."

500 Copyright "1989 Terra-Mar Resource Information Services and Eosat."

500 Includes text.

500 "SAT501."

651 0 Miami Region (Fla.) |v Remote-sensing maps.

691 9 Miami Region (Fla.) |v Remote-sensing maps |y 1988.

691 9 Florida |z Miami Region |v Remote-sensing maps |y 1988.

655 7 Remote-sensing maps |z Florida |z Miami Region |2 lcsh.

700 1 Tanaka, Stephanie M.

710 2 Portal Publications Ltd.

710 2 Eosat.

740 01 Miami and south Florida.

This example is a digitally-processed remote-sensing map based on a LANDSAT image, and includes text. This map covers the same area and closely resembles the false-color composite remote-sensing LANDSAT image in Example #4. Although it is mounted on paper with a title and modified to look like a poster, it would be considered a map, not just an image, as it has text and place-names on the face of the image. The MARC 300 and 650 fields would all be recorded for a map and not a remote-sensing image. Common U.S. policy is to use only the one 007 fixed field when cataloging a remote-sensing image in a cartographic format. There are situations in which extensive information such as contour lines has been added to the image which can not be coded in the remote-sensing image. To more correctly describe the item one could include two 007s such as the Canadian national libraries do. It would appear as "007   a |b j |d c |e a |f n |g z |h n". It is never incorrect to include two 007 fields for those items when the item being cataloged straddles the demarcation line between an image and a map. At the University of Florida the online catalog allows search limiting based on some elements in various fixed fields including the 007, giving functional advantage to multiple 007 fields in certain cases.

There is extensive documentation that the base is a LANDSAT product, but that it is a commercial product digitally processed and published by Terra-Mar Resource Information. The compilation data would be included in the note fields.

Example 7. Remote-Sensing Image of Miami (Poster) (See Figure G.)

OCLC: 27797422   Rec stat: c

Entered: 19930325 Replaced: 19980119 Used: 19970909

Type: e ELvl: I Srce: d Relf: z Ctrl: Lang: eng

BLvl: m SpFm: GPub: Prme: MRec: Ctry: mdu

CrTp: a Indx: 0 Proj: DtSt: s Dates: 1988,

Desc: a

040 FUG |c FUG |d OCL

007 r |b u |d c |f u |g f |h c |I b |j uu

034 0 a

052 3934 |b M5

090 G3934.M5A4 1988 |b .S2 RSP

049 FUGG

110 2 Satellite Snaps, Inc.

245 10 Miami from LANDSAT 5 / |c this print produced and distributed by

Satellite Snaps, Inc. ; image enhancement by KRS Remote Sensing ; LAND-SAT data distributed by the Earth Observation Satellite Company.

255 Scale not given.

260 [Ridgely, Md.] : |b Satellite Snaps, Inc., |c c1988.

300 1 remote-sensing image: |b col. ; |c 90 x 46 cm.

500 LANDSAT satellite image, processed to simulate natural color.

500 Shows Miami metropolitan area north from Fort Lauderdale, south to Homestead.

500 Image encapsulated in plastic.

651 0 Miami Region (Fla.) |v Remote-sensing images.

651 0 Miami (Fla.) |v Remote-sensing images.

655 7 Remote-sensing images |z Florida |z Miami Region |2 lcsh.

691 9 Miami Region (Fla.) |v Remote-sensing images.

691 9 Miami (Fla.) |v Remote-sensing images |y 1988.

710 2 KRS Remote Sensing (Firm)

710 2 Eosat.

This image closely resembles the previously discussed images and re-mote-sensing map. It has been included as an example because it is typical of images resembling posters being produced for commercial purposes by private firms. The image has no information added to it and there is very little compilation data available in the margin. Despite its presentation, it is considered a remote-sensing image and would be cataloged as such. Common U.S. policy is to use only the one 007 fixed field when cataloging a remote-sensing image in a cartographic format; this is a case where we would use a local option and include two 007s. At the University of Florida the online catalog allows search limiting based on some elements in various fixed fields including the 007, giving functional advantage to multiple 007 fields in some cases.

Example 8. Landsat Image of Coquimbo Region of Chile on CD-ROM (See Figure H.)

OCLC: 39885666   Rec stat: n

Entered: 19980917   Replaced: 19980917   Used: 19980917

Type: e ELvl: I Srce: d Relf: Ctrl: Lang: und

BLvl: m SpFm: z GPub: f Prme: MRec: Ctry: sdu

FIGURE F. *Miami & South Florida.* Reproduced with permission of the Map-Factory, Inc.

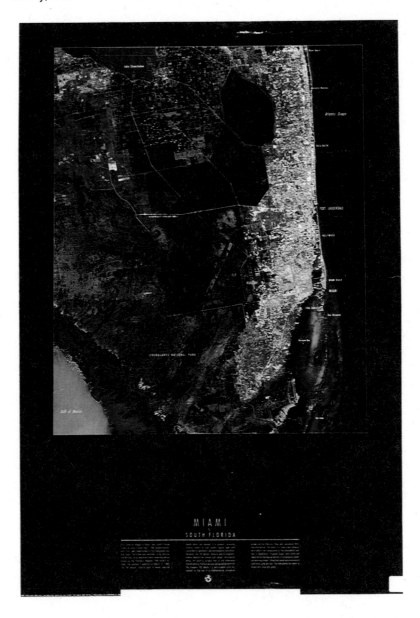

FIGURE G. *Miami from LANDSAT 5.* Reproduced with permission of Satellite Snaps, Inc.

FIGURE H. Landsat Image of Coquimbo Region of Chile on CD-ROM

FIGURE I. SPOT Image of the Pinhook Swamp and Osceola National Forest on CD-ROM

CrTp: a   Indx: 0   Proj: DtSt: s   Dates: 1986,

Desc: a

040 FUG |c FUG

007 r |b u |d c |e c |f u |g f |h b |i b |j aa

007 c |b o |d u |e g

034 0 z |d W0730100 |e W0713800 |f S0310500 |g S0324800

035 (FU)ALY7143

052 5332 |b C66

090 G5332.C66A4 1986 |b .E7

049 FUGG

110 2 EROS Data Center.

245 10 [Coquimbo Region, Chile, LANDSAT satellite image. |n Scene #LT5001082008624310] |h [computer file]

246 3 Chile, Coquimbo region

246 3 LANDSAT satellite image, Coquimbo region, Chile

255 Scale indeterminable |c (W 73 01'-W 71 38'/S 31 05'-S 32 48').

260 [Sioux Falls, S.D.] : |b USGS, EROS Data Center, |c [1997]

300 1 computer optical disc ; |c 4 3/4 in.

500 LANDSAT 5 (NASA satellite)

500 Path 01, Row 82.

500 Date of image: August 31, 1986.

500 "0119711060073 1 NLP 1/1"-Label.

538 System requirements: computer with CDROM reader; image-processing software and hardware capable of reading and displaying LANDSAT data which are Band Interleaved by pixel pairs-Goddard-formatted files on cartridges.

651 0 Coquimbo (Chile: Region) |v Remote-sensing images.

655 7 Remote-sensing images |2 lcsh

691 9 Coquimbo (Chile: Region) |v Remote-sensing images.

691 9 Chile |z Coquimbo (Region) |v Remote-sensing images.

710 1 United States. |b National Aeronautics and Space Administration.

There is no title and little other bibliographic information on this CD-ROM. According to the label on the face of the CD, it is from the USGS EROS Data Center, Sioux Falls, SD. Some other numbers are also given there, one of which begins with LT and is 18 characters long. The container label repeats these numbers along with other mysterious abbreviations, codes and numbers and a date: 11/09/97. Handwritten on this label is "Chile-8.31.86." For further clues towards describing this item, we must put it into a computer and print out the nine-page file summary that is equivalent to the header on a magnetic tape. Some of the additional data revealed here are "satellite = 5," and an indication that the image was recorded from NASA's LANDSAT 5 satellite. Further into the file we find that the long number beginning with "LT" is the "Scene Id List," the most important unique identifier for the item; and the date of recording is 08/31/86. The more recent date is a production date for the CDROM product. Near the end of the file printout are latitude and longitude for the center of the image, and for each of the four corners of the image. Path and row are also given on the file summary printout.

To ascertain coverage for purposes of classification and assignment of a subject, we must go to a map which provides coordinates and place names. There we learn that these coordinates describe a rectangle on the southern coast of Chile in the Coquimbo region. Classification of a region in Chile (including a Cutter for Coquimbo region and an A4 Cutter to represent the remote-sensing aspect of the item, the date of recording (1986) and a Cutter for the main entry) are accomplished by consulting the G schedule. Coquimbo (Chile: Region) is easily found in LCSH as a geographic subject term. The form subdivision, "Remote-sensing images," is required. We also choose the option of adding local reverse geographic subject headings.

The next step is to translate the information we have just derived into a bibliographic description and USMARC format. We begin by using a map format record. Since we must assign a title, we call it, "[Coquimbo Region, Chile, LANDSAT satellite image] Scene #LT. . . ," using delimiter n to set off the number, then adding a general material designator term, "[computer file]," in a delimiter h. Scale is indeterminable, but coordinates are known and should be recorded in the 255 and 034 fields. The place of publication, publisher, and date published are available from the labels. The physical description is "1 computer optical disc ; 4 3/4 in." Notes are needed for "LANDSAT 5," "Path 01, Row 82," and the other numbers found on the face of the CD. A systems-requirement note (538) tells that a computer with a CD-ROM reader and appropriate software are needed. At the top of the record, fixed fields are now routinely filled in, based on the information we have recorded in the variable fields. Two separate 007 fields are required, one for the remote-sensing properties and another for the computer-file properties of the item. This is an example of previous MARC rules; currently, a 006 is used for the latter properties.

Example 9. SPOT Image of the Pinhook Swamp and Osceola National Forest on CD-ROM (See Figure I.)

OCLC: NEW   Rec stat: n

 Entered: 19980903   Replaced: 19980903   Used: 19980903

 Type: e ELvl: I Srce: d Relf: Ctrl: Lang: eng

 BLvl: m SpFm: GPub: f Prme: MRec: Ctry: flu

 CrTp: a Indx: 0   Proj: DtSt: r Dates: 1998,1988

 Desc: a

040 FUG |c FUG

007 c |b o |d u |e g

007 r |b u |d c |e c |f u |g f |h c |i b |j zz

034 0 a |b 500000 |d W0829608 |e W0820036 |f N0306447 |g N0299761

052 3932 |b O8

090 G3932.O8 A4 1988 |b G41

110 2 SPOT Image Corporation.

245 10 Pinhook Swamp, Osceola National Forest, Florida |h [computer file] |b scene id 12 x 617289881025163723.

255 Scale varies.

256 Computer data and programs.

260 Paris : |b SPOT, |c 1988.

300 1 computer tape reel : |b 8 track.

538 Systems requirements: IBM compatible computer with MS-DOS version 3.1 or later; 640K of RAM; a CD-ROM reader using Microsoft Extensions, version 2.0 or later.

500 Title from leader.

516 JPEG compressed data image files.

500 Mastered 9/98.

533 Computer optical disc. |b Gainesville, Fla. : |c University of Florida, |d 1998. |3 5 1/4 in. ·

651 0 Pinhook Swamp (Fla.) |v Remote-sensing images |v Databases.

651 0 Osceola National Forest (Fla.) |v Remote-sensing images |v Databases.

691 9 Pinhook Swamp (Fla.) |v Remote-sensing images |v Databases |y 1988.

691 9 Osceola National Forest (Fla.) |v Remote-sensing images |v Databases |y 1988.

691 9 Florida |z Pinhook Swamp |v Remote-sensing images |v Databases |y 1988.

691 9 Florida |z Osceola National Forest |v Remote-sensing images |v Databases |y 1988.

655 7 Remote-sensing images |v Databases |2 lcsh.

This remote-sensing map recorded in 1988 from the SPOT satellite was held in the library for years on magnetic tape. It was recently reformatted to CD-ROM because of degradation to the tape. Since most current computers have CD input options, this is more convenient to use than magnetic tape. The original record was modified without deleting reference to the original format as it is still held by the library and provides compilation background. Again, two 007 fields were used to code the computer file aspects as well as those relating to remote sensing.

C. Internet Resource Composed of Remote-Sensing Images Linked to Maps and Place Names

Example 10. TerraServer Internet Database.

OCLC: 40107568   Rec stat: n

Entered: 19981014   Replaced: 19981014   Used: 19981014

Type: e ELvl: I Srce: d Relf: Ctrl: Lang: eng

BLvl: m SpFm: GPub: Prme: MRec: Ctry: xxu

CrTp: e Indx: 1 Proj: DtSt: m Dates: 1998,9999

Desc: a

040 FUG |c FUG

006 [aab c   001 0]

007 c |b r |d c |e n |f u

007 r |b u |d z |e c |f u |g z |h b |I b |j aa

034 0 a

052 3200

090 G3201.A4

049 FUGG

110 2 Microsoft Corporation.

245 10 TerraServer |h [computer file] / |c Microsoft.

246 30 Terra Server

246 30 Microsoft TerraServer

255 Scales indeterminate.

256 Computer data and programs, database size: 1.01 terabytes.

260 [S.l.] : |b Microsoft, |c c1998-

500 Description based on title screens October 14, 1998, dates of the images vary.

500 Additional contributors: Compaq Storage Works, USGS, SPIN-2.

500 "TerraServer uses Microsoft SQL Server 7.0 Enterprise Edition to host the world's largest online database."--From Website.

500 Includes indexes.

520 Includes USGS digital orthophotoquad aerial photographs, satellite images from SPIN-2, and maps with place names from ENCARTA for many places in the world with greater concentration on North America and Europe.

538 Remote access through the Internet using a browser, best viewed with color monitor.

530 Also available for purchase.

650 0 World maps |v Databases.

651 0 North America |v Remote-sensing maps.

651 0 Europe |v Remote-sensing maps.

651 0 North America |v Maps.

651 0 Europe |v Maps.

710 2 Geological Survey (U.S.)

710 2 SPIN-2.

710 2 Compaq Computer Corporation.

856 4 |3 Online maps and remote-sensing images: |u
http://terraserver.microsoft.com

This free Internet resource composed of remote-sensing maps, other maps, and a gazetteer became available recently, and was selected for access in our online catalog. A map format was chosen for the record, with computer-file aspects accounted for by added fields in the record. Since the images are associated with coordinates and maps showing place-names, the form subheading "remote-sensing maps" is used instead of "remote-sensing images." It is "published" by Microsoft, with the help of three other acknowledged contributors. Microsoft is given as the main entry, and the others are recognized in a note and with corporate body added entries (710 fields). The Internet address is given in an 856 field which both displays and functions as an automatic link in our catalog. A summary note (520 field) is used to describe the resource. The 245 field includes a general material designator (subfield h) for computer file, and there are 007 fields for both the remote-sensing and computer-file aspects of the item. Since we do not have a physical item, there is no physical description field (300). While we did not find an explicit statement regarding updating, we assumed that it would occur and left the dates in the fixed field and in the publisher statement open-ended.

VI. CONCLUSION

Cataloging remote-sensing images for a library can be a challenging as well as rewarding experience. Remote-sensing images have many characteristics in common with maps as both show the geographic relationships of features by a graphic representation. To catalog images using the USMARC Maps Format, both the similarities and differences need to be identified. The intent of this article was to help catalogers to better understand and describe remote-sensing images. The authors hope they have provided encouragement to the timid and solace to the overwhelmed, that these are not unfathomable materials. There are some methods for attacking the larger backlogs of remote-sensing images, which were examined in the case studies.

The characteristics of remote-sensing materials and scope of the materials were considered to provide a foundation for cataloging the items. A primary issue was the difference between a map and a remote-sensing image. The map is a compilation of data manipulated by a cartographer to convey information by showing spatial relationships. Because of space and scale restrictions only selected portions of the entire universe are portrayed. A remote-sensing image records all information as it appears at one specific time with no interpretation or analysis included. When additional information is placed on the face of the item it becomes a remote-sensing map.

The elements of remote-sensing cataloging were identified to assist the cataloger in interpreting different Anglo-American cataloging rules. The rules governing the expression of the different items in cataloging records were discussed. The various satellite and aerial photography platforms and programs plus their associated codes were explained. Understanding the nature of these platforms and sensors is essential for cataloging remote-sensing materials with such tools as USMARC, LCSH and LC classification. The discussions were conducted with reference to the MARC fields with particular attention given to the explanation of the new 007 field for remote-sensing.

To aid in further understanding how to catalog these items, examples were given with their catalog records for various types of remote-sensing images and maps. These examples showed that while remote-sensing materials may appear highly technical and even esoteric, there are logical explanations behind their creation and it is not necessary to be a rocket scientist to catalog them.

Remote-sensing images differ from polished, constructed maps in that they have the clean, unspoiled flavor of raw scientific data. These images may form the basis of further scientific exploration, which profoundly affects the universe. Catalogers can be instrumental in facilitating the researcher's access to these materials by thorough and accurate description in the catalog record.

VII. SELECTED REFERENCES

American Library Association, Map & Geography Round Table. *Remote-sensing Imagery: Identification, Control, and Utilization.* (Unpublished Pre-Conference Notebook) Presenters: Armstrong, HelenJane and Elizabeth Mangan with Practicum by Mary Larsgaard. Chicago, June 22, 1990.

American Society of Photogrammetry. *Multilingual Dictionary of Remote Sensing and Photogrammetry.* Falls Church, Va.: The Society, 1984.

Andrew, Paige. "Glossary." (Draft prepared for 2nd ed. of *Anglo-American Cataloging and Classification of Cartographic Materials*, not yet published).

Anglo-American Cataloguing Committee for Cartographic Materials. *Cartographic Materials; A Manual of Interpretation for AACR2.* Chicago: American Library Association; Ottawa: Canadian Library Association; London: The Library Association, 1982.

Anglo-American Cataloging Rules. 2nd ed. rev. Chicago: American Library Association, 1988.

Avery, Thomas Eugene and Graydon Lennis Berlin. *Interpretation of Aerial Photographs.* 4th ed. Minneapolis: Burgess Publishing Co., 1985.

Bibliographic Formats and Standards. Dublin, Ohio: OCLC, 1998.

Campbell, James B. *Introduction to Remote Sensing.* 2nd ed. New York: Guilford Press, 1996.

Centre Spatial de Toulouse and SPOT IMAGE. *SPOT User's Handbook.* Toulouse, France, 1988.

Library of Congress. Cataloging Distribution Service. *USMARC Format for Biblio-graphic Data*. 1994 ed. Washington, D.C.: The Service, June 1997 Update.

Library of Congress. Geography and Map Division. *Map Cataloging Manual*. Washington, D.C.: Cataloging Distribution Service, Library of Congress, 1991.

Library of Congress. Cataloging Policy and Support Office. *Subject Cataloging Manual. Subject Headings*. 5th ed. Washington, D.C.: Cataloging Distribution Service, Library of Congress, 1996-

Library of Congress. Subject Cataloging Division. *Classification: Class G, Geography, Maps, Anthropology, Recreation*. Washington, D.C.: Cataloging Distribution Service, Library of Congress, issued 1997.

Library of Congress. Cataloging Distribution Service. *Library of Congress Subject Headings*. 21st ed. Washington, D.C.: The Service, 1998.

Lillesand, Thomas and Ralph Kiefer. *Remote Sensing and Image Interpretation*. 2nd ed. New York: John Wiley, 1987.

Short, Nicholas M. *The Landsat Tutorial Workbook; Basics of Satellite Remote Sensing*. Washington, D.C.: National Aeronautics and Space Administration, 1982.

Strawn, Gary. *CLARR, the catalogers toolkit*. (Unpublished computer file), 1998.

APPENDIX I

Landsat and SPOT Spectral-Band Information

1. Landsat

   a. MSS (Multispectral Scanner)–in microns; first Band number given is for Landsats 4-6, second is for Landsats 1-3

      i.   Band 1/5: 0.5-0.6

      ii. Band 2/6: 0.6-0.7

      iii. Band 3/7: 0.7-0.8

      iv. Band 3/8: 0.-1.1

   b. TM (Thematic Mapper)–in microns

      i.   Band 1: 0.45-0.52

      ii. Band 2: 0.52-0.60

      iii. Band 3: 0.63-0.69

      iv. Band 4: 0.76-0.90

      v.   Band 5: 1.55-1.75

      vi. Band 6: 10.4-12.5

      vii. Band 7: 2.08-2.35

(From Campbell, pp. 431, 432)

2. SPOT–in microns

   a. SPOT 1, 2, 3: HRV (High Resolution Visible) systems

      i. panchromatic: 0.51-0.72

      ii. multispectral (color infrared): 0.50-0.59; 0.61-0.68; 0.79-0.89

   b. SPOT 4, 5

i.   0.49-0.73 band is replaced by 0.61-0.68 band

ii.   added band: 1.58-1.75

iii.   VMI (Vegetation Monitoring Instrument)

      a.   0.43-0.47

      b.   0.50-0.59

      c.   0.61-0.68

      d.   0.79-0.89

      e.   1.58-1.75

(From Campbell, pp. 496-97)

APPENDIX II

USMARC Field 007 for Remote-Sensing Images

007–REMOTE-SENSING IMAGE

Indicates that the item is a remote-sensing image which is defined as an image produced by a recording device that is not in physical or intimate contact with the object under study. This may be a map or other image that is obtained through various remote-sensing devices such as cameras, computers, lasers, radio frequency receivers, radar systems, sonar, seismographs, gravimeters, magnetometers, and scintillation counters.

00-Category of material

    r-Remote-sensing image

01-Specific material designation

    A one-character code that indicates the special class of material, usually the class of physical object, to which an item belongs.

#-No type specified

02-Undefined; contains a blank (#) or a fill character (|).

03-Altitute of sensor

Indicates the general position of the sensor relative to the object under study.

a-Surface

b-Airborne

c-Spaceborne

n-Not applicable

u-Unknown

z-Other

04-Attitude of sensor

Indicates the general angle of the device from which a remote-sensing image is made.

a-Low oblique

b-High oblique

c-Vertical

n-Not applicable

u-Unknown

05-Cloud cover

Indicates the amount of cloud cover that was present when a remote-sensing image was made.

0-0-9%

1-10-19%

2-20-29%

3-30-39%

4-40-49%

5-50-59%

6-60-69%

7-70-79%

8-80-89%

9-90-100%

n-Not applicable

u-Unknown

06-Platform construction type

Indicates the type of construction of the platform serving as the base for the remote-sensing device. For the purposes of this data element, "platform" refers to any structure that serves as a base, not only flat surfaces.

a-Balloon

b-Aircraft–low altitude

c-Aircraft–medium altitude

d-Aircraft–high altitude

e-Manned spacecraft

f-Unmanned spacecraft

g-Land-based remote-sensing device

h-Water surface-based remote-sensing device

i-Submersible remote-sensing device

n-Not applicable

u-Unknown

z-Other

07-Platform use category

Indicates the primary use intended for the platform specified in 007/06 (Platform construction type).

a-Meteorological

b-Surface observing

c-Space observing

m-Mixed uses

n-Not applicable

u-Unknown

z-Other

08-Sensor type

Indicates the recording mode of the remote-sensing device, specifically, whether the sensor is involved in the creation of the transmission it eventually measures.

a-Active

b-Passive

u-Unknown

z-Other

09-10-Data type

A two-character code that indicates the spectral, acoustic, or magnetic characteristics of the data received by the device producing the remote-sensing image. It can be used to indicate both the wave length of radiation measured and the type of sensor used to measure it.

aa-Visible light

da-Near infrared

db-Middle infrared

dc-Far infrared

dd-Thermal infrared

de-Shortwave infrared (SWIR)

df-Reflective infrared

dv-Combinations

dz-Other infrared data

ga-Sidelooking airborne radar (SLAR)

gb-Synthetic aperture radar (SAR)-Single frequency

gc-SAR-multi-frequency (multichannel)

gd-SAR-like polarization

ge-SAR-cross polarization

gf-Infometric SAR

gg-polarmetric SAR

gu-Passive microwave mapping

gz-Other microwave data

ja-Far ultraviolet

jb-Middle ultraviolet

jc-Near ultraviolet

jv-Ultraviolet combinations

jz-Other ultraviolet data

ma-Multi-spectral, multidata

mb-Multi-temporal

mm-Combination of various data types

nn-Not applicable

pa-Sonar–water depth

pb-Sonar–bottom topography images, sidescan

pc-Sonar–bottom topography, near surface

pd-Sonar–bottom topography, near bottom

pe-Seismic surveys

pz-Other acoustical data

ra-Gravity anomalies (general)

rb-Free-air

rc-Bouger

rd-Isostatic

sa-Magnetic field

ta-Radiometric surveys

uu-Unknown

zz-Other

From the USMARC homepage, http://www.loc.gov/marc/

HANDLING EARLY CARTOGRAPHIC MATERIAL

Cataloging Early Printed Maps

Nancy A. Kandoian

SUMMARY. In the context of machine readable cataloging for national bibliographic databases, this paper describes the cataloging of mono-

Nancy A. Kandoian is Map Cataloger in the Map Division of The New York Public Library. She holds a BA in geography from Mount Holyoke College and an MLS from Rutgers University.

Address correspondence to: Nancy A. Kandoian, Map Division, Room 117, The New York Public Library, 5th Avenue and 42nd Street, New York, NY 10018-2788.

The author wishes to acknowledge, in addition to the authors of the publications cited herein, Alice Hudson for all the sharing of her enthusiasm for antique maps (with special attention to decorative cartouches and coats-of-arms!), Albert Howard for a recent discussion at the Osher Map Library about cataloging antiquarian materials, Arlyn Sherwood for a session and handouts at a workshop on cataloging special materials in Atlanta back in 1987, and as mentioned elsewhere, Ellen Caplan and Barbara Story for their recent consultations. Thanks also to editors Paige Andrew and Mary Larsgaard for their encouragement, constructive suggestions, and answers to questions, and to Velma Parker for her reading of the manuscript and her comments that helped me to see some issues in a new light.

[Haworth co-indexing entry note]: "Cataloging Early Printed Maps." Kandoian, Nancy A. Co-published simultaneously in *Cataloging & Classification Quarterly* (The Haworth Information Press, an imprint of The Haworth Press, Inc.) Vol. 27, No. 3/4, 1999, pp. 229-264; and: *Maps and Related Cartographic Materials: Cataloging, Classification, and Bibliographic Control* (ed: Paige G. Andrew, and Mary Lynette Larsgaard) The Haworth Information Press, an imprint of The Haworth Press, Inc., 1999, pp. 229-264. Single or multiple copies of this article are available for a fee from The Haworth Document Delivery Service [1-800-342-9678, 9:00 a.m. - 5:00 p.m. (EST). E-mail address: getinfo@haworthpressinc.com].

229

graphic early printed maps whether published separately or extracted from other publications. It deals with description and access to capture the essence of a rare or "antique" map to create a useful surrogate. The step-by-step approach, rather than breaking new ground, integrates rules and guidance from multiple sources, both cataloging tools and supplementary materials, in a narrative fashion, with reference to the sources, their specific rules, and stated policies. Reference is made throughout the text to ten sample catalog records, with MARC 21 tagging, that are appended to the article. *[Article copies available for a fee from The Haworth Document Delivery Service: 1-800-342-9678. E-mail address: getinfo@haworthpressinc.com <Website: http://www.haworthpressinc.com>]*

KEYWORDS. Early printed maps, antique printed maps, rare printed maps, map cataloging, descriptive cataloging, cataloging early maps, early cartographic items, rare materials cataloging

INTRODUCTION

Cataloging early maps to create standard, consistent records for library catalogs and national bibliographic databases involves, like all library cataloging, not simply following a set of rules but also making a series of decisions. One makes choices while interpreting and applying rules, while selecting information to describe the items in question and while providing access to them. The antique, or early, map cataloger's choices affect the balance of conflicting factors:

- bringing out information content vs. capturing characteristics of the artifact;
- describing the uniqueness of an early printed map vs. matching the description with a single master record in an automated bibliographic database;
- reconciling rules for cataloging cartographic materials in general with special rules and applications for early cartographic items;
- reconciling rules for early maps with those for rare books that might apply to equivalent fields;
- striving for the ideal of the complete, well-researched record vs. achieving a larger quantity of cataloging in a fixed amount of time by recording only bare essentials;
- working towards expansion and clarification of rules that will reduce time-consuming decision-making by the cataloger vs. leaving room for flexibility and judgment.

While weighing the decisions of cataloging, the librarian working with early printed maps must be conscious of how the rarity and value of the materials intensify the purposes of cataloging. Preservation needs dictate minimal handling of rare maps and the use of a surrogate (e.g., a catalog record) for preliminary selection of materials for research needs. The expensive and deliberate decisions to acquire early maps for the collection must be made with knowledge of one's existing holdings. Awareness of loss and theft requires that same knowledge of existing holdings, and the need-to-know is heightened where valuable and difficult-to-replace special materials are concerned.[1]

Noting the variety of institutions whose records for early maps appear in national bibliographic databases such as the OCLC Online Union Catalog (OCLC) and the Research Libraries Information Network (RLIN) makes it clear that many librarians deal with these conflicting factors and put their needs in the balance to accomplish some level of control. Undaunted by the challenges, United States institutions as diverse as Central State University in Edmond, Oklahoma, the University of Texas at Arlington, Knox College in Galesburg, Illinois, the Minnesota Historical Society, the State Library of Florida, the University of Illinois, Duke University, the Philadelphia Maritime Museum Library, Watertown Free Public Library in Massachusetts, the New York Public Library, and of course the Library of Congress, to name a few, are creating and entering records for early printed maps in national bibliographic databases. Whether or not the quality and completeness of the records can be improved, it is gratifying that records for early maps are appearing widely from various sources; the maps are being described, access is being provided, knowledge of their existence is being shared.

This author's experience at the New York Public Library in beginning to catalog the maps of the Lawrence H. Slaughter Collection reveals that the whole process of cataloging a collection of early maps is a learning experience achieved with time.[2] It often seems that each map presents new situations and choices to be made and that the rules do not address every situation. One constantly discovers and re-discovers rules as one goes along, realizing what has been done incorrectly or inconsistently in the past.

For rules, interpretation, and examples of their application, the early map cataloger will refer most heavily, beyond standard cataloging tools, to four manuals: *Cartographic Materials: A Manual of Interpretation for AACR2,* the Library of Congress Geography and Map Division's *Map Cataloging Manual,* Robert Karrow's *Manual for the Cataloging of Antiquarian Cartographic Materials,* and *Descriptive Cataloging of Rare Books.*[3] The basis for these manuals, of course, and the basis for creating complete, accurate, and consistent descriptions comes from *Anglo-American Cataloguing Rules, Second Edition (AACR2), Anglo-American Cataloguing Rules 2nd ed., 1988*

Revision (AACR2R), and *ISBD(CM): International Standard Bibliographic Description for Cartographic Materials (ISBD(CM)).*[4]

Cartographic Materials (CM) interprets AACR2 for maps, putting together in one progression the Chapter 1 "General rules for description" that apply to maps and the rules of Chapter 3, "Cartographic materials." Interspersed throughout are bold headings for "Early Cartographic Items," followed by rules, applications, and options, in some of the situations where antique maps merit different treatment or where they may present a different set of circumstances to recognize and handle. The priority of the committee creating the manual, to provide basic assistance without delay to map catalogers in using AACR2, did not leave it time to consider documents related to rare materials (published just within the year or two before CM's publication) that would have been necessary "to develop guidelines for the comprehensive treatment of early materials for the manual."[5] The revision of CM is in process, and a more expanded consideration of early cartographic materials cataloging issues is planned.[6]

The *Map Cataloging Manual* (MCM) sets down various map cataloging policies that have developed over the years in the Geography and Map Division of the Library of Congress (LC). It does not highlight or segregate applications for early printed maps, and it lacks an index entry for them as a class of materials, but it does include practices related to their treatment throughout its pages.

While Karrow's *Manual* predates the publication and adoption of *AACR2,* it does incorporate changes ordained by *ISBD(CM),* which formed a new "framework" for rules of description such as *AACR2.*[7] It does not correspond in all respects with the notes on "Early Cartographic Items" in *CM,* or policies put forth in the *MCM,* but it provides a unified and comprehensive approach to the cataloging of early maps and a realistic philosophy of what early map cataloging can and should accomplish. Moreover, it provides useful supporting advice and examples, and deals with some specifics that are not addressed in the other sources.

Descriptive Cataloging of Rare Books (DCRB), while addressing rare books, is worth reviewing for its treatment of elements that maps share with books: title, statement of responsibility, place of publication, publisher, and date of publication. In an article on "Cataloging Rare Maps," which is a useful supplement to the major manuals cited here, Nancy Vick and Nancy Romero thoughtfully point out areas of conflict between DCRB's predecessor and CM, where early map catalogers might do well to defer to their rare books counterparts.[8] Perhaps there will be reconciling of differences to standardize treatment, without compromising map elements of description, between the new CM edition and the latest DCRB. In the meantime, exposure to DCRB can serve as a kind of elaboration of the principles of cataloging early printed materials; in some cases it may turn out to have applicable advice for clarification, and in other cases, conflicting advice that either confounds the decision-making or serves the purposes at hand better.

The following pages contain a review of the relevant rules, interpretations, applications, policies, practices, and advice for the descriptive cataloging of antique printed maps, gleaned and integrated primarily from these manuals. The author's experience has been, and therefore the examples set forth here will be, related to what might be considered the most common among rare maps for Anglo-American institutional collections: European and American imprints of the 17th, 18th, and 19th centuries. Each major element of description is addressed below in the standard order presented in catalog records. A few principles and practices relevant to subject and descriptive access for early maps will also be considered. Some sample cataloging records referred to in the text are appended to this review.

TITLE

In transcribing the title, as well as the statement of responsibility, of an antique map, one does not give it special treatment such as indicating line breaks or showing the use of upper case letters in the source. The cataloger should follow the *AACR2* rules for capitalization that apply to both modern and early printed materials.[9] For punctuation, however, the cataloger has the choice of recording all punctuation in the source, or of following modern punctuation usage. Prescribed punctuation for catalog records, nevertheless, must be included in either case, even if double punctuation results. However, when the original and prescribed punctuation marks are the same, one need only give the prescribed punctuation mark(s).[10] For example, consider a title punctuated on a map in this way: *Map of the United States in North America: with the British, French and Spanish dominions adjoining, according to the Treaty of 1783.* If the original punctuation is to be maintained, and the first eight words are considered the title proper, a double colon can be avoided by simply including the prescribed space-colon-space to divide the title proper from other title information, thus: *Map of the United States in North America : with the British . . .* However, if the punctuation is to be maintained, but the first sixteen words are considered the title proper, the title will be punctuated in the cataloging record this way: *Map of the United States in North America: with the British, French and Spanish dominions adjoining, : according to the Treaty of 1783.* Note the double punctuation (but not repetition of the same punctuation mark) of the comma-space-colon-space.

For works published before 1801, *CM* instructs the cataloger not to supply diacritical marks to the transcription that are lacking in the item being cataloged. In the same rule (1B16, pp. 27-28), *CM* also provides guidelines for transcribing earlier forms of letters which are no longer in use or no longer used the same way, such as early forms of "s," ligatures, and interchanged uses of i/j and u/v. *DCRB* gives more detail and background information on this issue. For example, it elaborates to the point of instructing the cataloger to separate the components of a Latin "æ," but maintain the Scandinavian "æ" as is.[11] See Figure 1 and its accompanying cataloging record, Example 1, field 245 (figures and examples are found in the Appendix). A principle to fall back on in *CM* is that "[w]hen there is any doubt as to the correct conversion of elements to modern form, transcribe them from the source as exactly as possible."[12]

For modern maps, catalogers are accustomed to inserting "[*sic*]" to follow a word spelled improperly to alert the reader that the word was transcribed exactly from the source and that the error was made on the item being cataloged. While misprints should have that treatment, or a correction following "i.e." in square brackets, *DCRB* explains that "[*sic*]" should not be used when transcribing "words spelled according to older or non-standard orthographic conventions."[13] Therefore, placenames such as "New Iarsey" and "Pensilvania" should be transcribed as they are with no addition or explanation. See the 245 field of Example 3. As Karrow writes, "[i]diosyncratic spellings are recorded without comment."[14]

Cartographic Materials rules 1B13, 1B14, and 1B15 (pp. 26-27) give some guidance about what constitutes a title proper, and how to abridge it if that is desired. One must use the mark of omission to indicate where words or groups of words have been omitted from the source in the transcription. The cataloger must also refer back to rule 1B4 (p. 22), not under the heading of "Early cartographic items," for the reminder that the first five words of the title proper should never be omitted. The application to this rule elaborates that when a dedication "forms an integral part of the title proper and precedes the title proper" include the dedication in the title transcription and "do not omit the first five words of the dedication."[15] So, in the example of Bernard Ratzer's plan of New York, with the title-encompassing dedication: *To His Excellency Sr. Henry Moore, Bart. Captain General and Governour in Chief in & over the Province of New York & the Territories depending thereon in America Chancellor & Vice Admiral of the same, This Plan of the City of New York, Is most Humbly Inscribed, by His Excellency's Most Obedient Servant, Bern'd Ratzen, Lieut't in the 60th Reg't*, the dedication-title-statement of responsibility might be abridged in the cataloging record in the following way: *To His Excellency Sr. Henry Moore . . . this plan of the city of New York is most humbly inscribed / by His Excellency's most obedient servant, Bern'd Ratzen* [sic], *lieut't in the 60th Reg't*. Mottoes, quotations, and dedications that appear on the map and that are separate from the title are omitted without the mark of omission. They may be quoted in a note. (See under NOTES below.)

Vick and Romero's article gives examples highlighting the difficulties inherent in assessing what constitutes the title proper on many old maps, and the need to consider a break for the title proper that recognizes not only the mapmaker's intent shown in the type, but also grammatical constructions and distinctiveness for easier machine retrieval. They support the goal of avoiding title abridgement that may prevent distinguishing a map from its variants.[16] The title used as an example in the context of punctuation above shows the cataloger's discretion coming into play in the designation of the title proper, set off from other title information.

STATEMENT OF RESPONSIBILITY

When transcribing the statement(s) of responsibility from an antique map, in contrast to the practice for most modern maps, the cataloger should include titles of honor and distinction that appear with names on the map.[17] This is evident in the statement of responsibility for the Ratzer map, quoted above, and in the cataloging record, Example 2 (field 245), to accompany Figure 2.

The cataloger should cite *virtually all* persons named who played a role in the creation of the map. Karrow's guideline is to cite those who "have affected the depiction of reality on the finished map," i.e., "the surveyor, cartographer, draftsman, and engraver."[18] Karrow, as well as the *Map Cataloging Manual,* offer some commonly used Latin terms and contractions of those terms that appear on old maps to denote the role played in the creation of the map: e.g., *del.* or *delineavit* for the cartographer or drafter, *sculp.* or *sculpsit* for the engraver, *exc.* or *excudit* for the printer.[19] See, for example, field 245 subfield "c" in Example 4.

An area of potential difficulty on some old maps is the interpretation of where "other title information" concludes and statement of responsibility begins. When one examines the examples for the rules governing "other title information [that] includes a statement of responsibility" (*CM* rule 1E4, p. 32), "a noun phrase occurring in conjunction with a statement of responsibility [that should be treated] as other title information if it is indicative of the nature of the work" (*CM* rule 1F12, p. 39), and a statement of responsibility that names no person or body (*CM* rule 1F14, p. 40; *DCRB* rule 1G12, p. 19), one realizes the perplexing situations that call forth individual interpretations and decision making. In Examples 1 and 2, accompanying Figures 1 and 2, statements of responsibility (field 245, subfield "c") might arguably have begun at a later point in each transcription. The "in case of doubt" phrases (e.g., *CM* rule 1F12, p. 40, "In case of doubt, treat the noun or noun phrase as part of the statement of responsibility") help ease the problems of weighing the decisions.

EDITION

The edition statement on antique maps may be handled just as it is done for modern maps, using standard abbreviations, or, in contrast, it may be transcribed exactly as it appears on the map (*CM* rule 2B9, p. 47). If the edition statement is entwined in the title or statement of responsibility, it is included in that element of the description and is not repeated as a separate edition statement (*CM* rule 2B8, p. 47).

MATHEMATICAL DATA

In order to express the map scale as a representative fraction, which is how it is rarely expressed on an early map, it is useful to have a Map Scale Indicator to convert measures on a bar scale or along the marginal grid of latitude. Modern equivalents of some units of measure employed on the bar scales of early maps can be found in *CM*'s Appendix B.3, Table 1 (p. 172). A footnote to Appendix B.2B cites two sources for additional conversion figures (p. 167). Another printed source is *Webster's New International Dictionary of the English Language* (2nd ed., unabridged)[20] in the table under "Measure." Conversion figures can be used with measurements taken along the bar scale to calculate a representative fraction scale. An Internet source for modern equivalents of various units of measure is "Mathematical Data for Bibliographic Descriptions," prepared by Jan Smits, Map Curator, Koninklijke Bibliotheek, the Netherlands, at URL <http://www.konbib.nl/kb/skd/skd/mathemat.htm>. Smits suggests adding a note to the scale statement to reveal measurements taken along the bar scale, e.g., "Scale [ca. 1:7,900], measurement derived from scale bar (900 rods = 33 mm)." With no rule in *CM* calling for such a phrase in the scale statement, such an addition, while informative as a justification for the calculated scale fraction, might more properly be placed in a note field of the description.

One must remember to express longitude coordinates in the mathematical data area of a record using Greenwich as the prime meridian, no matter what serves as prime meridian for longitude on the map itself. The origin for the designation of meridians varies widely on old maps, and longitude may be measured from 0 to 360 degrees or according to a metric centesimal system, rather than from 0 to 180 degrees east and west of a prime meridian. (See the longitude scale along the top margin of Figure 1 for an example of the 360-degree measure.) One may specify in the notes what serves as prime meridian on the map. In practice, many catalogers opt not to include coordinates when cataloging early maps, whether the map shows a marginal grid or not. The longitude in particular, besides being measured from a different origin, was imprecisely calculated in former times, and one is obliged to compare the antique map with a modern one to approximate the area covered–a time-consuming endeavor. For very early maps, it may seem a virtually meaningless task.

On the one hand, one might consider that map records are not yet general-

ly going into databases that are searchable by coordinates, so the approximation, even if a somewhat meaningful expression of reality, may be judged not
worth the time and effort. On the other hand, if map areas are represented by
title and subject terms that only vaguely convey their areas of coverage for
modern interpretation, approximated coordinates are a way to describe in a
universal language the map's geographical coverage. Moreover, it is becoming possible to put map records into coordinate-searchable databases of digital geospatial metadata.[21] As that integration of records becomes increasingly
desirable, the inclusion of coordinates in map records will be of added importance.

PUBLICATION, DISTRIBUTION, ETC., AREA

DCRB instructs that the place of publication for rare books be transcribed
along with any prepositions appearing before the name of the place (and with
other terms appearing in conjunction with the placename as well).[22] For
example:

À Paris

Cartographic Materials does not specifically address this point for early
cartographic items, but rule 4C8 and its application could be interpreted to
favor such transcription. "Give the place of publication, etc., as it is found in
the item . . . Give the place of publication in the orthographic form and the
grammatical case in which it appears in the source . . ." (pp. 70-71). *CM*
explicitly tells the cataloger of modern maps to drop such prepositions unless
case endings would be affected (rule 4B4, p. 67).

In standard modern map cataloging, *CM* instructs the cataloger, when
supplying an uncertain place of publication in the absence of a place in the
prescribed sources of information, to supply the name *in the language of the
chief source of information,* with a question mark and in square brackets (rule
4C6, p. 69). In contrast, but for subtly different conditions for early cartographic items, *CM*'s rule is to supply the name *in English* in brackets when
only an address or sign of the publisher, but not a city, appears on the
publication (rule 4C10, p. 71). When the place of publication does appear on
the item but its language or form may make it incomprehensible to many
catalog users, the cataloger may include the modern placename in brackets to
aid understanding of the transcription (rule 4C8, p. 70). For example:

Amstelodami [Amsterdam]

CM's rule 4C9 (p. 71) says, "[i]f the full address or sign of the publisher,
etc., appears in the prescribed sources of information, add it to the place if it

aids in identifying or dating the item." If there is no date on the map, the address or sign may be a primary clue to help establish the date or possible range of dates.[23] It may also distinguish a map from a variant state or edition. Rule 4C9, with its examples, guides the cataloger to include the address or sign of the publisher in the subfield with the place of publication. (See the 260 fields of Examples 3 and 5.) But the very next rule, 4C10 (p. 71), says to "[r]ecord the address or sign *within the publisher statement*" (emphasis added) when only an address or sign appears in the publication and the placename is being supplied. An example of this situation is illustrated by Figure 2, and its accompanying Example 2. *DCRB*, in rules 4B11 (worded almost precisely the same as *CM*'s 4C10), 4C2, and 4C4, seems to be instructing that any publisher address deemed necessary for inclusion in the record be put with the publisher statement (pp. 30, 31-32). But rule 4B2 (p. 29) might be interpreted as allowing its placement with the "principal place of publication" in circumstances where it is phrased directly with the place, because it ordains the inclusion of "accompanying words or phrases associated with the [place]name" in the place of publication subfield.[24]

If a publisher as such is not identified on the map, but a bookseller or printer is named, or the map has been "printed for" an identified person, group, or firm, these are transcribed in the field for publisher. See Figure 2 and its accompanying Example 2, and see also Example 6, field 260. If a distinction is made, however, between printer on the one hand, and the publisher or bookseller on the other, then *CM* instructs the cataloger to give the place and name of printer as instructed in the rules for maps in general, i.e., in parentheses following the publication date (rule 4G5, p. 84). For any details considered too lengthy to include, or judged not important for identifying the map, use the mark of omission (. . .) to indicate that some publisher details appearing on the map have been left out (rule 4D9, p. 75). An "[etc.]" rather than a mark of omission is called for when separate publisher statements beyond the initial statement (rather than details pertaining to one) are omitted (rule 4D10, p. 76). The same is done when more than one place of publication appears on the item, and the cataloger chooses to record only the first one (rule 4C11, p. 71). See the 260 fields of Examples 7 and 8.

When the date of publication appears on the map in Roman numerals, transcribe it in the imprint area in Arabic numerals, and do not put it in brackets unless this is called for for another reason (*MCM*, 2.10). Include the publication day and month with the year, as they appear on the map, if they are specified (*CM* rule 4F10, pp. 82-83). See field 260 subfield "c" of Example 6.

The Library of Congress also chooses options to convert dates to modern chronology (to the Gregorian or Julian calendar) and to "formalize" (i.e., express in shortened, traditional numerical form) very long dates, such as ones written out in Latin words. See the examples in *CM* under rule 4F10 (pp. 82-83).

The *Map Cataloging Manual* explains that dates of registration and deposit, included in such phrases as "Entered according to Act of Congress, Novr. 11th in the year 1835 . . ." are not copyright or publication dates (2.10). The manual refers to these phrases as "pre-copyright statements" (*MCM,* 3.3, 3.18). Publication dates may be inferred from dates in such statements, and therefore the dates are included in the imprint statement of the record in square brackets in the absence of a specified publication date. The pre-copyright statement may be quoted in a note (see NOTES section below).

Many of the early maps waiting to be cataloged in libraries have been taken out of atlases or other publications. When one is able to determine fairly certainly the publication from which a map has come (about which see more below, under NOTES), the publication date of that atlas or periodical or monograph should be considered the publication date of the map, despite any date that may appear on the map itself, according to the *MCM* (2.11). That source-document publication date should be entered in the publication, distribution, etc., area of the map's catalog record in square brackets. It is nevertheless important to find a way to include the date that appears on the map itself in the catalog record also. Depending on its context, it may find a place in the title or statement of responsibility area (see Example 4, field 245 subfield "c"), or in a note on the nature and scope of the item if it conveys the date of situation of the map and does not fit elsewhere in the description. If the date on the map itself is presented as a publication date (e.g., in bottom margin of map, "Published . . . New York, 1795") and the map is known to come from a publication of another date (say, 1796), include both dates in the publication, distribution, etc. area, with the latter in square brackets as a correction of the former ("1795 [i.e., 1796]").[25] Notes, including a quoted publication statement from the map and the "From . . ." note telling the source of the map, can explain this correction.

When, on the other hand, the cataloger is able to determine that a map in question has appeared in one or more particular publications, but it is not clear (from numbering, format, condition, documented bibliographic history, or any other available evidence) from which publication this particular item in hand actually came, a date on the map itself is preferred for publication date or inferring publication date, rather than the date of a publication from which it might or might not have come (*MCM,* 2.10). See Example 6.

In the absence of a date on the map itself, the date of a known publication in which the map appeared is used for inferring publication date (*MCM,* 2.10). In this case, the date will be transcribed followed by a question mark, in square brackets. The *Map Cataloging Manual* advises that if it is determined that a map appeared in several editions of a work (or, by extension, in several works of varying dates), the edition and history note should cite the earliest known edition (or work) that includes the map (3.21). (See the NOTES section below.) Extending that principle back to the inferred date of publication, one might use the publication date of the earliest known edition/ work that includes the map to infer the publication date of the map, and cite it with a question mark and in square brackets. See Examples 2 and 5. Alternately, the cataloger might cite a range of years of possible publication (but limited to a range of less than 20 years by the example of *CM* rule 4F7, pp. 80-81), based on the earliest and latest dates of imprint of the sources in which it is known to have appeared. See an example in Vick and Romero's article (p. 13), regarding a map from "[between 1763 and 1777]." This latter method would require notes of justification that cite both the earlier and later appearance of the map.

If a map is undated and a dated source is not determined, further searching in cartobibliographies and other reference books may reveal a scholar's approximation of the date, or may suggest a possible range of years based on the mapmaker's lifespan, the publisher's address, watermarks, style, or geopolitical content.[26] *CM* suggests all sorts of ways to express approximated dates, decades, and spans of years, and leaves no doubt that an approximated date or range must be supplied in the absence of a publication date (rules 4F7 and 4F11, pp. 80-81, 83).

PHYSICAL DESCRIPTION AREA

A basic principle of physical description in cataloging, expressed as an application under rule 5B2a of "Extent of item" in *Cartographic Materials,* prescribes the following: "Describe the physical state of the item in hand at the time of cataloguing . . . regardless of how it was issued by the publisher/ printer . . ." (p. 93). One is referred back to the "cardinal principle" of *AACR2* that "the starting point for description is the physical form of the item in hand, not the original or any previous form in which the work has been published" (*CM,* pp. 5-6). While many librarians rub elbows with this principle in the context of microform reproductions, the cataloger of early printed maps might well be faced with another twist on its implications when examining antique maps. If a multi-sheet map was joined after the issue of the map, if a map user dissected a single sheet and then mounted it on cloth for folding, if a map was published in its uncolored state and subsequently colored, the principle quoted above is saying that the physical description area of the catalog record should reflect the joined map, the dissected and mounted map, the colored map, respectively, i.e., the map in its current state, rather than the map in its original condition at the time of issue. This presents a dilemma for the cataloger whose automated environment, e.g., OCLC, pays heed to the concept of the master record. Libraries make use of one record in the database to represent a particular bibliographic entity, they download the record for local use, and they attach their holdings symbol to the record, rather than creating a unique record for each copy of that entity.

In regard to the issue of color, *CM,* in the application to rule 5C3 of the "Other physical details" subfield of physical description, states that "if a printed item is hand coloured, state this as 'hand col.' Do not differentiate between colour and partial colour. Details . . . may be recorded in a note" (p. 106). The *MCM* elaborates on the use of the phrase "hand col." by directing catalogers not to use that description "for otherwise uncolored maps that have colored annotations" (2.15). It goes on to suggest that the distinction between the two situations is assessed by the cataloger with a judgment of how the work was issued. A map thought to be colored before it was issued is described as "hand col.," while one that was issued and "then colored in a manner unique to the copy, usually for the owner's use, is *annotated*" (2.15). The information about color annotations would be included in a "copy being described" or "copy specific" note (3.50-3.51; see more about this in the NOTES section below), not in the physical description field. Though not spelled out in the *MCM,* annotations are understood to be additions by hand that note or highlight certain features for a specific purpose, for example showing a traveler's route or directions to a site, or pinpointing some locations of interest to the map user.

In dealing with this issue of hand color and annotations the manual indirectly raises another issue, without actually addressing it: what about an antique map that has been colored–with general wash or outline color, not with specific annotations–since its publication? There is the not uncommon situation on the antique map market of uncolored maps that have been colored prior to re-sale. Does one simply describe this as "hand col." under "Other physical details"? The *AACR2* principle cited above would seem to answer yes. After all, early printed maps by their nature–with changes in plates, paper, ink, and their conditions of use and storage over long periods of time–would each seem to be unique artifacts, products of the conditions under which they were produced and of their individual histories. Do they not each deserve their own unique descriptions that take account of their own unique characteristics? In addition, map catalogers are often faced with antique maps with no information readily available about their original condition. On the other hand, though it may seem like a surrender to automated system needs to balance the scale in the other direction, there would seem to be some virtue in co-locating records via the master record for maps that were known to be basically the same at the time of issue. In a certain way, the result facilitates comparative studies of differing states and editions of a map: by playing down the differences between same-state maps, it allows for accentuating the differences between variants.[27]

The Library of Congress and its machine readable cataloging system (MARC) enable a compromise solution with their allowance for "copy being described" and "copy specific" notes. So, if in the course of cataloging research, or through experience looking at old maps one realizes that a map has modern color (e.g., it becomes clear that maps published in the *London Magazine* were uncolored, and one of its maps being cataloged is hand colored) the physical description area may reflect the original conditions, and a note (see more below) will follow to convey the copy-specific details. On the other hand, if it is more important for a local catalog to describe concisely, with minimal notes and with minimal research, the library's holdings in their current form, or to consider and note the unique characteristics of each artifact in the collection, the solution will be to follow *AACR2* principles strictly to create master records with the maps' current conditions reflected in the physical description area.[28]

The final subfield of physical description includes the dimensions of the map. Rule 5D1a of *CM* presents the option of expressing the size of early cartographic items with dimensions to the nearest millimeter (p. 107). As *CM* indicates, LC does not apply this option, nor does it seem to be in evidence in other United States libraries' antique map records. As the neat lines of early printed maps are often not straight, it is not useful to offer measurements down to the level of detail of millimeters.

NOTES

Follow the order of notes prescribed for map cataloging in general, as specified in both *Cartographic Materials* (pp. 131-150) and the *Map Cataloging Manual* (3.4-3.6). One should begin with any needed expressions of the nature and scope of the item, including how relief is shown, and indicate the languages of a multilingual map if the title does not make them clear. Then one makes notes that relate to each field of the description in turn as needed. Imperative are notes that justify supplied information, "all added entries and subject headings, all aspects of the call number, and any internal peculiarities or seeming contradictions in the bibliographic description" that are not otherwise explained by the title and other elements of description (*MCM*, 3.2). Following are some "note"-worthy topics that often arise for early printed maps.

Source of Title Proper

Multiple titles do not seem to be as much of a problem on early cartographic items as on modern commercial folded maps. But a situation that occasionally surfaces is a map with one title in a cartouche, within the neat line, and another title in the margin. Vick and Romero cite an example of this (p. 7). In such a situation, the cataloger must be sure, as for any map, to indicate in the notes the source of the title proper, and give a varying form of title entry to the other title, preceded by a brief phrase of explanation, in the 246 field.

Dedication

When a map includes a dedication that is separate from the title and statement of responsibility, and therefore not transcribed in those fields, it may be quoted in a note (*CM* rule 7B5, p. 136). For example, see the dedication in the notes of Example 4.

Statement of Responsibility

Notes that relate to the statement of responsibility may include information regarding the responsibility for creation of the map that was not expressed in the statement of responsibility. This is the place to quote a "pre-copyright statement," such as "Entered according to Act of Congress, Novr. 11th in the year 1835 by Thomas Illman . . ." This will be justification for entering, by main or added entry, the map under this person's name. It may also serve as a justification note for an inferred date of publication. The *MCM* also points out that the statement should be quoted to justify a call number date or if the stated date differs from the date in the publication, distribution, etc., area (3.18-3.19).

Statements such as "Cum privilegio" also appear on some early maps as a sort of pre-copyright indicator, an indication that the mapmaker has been granted the privilege by the powers that be to publish the map. In effect, it lends authority to the publication. Karrow (sections 3.4136 and 3.74) recommends quoting this in a note. The phrase may even turn out to be a distinguishing characteristic for a state of a map, as in the case of Homann's *Regni Mexicani seu Novae Hispaniae, Ludovicianae . . .* , which exists both with and without the statement.[29]

Source, Edition, History

Researching an antique map in cartobibliographies to help provide lacking publication information for the map's basic description may reveal facts about the source or edition of the map. Even if the map apparently does not lack the specifics of its publication it is worthwhile to search for more information about the context of its publication so that the catalog record not only describes, but also identifies the map, distinguishing it from other similar maps.

Because of the nature of maps produced from copperplate engravings, which were frequently modified and printed with changes over time,[30] and the situation of maps being removed from atlases and other publications, cartographic historians and bibliographers have discovered and noted edition and source information and published comparative studies and annotated bibliographies revealing their findings.[31] The early map cataloger, to the extent possible, considering the resources of time and reference materials available, should seek out this information. It may be a one-step process of consulting a general bibliography (library collection-oriented, such as the British Museum map catalog,[32] or place-of-imprint-oriented, such as Wheat and Brun)[33] or a regional bibliography (such as Cumming).[34] These sources may reveal the publication from which a map came. Vick and Romero give an example of searching in a two-step process, where *Tooley's Dictionary of Mapmakers*[35] leads them to a title which they find in Phillips' *List of Geographical Atlases*.[36] There, specific map plates are listed under the atlas entry, and one of them is matched with the map in hand. Appended to their article is a basic bibliography of cartobibliographic reference sources to support the antique map cataloger in this type of research.[37]

Consulting comparative studies, such as Stevens and Tree's "Comparative Cartography,"[38] or Coolie Verner's *Carto-bibliographical Study of The English Pilot, The Fourth Book,*[39] may help to distinguish one map from another in a series printed from the same copperplate, or in a series of varied editions.

These examples of source, edition, and history research would result in notes beginning with such words and phrases as:

From . . .

Appears in . . .

Earlier ed. appears in . . .

Based on . . .

Differs from . . .

Examples of notes like these appear in many of the sample records appended to this article. The *MCM* (3.20-3.21) gives guidelines for making judgments about whether a map has actually been extracted from a particular publication, or whether one can make no conclusion beyond the fact that it has appeared in that publication (perhaps in addition to other publications). The manual also demonstrates how to construct notes that distinguish two otherwise identical maps by bringing out in each catalog record how one differs from the other (3.19). See the "Differs from . . ." note in Example 9.

Confusion may arise with terms to use in referring to varying versions of maps, i.e., states, issues, impressions, or editions. Moreover, lack of complete information may prevent a firm designation. These terms may apply somewhat differently to maps than to books, and some difference in the use of the terms may be noticed among the works of different cartobibliographers. Some background material with suggested definitions can be found in the writings of, for example, Coolie Verner, Lloyd Brown, and Francis J. Manasek.[40] Use of a more generic term, like "variant," may be a solution when the situation is not clear or when there is disagreement among cartobibliographies about what is involved. "Variant" is defined in both *CM*'s and *DCRB*'s glossary as "a copy showing any bibliographically significant difference from one or more other copies of the same edition. The term may refer to an impression, issue, or state."[41]

Making a note that one map is "based on" another may be informative, but the *MCM* advises that such a note is only included "on the basis of actual knowledge, not conjecture." LC map catalogers do "no searching . . . to determine the work upon which a map is based" (3.19).

Donor, Source, Previous Owners

In many cases a library will want to preserve the name of the donor of antique maps, or identify in whose possession or collection the maps formerly existed, if the individual or group does not object to being identified in a public catalog. The *MCM* suggests a method using a "copy specific note" (in a MARC 500 field with a subfield 5 plus the library's *National Union Catalog (NUC)* symbol tacked on to the end of the note) and access via an added entry for the personal or institutional donor (MARC 700 or 710 field, also with a subfield 5) (3.22).

An example might be:

500 Gift of the Estate of Lawrence H. Slaughter, 1997. $5 NN

700 1 Slaughter, Lawrence H. $5 NN[42]

Mathematical and Other Cartographic Data

The *MCM* advises the cataloger to "record the prime meridian if stated on the item and if other than Greenwich" (3.22). The manual explains that prior to 1884, when Greenwich was almost universally adopted as the prime meridian, maps have varied prime meridians. Table 3 of *CM*'s Appendix B.3 (p. 173) lists many locations used for prime meridians, with their differences in longitude from Greenwich. Some maps even show marginal grids using multiple prime meridians, as for example degrees from Ferro along the northern edge of the map, and from London along the southern edge of the map. See the note in Example 8.

Orientation of early maps shows much variation. There are not only the medieval mappae mundi with their eastern orientations, but common examples for early maps of North America are western oriented maps of Virginia and New England, in effect showing the explorers' approach to the New World.[43] As the *MCM* explains, "whenever north is positioned at an angle of 45° or more from the top of the map, a note is included to indicate the orientation" (3.23). The manual suggests wording for the note, expressing orientation in terms of north. An example would be:

Oriented with north to the right.

as in the notes found in Example 5.

Physical Description

The cataloger may note watermarks appearing in the paper of a map by examining the map on a light table or holding it up to light. The *MCM* advises noting their presence on all items produced before 1900. A general note (e.g., "Has watermarks") will suffice if the watermarks defy simple description or are unintelligible. A clear and simple watermark may be described explicitly (*MCM*, 3.25). For example:

Watermark: J Whatman.

Watermark includes fleur-de-lis.

This information enhances the physical description of the artifact and may help to date an undated map.[44] Many of the sample appended records include notes on watermarks.

Contents Notes

Library of Congress Geography and Map Division policy is to note or describe decorative cartouches and compass roses only in exceptional circumstances, i.e., those "judged to be of artistic or reference value."[45] Use of a map collection by designers and visual artists may heighten the reference need for such notes. Coats of arms, according to LC policy, are usually mentioned as a separate type of illustration only when they are numerous or the only type of illustration present, but even then the mention is optional.[46] It should be noted, however, that they might help to place the map in a political and historical setting or indicate the allegiance or patronage of the mapmaker.

Reference to Published Descriptions

A ubiquitous note for all kinds of rare library materials is the brief citation known as the "reference to published descriptions" note, coded in MARC records as a 510 field. It is useful to consult *CM* rule 7B15 (p. 145), the *MCM* pp. 3.28-3.30, and *DCRB* rule 7C14 (pp. 62-63), to get a sense of the purpose of the note and to see varied examples. The note tells in an abbreviated form where one can find reference(s) in a standard published list, bibliography, or other source, to the item being cataloged. *DCRB*'s guidelines are worth following: cite a list or bibliography "when it would serve to distinguish an edition (or variant) from similar editions (or variants), when it would substantiate information provided by the cataloger, or when it would provide a more detailed description of the publication being cataloged" (p. 63). *CM* includes an example that suggests that the reference note might be used even when the cited work is not an exact match but an informative entry for a variant of the item being cataloged: "Reference: Similar to Phillips 4195" (p. 145). The *MCM* includes a list of standard citation forms for some major cartobibliographies, and instructs LC map catalogers to submit citation forms of additional sources for approval (3.28-3.29). Other libraries wanting to create brief citation forms of additional sources should consult *Standard Citation Forms for Published Bibliographies and Catalogs Used in Rare Book Cataloging*, prepared by Peter M. VanWingen and Belinda D. Urquiza.[47] This volume lists standard citations for major published bibliographies (including only a few major cartobibliographies: the British Museum Map Room's *Catalogue of Printed Maps*, Koeman's *Atlantes Neerlandici*,[48] and the Library of Congress Map Division's *List of Geographical Atlases*)[49] in addition to providing guidelines ("Working principles," xv-xxvii) for formulating in a consistent way brief citation forms for other sources. One should keep a list, in effect an authority file, of such citation forms created for one's own library's continued use, and it will serve also as an interpretive key for those not familiar with the sources cited. VanWingen and Urquiza suggest that even when a bibliography is cited in a general note to justify or amplify part of the bibliographic description of the catalog record, it should also be cited in a reference note in the standard form, despite the repetition that results (xxvi). See notes in Example 4 and Example 5.

The *MCM* explains that "References to non-standard, infrequently-cited published descriptions are made in a longer, more complete form as a general note (tag 500). Usually the phrase *Described in*: is used to introduce the note" (3.30). This type of note appears in Example 7.

Numbers

Notes for miscellaneous numbers that appear on maps that may help to identify them (see *CM* rule 7B19, p. 148) can be useful for antiquarian items as clues to atlases or other publications from which they have come. Frequently such numbers appear in the margins of old maps. See evidence of this in Example 10.

Copy Being Described and Copy Specific Notes

Under PHYSICAL DESCRIPTION AREA above, the possibility of notes pertaining to the particular copy of a printed map being cataloged was introduced. These notes may convey such information as unique or modern coloring, manuscript annotations, the physical state, and imperfections in the map. *MCM* states that for "antiquarian or rare materials, even minor imperfections may be noted" (3.51).

The "copy being described" note, as described by the *MCM,* is coded as a local note (MARC field 590) and is introduced by a phrase identifying the holding library (3.50). For example:

NYPL copy has ms. notation in ink on verso: North America.

NYPL copy imperfect: surface mutilation on center of sheet.

NYPL copy dissected and mounted on cloth.

The "copy specific" note is used if the "information is needed as justification for another part of the bibliographic description" (*MCM,* 3.51) or where there is reference value to the information which would benefit other libraries.[50] Example:

Annotated in ink to show changes in house numbers. $5 NN

Hand coloring added at later date. $5 NN

The "copy specific" note has coding as MARC 21 field 500, with the addition of subfield 5 and the holding library's *NUC* symbol. The difference in

MARC 21 coding means that 590 local notes will not appear in master records, while 500 notes will. See how the 590 field is used in Examples 7 and 8, and the 500 field with subfield 5 in Example 5.

ACCESS

Subject and descriptive access follow basically the same principles for early materials as for modern materials, with the same needs to adhere to consistent authorities. But it is worthwhile to address a few points that may arise in the context of early maps.

Subject Access

Items which were produced prior to 1800 receive the subheading "Early works to 1800" after the geographical or topical heading and any "Maps" form subdivision present. (This subheading receives the MARC 21 subfield coding "v".)[51] Examples:

New Jersey–Maps–Early works to 1800.

Great Britain–Road maps–Early works to 1800.

Nautical charts–Chesapeake Bay (Md. and Va.)–Early works to 1800.

"History" and "Historical geography" are *not* used as subject subdivisions for early maps, unless the maps, at the time they were produced, showed historic sites and events or conditions for a previous period of time. The fact that early maps are documents of history does not mean that they were created to show history as a topic. An example of appropriate use of such a heading is:

- Title: Carte géographique de France avec des remarques curieuses sur l'ancienne et la nouvelle géographie. [1705?]
- Note: Inset, with indexes: Petite carte de la France ancienne ou des Gaules.
- Subject headings: France–Maps–Early works to 1800.
    France–Historical geography–Maps–Early works to 1800.

If this were simply a "Carte géographique de France" from 1705, the former subject heading would be used, but not the latter.

The Library of Congress's *Subject Cataloging Manual* makes clear that "[i]t is subject cataloging policy to assign as a subject heading or as a geographic subdivision, only the latest name of a political jurisdiction that has had one or more earlier names, as long as the territorial identity remains essentially unchanged."[52] This is echoed in the *Map Cataloging Manual* on page 4.6 where it more generically refers to "places"–not just political jurisdictions but also geographic features. Example:

- Title: Map of Ceylon, the key to the eastern empire.
- Date: 1850.
- Subject heading: Sri Lanka–Maps.

Confusion may become apparent when dealing with changing territorial identities and names that remain the same. A simple example involves Queens County, New York. J.B. Beers & Co.'s *New Map of Kings and Queens Counties, New York* (New York, 1886) covers a pre-1899 Queens County that encompassed the Queens and Nassau counties of today. The necessary subject headings are:

Kings County (N.Y.)–Maps.

Queens County (N.Y.)–Maps.

Nassau County (N.Y.)–Maps.

A justification note can explain the former coverage of Queens County. A more complex example involves the eighteenth century Louisiana of the French, which covered a large swath of North America, many times the area of today's state of Louisiana. The Library of Congress has assigned maps such as Jacques Nicolas Bellin's *Carte de la Louisiane et des Pays Voisins* ([Paris?, between 1755 and 1762]) the subject headings:

North America–Maps–Early works to 1800.

and Louisiana–Maps–Early works to 1800.[53]

There is a rationale there, but it is difficult to find ready documentation for it. These examples just scratch the surface of name and area changes over time. In the absence of clear documentation such as subject scope and usage notes in LC's subject and name authority files, as one comes upon these dilemmas one may have to fall back on the time-honored method of examining existing records, trying to understand their logic, trying to get a sense of "geographic names with which most readers are likely to be familiar, hence the ones to which they will resort,"[54] and proceed consistently from there.

Descriptive Access

The descriptive cataloger, as mentioned above under STATEMENT OF RESPONSIBILITY, will make every effort to cite the names of all persons and corporate bodies involved in the creation and production of an early map. These names all should be traced for access as the *Map Cataloging Manual* advises in its guidelines regarding pre-1800 maps. "This includes engravers, delineators, publishers, printers, etc., but does not usually include individuals to whom a work is dedicated."[55] Not only will the effort be much appreciated by cartographic historians for years to come, but it will also yield dividends for the cataloger days hence, working on other seemingly unidentifiable early maps. For materials published between 1800 and 1900, entries are made for those who had "a major part in the production of an item"; printers and those who engraved only the illustrations on the map do not usually receive added entries. But flexibility allows for the decision to be made on a case-by-case basis. The general philosophy is, when in doubt, make an added entry.[56]

The choice of a main entry, as Karrow says, is "sometimes rather arbitrary . . . Nevertheless, a decision must be made . . ." Recognizing the difficulty facing early map catalogers in making that choice when there are many named contributors or responsible parties on a map to choose from, Karrow offers a useful order of preference in the absence of a "name . . . stressed by size or location of lettering or by inclusion in the title proper." This is his ordered list:

a. the surveyor whose work provided the basis for the map,
b. the cartographer or draftsman who drew the map,
c. the compiler or editor who assembled the map from other sources,
d. the engraver or woodcutter who worked the plate or block from which the map was printed,
e. the printer or publisher of the map,
f. the claimant of copyright, or
g. the person in whose book the map is found.[57]

When it is determined that a map was extracted from another publication and the publication is identified, an added entry may be made to that work as a "related work." The *Map Cataloging Manual* says that a "simple entry," as opposed to an analytical one, is made in such cases, and likewise may be made when there is an identified work "upon which the map being cataloged is based."[58] "[T]he heading is that of the person or corporate body or the title under which the related work is, or would be, entered. If that heading is for a person or body, and the title of the related work differs from that of the work being catalogued, add the title of the related work to the heading to form a name-title added entry heading."[59] See related work entries in 700 or 740 fields (with justification in 500 notes) in several of the appended examples.

IN CONCLUSION

A good body of literature exists now to guide the work of the early printed map cataloger. To further guide this work, there is a need for:

- an ironing out of the differences between the mutually applicable rules for early cartographic items and rare books; and
- a widely accessible, updated manual that addresses the cataloging of early printed maps in a comprehensive manner, in basic harmony with *AACR2R*.

While the committee revising *CM*[60] brings to fruition the second point above, it is no doubt dealing with the first one.

Beyond the needs for *guiding* antique map cataloging, *support* of this work requires something more:

- access to basic reference materials of all kinds that provide a foundation for the interpretation and description of antique maps;[61] and
- the cartobibliographic research and access to it that are foundations for identifying early maps.[62]

The research, identification, and cataloging will continue to build on each other.

One more necessary ingredient for guidance *and* support is:

- enhanced communication among early map catalogers, through Internet discussions and meetings at professional association conferences.

This depends on the outreach and vociferousness of individual early map catalogers. It is an ongoing need to promote mutual input on the weighing of cataloging decisions, a common understanding of how we deal with issues that arise, and on sharing information about reference materials that are available to support the description and identification of early printed maps.

ENDNOTES

1. Mary L. Larsgaard, *Map Librarianship: An Introduction*, 3rd ed. (Englewood, Colo.: Libraries Unlimited, 1998), 197; and Mary L. Larsgaard and Katherine Rankin, "Helpful Hints for Small Map Collections: Cataloging Those Maps–Why Do It?" *MAGERT Electronic Publication No. 1* at URL <http://www.sunysb.edu/libmap/larsg.htm> rev. by Stephen Rogers 6/97.

2. Nancy Kandoian, "The Lawrence H. Slaughter Collection: The Cataloger's Viewpoint," *Meridian* 13 (1998): 33-40.

3. *Cartographic Materials: A Manual of Interpretation for AACR2.* (Chicago: American Library Association, 1982); Library of Congress, Geography and Map Division; *Map Cataloging Manual* (Washington, D.C.: Cataloging Distribution Service, Library of Congress, 1991); Robert W. Karrow, Jr. *Manual for the Cataloging of Antiquarian Cartographic Materials*, 2nd draft (Chicago: Newberry Library, 1977); and *Descriptive Cataloging of Rare Books*, 2nd ed. (Washington, D.C.: Cataloging Distribution Service, Library of Congress, 1991).

4. *Anglo-American Cataloguing Rules*, 2nd ed. (Chicago: American Library Association, 1978); *Anglo-American Cataloguing Rules*, 2nd ed., 1988 revision (Chicago: American Library Association, 1988); and International Federation of Library Associations and Institutions, Joint Working Group on the International Standard Bibliographic Description for Cartographic Materials, *ISBD(CM): International Standard Bibliographic Description for Cartographic Materials* (London: IFLA International Office for UBC, 1987).

5. *Cartographic Materials*, viii-ix.

6. Larsgaard, *Map Librarianship*, 197.

7. Robert W. Karrow, Jr., "Innocent Pleasures: ISBD(CM), AACR2, and Map Cataloging," Special Libraries Association Geography and Map Division *Bulletin* 126 (Dec. 1981): 3.

8. *Bibliographic Description of Rare Books* (Washington, D.C.: Library of Congress, Office for Descriptive Cataloging Policy, 1981); and Nancy J. Vick and Nancy L. Romero, "Cataloging Rare Maps," *Cataloging & Classification Quarterly* 10, no. 4 (1990): 5, 10-12. The complete article is on pp. 3-18.

9. *AACR2* 1988 rev., Appendix A, 563-599.

10. *Cartographic Materials*, 12, rule 0C1.

11. *Descriptive Cataloging of Rare Books*, 6-7, rule 0H, and "Appendix B: Early Letter Forms,"69-70.

12. *Cartographic Materials*, 27-28, rule 1B16.

13. *Descriptive Cataloging of Rare Books*, 5, rule 0g.

14. Karrow, *Manual*, section 3.024.

15. *Cartographic Materials*, 22.

16. Vick and Romero, "Cataloging Rare Maps," 6-10.

17. *Cartographic Materials*, 39, rule 1F7.

18. Karrow, *Manual*, section 3.15.

19. Ibid., section 2.12; and *Map Cataloging Manual*, 2.2.

20. *Webster's New International Dictionary of the English Language*, 2nd ed., unabridged (Springfield, Mass.: G. & C. Merriam Co., 1934 and later issues).

21. This was discussed at "Building the National Spatial Data Infrastructure Metadata Catalog: The Librarian's Role," a program presented at the American Library Association (ALA) Annual Conference, Washington, D.C., June 29, 1998, by the ALA Map and Geography Round Table, Geographic Technologies Committee.

22. *Descriptive Cataloging of Rare Books*, 29, rule 4B2.

23. Katherine H. Weimer, Elka Tenner, and Richard Warner, "Methods for Determining the Date of an Undated Map," Western Association of Map Libraries *Infor-*

mation Bulletin 29, no. 1 (Nov. 1997): 8. Resources such as Sarah Tyacke's *London Map-sellers, 1660-1720* (Tring, England: Map Collector Publications, 1978) tell when certain publishers were operating at certain addresses.

24. Looking at *AACR2* as the basis for the rules, one sees in the original 1978 edition of *AACR2* that rule 2.16C was the basis for *CM*'s rule 4C9. But rule 2.16C is changed in *AACR2* rev. 1988, and no longer instructs that an address of publisher worthy of inclusion in the record be included with the place. *DCRB*, 1991, consistent with the *AACR2* revision, has no explicit instruction to include address with place, but rather is explicit about including it with publisher. Nevertheless, *AACR2* rev. 1988 does show an example, with rule 2.16D, where an address of sorts, phrased with the place name, is included in that subfield: "Enprynted at Westmyster in Caxtons hous: By me Wynken the Worde" (p. 88).

25. *Cartographic Materials*, 79, rule 4F2.

26. See Weimer, Tenner and Warner, "Methods for Determining the Date of an Undated Map," 6-13, for some concrete methods and references that will aid in determining the date of an undated map.

27. Becoming familiar with such works as Coolie Verner's "Carto-Bibliographical Description: The Analysis of Variants in Maps Printed from Copperplates," *American Cartographer* 1, no. 1 (April 1974): 77-87; Karrow's "Cartobibliography," *AB Bookman's Yearbook* (1976): 43-52; and his "Role of Cartobibliography in the History of Cartography," Association of Canadian Map Libraries *Bulletin* 57 (Dec. 1985): 1-13, may help to clarify one's thoughts on how best to apply the principles discussed here.

28. The author is indebted to Barbara Story of the Library of Congress and Ellen Caplan of OCLC in helping to work through these issues, but any confusion promoted here is the author's own doing.

29. Johann Baptist Homann, *Regni Mexicani seu Novae Hispaniae, Ludovicianae, N. Angliae, Carolinae, Virginiae, Pensylvaniae . . .* [map], [ca. 1:10,000,000] (Noribergae: [Homann Erben, between 1712 and 1730?]). See the description in Woodbury Lowery, *The Lowery Collection: A Descriptive List of Maps of the Spanish Possessions within the Present Limits of the United States, 1502-1820.* (Washington: Government Printing Office, 1912), 332, no. 473.

30. See Verner, "Carto-Bibliographical Description," 77-87.

31. Karrow "The Role of Cartobibliography," 1-13.

32. British Museum, Map Room, *Catalogue of Printed Maps, Charts and Plans*, 15 vols. (London: British Museum, 1967).

33. James Clements Wheat and Christian F. Brun, *Maps Published in America before 1800: A Bibliography* (New Haven: Yale University Press, 1969).

34. William P. Cumming, *The Southeast in Early Maps*, 3rd ed. (Chapel Hill: University of North Carolina Press, 1998).

35. Ronald Vere Tooley, *Tooley's Dictionary of Mapmakers* (New York: A.R. Liss, 1979).

36. Library of Congress, Map Division, *A List of Geographical Atlases in the Library of Congress, with Bibliographical Notes*, compiled under the direction of P. L. Phillips, 9 vols. (Washington, D.C.: Government Printing Office, 1909-1992).

37. Vick and Romero, "Cataloging Rare Maps," 12-13, 15-18.

38. Henry Stevens and Roland Tree, "Comparative Cartography," in *The Mapping of America*, comp. by R.V. Tooley (New York: R.B. Arkway, 1980), 41-107.

39. Coolie Verner, *A Carto-Bibliographical Study of The English Pilot, The Fourth Book* (Charlottesville, Va.: University of Virginia Press, 1960).

40. Verner, "Carto-Bibliographical Description"; Lloyd A. Brown, *Notes on the Care and Cataloguing of Old Maps* (Windham, Conn.: Hawthorn House, 1940), 83-89; and Francis J. Manasek, *Collecting Old Maps* (Norwich, Vt.: Terra Nova Press, 1998), 63-65.

41. *Cartographic Materials*, 237, and *Descriptive Cataloging of Rare Books*, 82.

42. The New York Public Library actually handles this in a different way, making a local added entry (799) for the collection resulting from the gift. This can be seen in many of the examples appended to this article.

43. See, for example, *Virginia, Discovered and Described by Captain John Smith*, graven by William Hole [map], [ca. 1:1,200,000], ([Oxford, 1612]) and Willem Janszoon Blaeu, *Nova Belgica e Anglia Nova* [map], [ca. 1:3,200,000], ([Amsterdam: Blaeu, 1635]).

44. Weiner, Tenner, and Warner, "Methods for Determining the Date of an Undated Map," 10.

45. *Map Cataloging Manual*, 3.33.

46. Ibid.

47. Peter M. VanWingen and Belinda D. Urquiza, *Standard Citation Forms for Published Bibliographies and Catalogs Used in Rare Book Cataloging*, 2nd ed. (Washington, D.C.: Library of Congress, Cataloging Distribution Service, 1996).

48. Cornelius Koeman, *Atlantes Neerlandici: Bibliography of Terrestrial, Maritime and Celestial Atlases and Pilot Books, Published in the Netherlands up to 1880*, 6 vols. (Amsterdam: Theatrum Orbis Terrarum, 1967-1985).

49. VanWingen and Urquiza, *Standard Citation Forms*, 23, 67, 73.

50. Phone conversation with Barbara Story of the Library of Congress Geography and Map Division, 1998.

51. This is a change in subheading that has occurred since the *Map Cataloging Manual* p. 4.14 was published in March 1991. See Library of Congress, Office of Subject Cataloging Policy, *Subject Cataloging Manual: Subject Headings*, 4th ed. (Washington, D.C.: Cataloging Distribution Service, Library of Congress, 1991-), H1576, Aug. 1995.

52. *Subject Cataloging Manual*, H708, 1, Feb. 1991.

53. Library of Congress control number 73-693831.

54. David Judson Haykin, *Subject Headings: A Practical Guide* (Washington, D.C.: U.S. Government Printing Office, 1951), 45.

55. *Map Cataloging Manual*, 5.1. Surveyors, also, should be added to that list.

56. Ibid., 5.2.

57. Karrow, *Manual*, section 2.11.

58. *Map Cataloging Manual*, 5.4. See also *AACR2*, 1988 revision, 351-354, rule 21.28.

59. *AACR2*, 1988 revision, 356, rule 21.30G1; see also *Library of Congress Rule Interpretations* (Washington, D.C.: Cataloging Distribution Service, Library of Congress, 1990-), rule 21.30M, Jan. 5, 1989.

60. The Anglo-American Cataloguing Committee for Cartographic Materials.

61. The Bibliographic Standards Committee of the Rare Books and Manuscripts Section of the Association of College and Research Libraries has developed a Website, "Resources for the Rare Materials Cataloger," with links to such potentially useful reference sites for the antique map cataloger as a Latin placenames directory, a calendar conversion system, and a watermark archive. The URL is <http://www. library.upenn.edu/ipc/index.html>

62. Karrow's "Role of Cartobibliography" and Vick and Romero's "Cataloging Rare Maps" both make articulate pleas for this work but from different angles, and perhaps even from opposite poles of spareness and detail.

BIBLIOGRAPHY

The sources cited below, with a few exceptions, have been cited within the text of this paper. They include the major tools for early map cataloging, supplementary related articles, and some supportive reference sources. While cartobibliographies form a major body of reference works to be consulted by the cataloger of early printed maps, they have not been included here. As mentioned earlier, Vick and Romero append a basic list of these sources to their paper. Pelletier, Ristow, and Wolf, cited below, can lead one to more such sources.

Association of College and Research Libraries. Rare Books and Manuscripts Section. Bibliographic Standards Committee. "Resources for the Rare Materials Cataloger." URL <http://www.library.upenn.edu/ipc/index.html>

Cartographic Materials: A Manual of Interpretation for AACR2. Chicago: American Library Association, 1982.

Descriptive Cataloging of Rare Books, 2nd ed. Washington, D.C.: Cataloging Distribution Service, Library of Congress, 1991.

Karrow, Robert W., Jr. "Cartobibliography." *AB Bookman's Yearbook* (1976): 43-52.

_____. *Manual for the Cataloging of Antiquarian Cartographic Materials*, 2nd draft. Chicago: Newberry Library, 1977.

_____ . "Role of Cartobibliography in the History of Cartography." Association of Canadian Map Libraries *Bulletin* 57 (Dec. 1985): 1-13.

Library of Congress. Geography and Map Division. *Map Cataloging Manual.* Washington, D.C.: Cataloging Distribution Service, Library of Congress, 1991. (Also available in electronic form as a part of *Cataloger's Desktop*)

Pelletier, Monique, ed. *How to Identify a Mapmaker.* Paris: Comité Français de Cartographie, 1996.

Ristow, Walter W., comp. *Guide to the History of Cartography: An Annotated List of References on the History of Maps and Mapmaking.* Washington, D.C.: Library of Congress, 1973.

Smits, Jan. "Mathematical Data for Bibliographic Descriptions." URL <http://www. konbib.nl/kb/skd/skd/mathemat.htm>

Tooley, Ronald Vere. *Tooley's Dictionary of Mapmakers.* New York: A.R. Liss, 1979.

VanWingen, Peter M. and Belinda D. Urquiza. *Standard Citation Forms for Published Bibliographies and Catalogs Used in Rare Book Cataloging*, 2nd ed. Washington, D.C.: Library of Congress, Cataloging Distribution Service, 1996.

Verner, Coolie. "Carto-Bibliographical Description: The Analysis of Variants in Maps Printed from Copperplates." *American Cartographer* 1, no. 1 (April 1974): 77-87.

Vick, Nancy J. "Analyzing Atlases." Western Association of Map Libraries *Information Bulletin* 19, no. 1 (Nov. 1987): 30-32.

Vick, Nancy J. and Nancy L. Romero. "Cataloging Rare Maps." *Cataloging & Classification Quarterly* 10, no. 4 (1990): 3-18.

Webster's New International Dictionary of the English Language, *2nd ed., unabridged. Springfield, Mass.: G. & C. Merriam Co., 1934 and later issues.*

Weimer, Katherine H., Elka Tenner, and Richard Warner. "Methods for Determining the Date of an Undated Map." Western Association of Map Libraries *Information Bulletin* 29, no. 1 (Nov. 1997): 6-13.

Wolf, Eric W., comp. and ed. *The History of Cartography: A Bibliography, 1981-1992*. Washington, D.C.: Washington Map Society in association with Fiat Lux, 1992.

APPENDIX

The following sample cataloging records are meant to serve as examples of points made in the text of this paper. References are made to these examples throughout the text. They represent the attempts of one map cataloger to meet the challenges of describing, identifying, and providing access to antique printed maps.

FIGURE 1.Cartouche and Part of Northwest Corner of Map Described in Example 1

Courtesy of Lawrence H. Slaughter Map Collection, Map Division, Humanities and Social Sciences Library, The New York Public Library, Astor, Lenox and Tilden Foundations.

FIGURE 2. Cartouche and Part of East Side of Map Described in Example 2

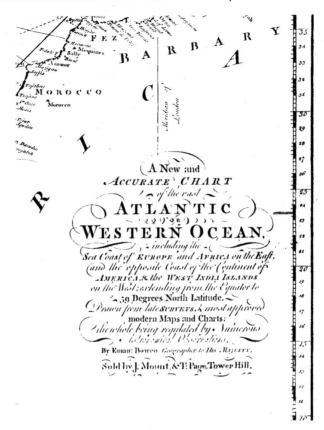

Courtesy of Lawrence H. Slaughter Map Collection, Map Division, Humanities and Social Sciences Library, The New York Public Library, Astor, Lenox and Tilden Foundations.

EXAMPLE 1

```
100 1    Homann, Johann Baptist, |d 1663-1724.
245 10   Regni Mexicani seu Novae Hispaniae, Floridae, Novae Angliae,
Carolinae, Virginiae et Pensylvaniae necnon insularum archipelagi Mexicani in
America Septentrionali / |c accurata tabula exhibita à Ioh. Baptista Homanno.
255      Scale [ca. 1:10,000,000] |c (W 110°--W 55°/N 45°--N 8°).
260      Noribergae |b [Homann Erben,|c 1759?]
300      1 map : |b hand col. ; |c 46 x 56 cm.
500      Relief shown pictorially.
500      Some references suggest earlier possible imprint date.
500      Includes notes and ill.
510 4    LC Maps of North America, 1750-1789, |c 82
510 4    Phillips, |c 3474, 5959
590      NYPL copy has ms. note in ink in upper right margin: 6.
651  0   North America |v Maps |v Early works to 1800.
651  0   New Spain |v Maps |v Early works to 1800.
710 2    Homann Erben (Firm)
799      Lawrence H. Slaughter Collection ; 452.
```

EXAMPLE 2

```
100 1    Bowen, Emanuel, |d d. 1767.
245 12   A new and accurate chart of the vast Atlantic or Western Ocean
: |b including the sea coast of Europe and Africa on the east, and the
opposite coast of the continent of America & the West India Islands on the
west; extending from the equator to 59 degrees north latitude / |c drawn from
the late surveys, & most approved modern maps and charts, the whole being
regulated by numerous astronomical observations by Eman. Bowen geographer to
his majesty.
255      Scale [ca. 1:12,000,000] |c (W 83°--E 9°/N 57°--N 0°).
260      [London] : |b Sold by J. Mount & T. Page, Tower Hill, |c [1778?]
300      1 map ; |c 59 x 78 cm.
500      Depths shown by soundings in the Grand Banks of Newfoundland. Shallow
areas shown by stippling.
500      Originally printed on 2 sheets.
500      Watermarks.
500      "Degrees of longitude from London."
500      Appears in The English pilot. The fourth book. 1778 to 1794 eds.--
Verner.
510 4    Verner, C. Bibliographical note. In The English pilot. The fourth
book, London, 1689. Amsterdam, 1967, |c p. xx
650  0   Nautical charts |z North Atlantic Ocean |v Early works to 1800.
651  0   North Atlantic Ocean |v Maps |v Early works to 1800.
710 2    Jno. Mount and Tho. Page.
740 0    English pilot. |n Fourth book.
799      Lawrence H. Slaughter Collection ; 472.
```

EXAMPLE 3

```
100 1    Lea, Philip, |d fl. 1683-1700.
245 12   A new map of New England, New York, New Iarsey, Pensilvania, Maryland,
and Virginia / |c by Philip Lea.
255      Scale [ca. 1:1,525,000].
260      London, at the Great Toy, Spectacle, Chinaware, and Print Shop, at ye
corner of Ludgate Street near St. Paul's : |b Sold by Geo. Willdey, :c
[between 1715 and 1720?]
300      1 map ; |c 45 x 54 cm.
500      Covers Massachusetts, Rhode Island, Connecticut, southeastern New
York, New Jersey, eastern Pennsylvania, Delaware, eastern Maryland, and
eastern Virginia.
500      Relief shown pictorially. Depths shown by soundings.
500      Watermark.
500      Includes inset of New York Harbor area.
510 4    Stevens & Tree, |c 35(d)
651  0   New England |v Maps |v Early works to 1800.
651  0   Middle Atlantic States |v Maps |v Early works to 1800.
700 1    Willdey, George.
799      Lawrence H. Slaughter Collection ; 490.
```

EXAMPLE 4

```
100 1    Senex, John, |d d. 1740.
245 12   A new map of the English empire in America : |b viz. Virginia,
Maryland, Carolina, New York, New Iarsey, New England, Pennsylvania,
Newfoundland, New France &c. / | crevis'd by Ion. Senex, 1719 ; I. Harris,
sculp.
255      Scale [ca. 1:6,500,000] |c (W 95°--W 53°/N 52°--N 25°).
260      [London : |b Printed for Daniel Browne ... , John Senex ... , |c 1721]
300      1 map : |b hand col. ; |c 50 x 59 cm.
500      Relief shown pictorially.
500      Dedication at head of title: Most humbly inscrib'd to Hewer Edgly
Hewer of Clapham Esqr: &c.
500      From the author's New general atlas. 1721.
500      Printed from the same plate as Robert Morden's map of the same title,
ca. 1695--Cumming, 119, 172.
500      Watermarks include fleur-de-lis and "H D."
500      Insets: The harbour of Boston or Mattathusetts Bay -- A generall map
of the coasts & isles of Europe, Africa and America.
510 4    Cumming |c 172
510 4    Karpinski, L.C. Printed maps of Michigan, |c XLVI
510 4    Stevens & Tree, |c 20
590      NYPL copy includes ms. notations in ink plotting sea route.
651  0   North America |v Maps |v Early works to 1800.
651  0   Great Britain |x Colonies |z America |v Maps |v Early works to 1800.
700 1    Harris, John, |d fl. 1680-1740.
700 1    Senex, John, |d d. 1740. |t New general atlas.
700 1    Morden, Robert, |d d. 1703. |t New map of the English empire in
America.
799      Lawrence H. Slaughter Collection ; 503.
```

EXAMPLE 5

```
100 1    Tiddeman, Mark.
245 12   A draught of New York from the hook to New York Town / |c by Mark
Tiddeman.
255      Scale [ca. 1:100,000].
260      London, upon Tower Hill : |b Printed for W. Mount & T. Page, |c
[1737?]
300      1 map ; |c 46 x 57 cm.
500      Covers Amboy to Jamaica Bay, Harlem to Shrewsbury River.
500      Relief shown by hachures and shading. Depths shown by soundings.
500      Oriented with north to the right.
500      From The English pilot. The fourth book. This state appears in eds.
from 1737-1773--Verner.
500      Hand coloring added at later date. |5 NN
510 4    Verner. Carto-bibliographical study of The English pilot, |c
p. 66-67.
650  0   Nautical charts |z New York Region |v Early works to 1800.
650  0   Harbors |z New York Region |v Maps |v Early works to 1800.
651  0   New York Region |v Maps |v Early works to 1800.
710 2    W. Mount & T. Page.
740 0    English pilot. |n Fourth book.
799      Lawrence H. Slaughter Collection ; 814.
```

EXAMPLE 6

```
245 00   Bowles's new pocket map of the Atlantic or Western Ocean : |b laid
down from the latest discoveries, and regulated by numerous astronomical
observations.
255      Scale [ca. 1:13,500,000] |c (W 85°--E 10°/N 52°--N 0°).
260      London : |b Printed for the proprietor Carington Bowles, |c
1 Jan. 1779.
300      1 map : |b hand col. ; |c 44 x 55 cm.
500      Shows sailing routes.
500      Depths shown by soundings in fishing banks off Newfoundland. Shallow
areas shown by stippling.
500      Appears in Palairet's Bowles's universal atlas. [1775-1780].
500      "Degrees of longitude from London" and "West long. in hours & minutes
of time from the meridian of London."
500      "Published as the act directs."
500      Watermarks include "J Whatman" and fleur-de-lis on shield.
510 4    Phillips |c 5988
650  0   Trade routes |z North Atlantic Ocean |v Maps |v Early works to 1800.
650  0   Nautical charts |z North Atlantic Ocean |v Early works to 1800.
651  0   North Atlantic Ocean |v Maps |v Early works to 1800.
651  0   Atlantic Ocean |v Maps |v Early works to 1800.
700 1    Bowles, Carington, |d 1724-1973.
700 1    Palairet, Jean, |d 1697-1774. |t Bowles's universal atlas.
799      Lawrence H. Slaughter Collection ; 517.
```

EXAMPLE 7

```
100 1    Faden, William, |d 1750?-1836.
245 14   The Province of New Jersey : |b divided into East and West, commonly
called the Jerseys / |c engraved & published by Wm. Faden, Charing Cross,
December 1st, 1777 ; Croisey, sculpt.
255      Scale [ca. 1:430,000] |c (W 76°00'--W 73°20'/N 41°30'--N 38°40').
260      Paris : |b Chez les Srs. Perrier et Verrier, eleves et successeurs de
M. Julien ... , |c [1778?]
300      1 map ; |c 74 x 55 cm.
500      Relief shown by hachures.
500      Prime meridian: Philadelphia.
500      Shows county boundaries and "Division line run in 1743 between East
New Jersey and West New Jersey."
500      "This map has been drawn from the survey made in 1769 ... by Bernard
Ratzer ... and from another large survey of the northern parts ... by Gerard
Banker."
500      Described in Imago mundi, v. 48 (1996), p. 168.
500      Includes table of "Astronomical observations."
546      Cartographic source note in English and French.
590      NYPL copy mounted on paper and cloth. Engraver note in lower left
margin mutilated. Ms. note in pencil in upper left margin: *ENO.
651  0   New Jersey |v Maps |v Early works to 1800.
651  0   New Jersey |x Administrative and political divisions |v Maps
|v Early works to 1800.
700 1    Croisey, P.
700 1    Ratzer, Bernard.
700 1    Bancker, Gerard, |d 1740-1799.
710 2    Perrier et Verrier.
```

EXAMPLE 8

```
100 1    Palairet, Jean, |d 1697-1774.
245 10   Carte des possessions angloises & françoises du continent de
l'Amérique septentrionale, 1755 / |c Thos. Kitchin, sculpt.
255      Scale [ca. 1:7,000,000] |c (W 100°--W 52°/N 53°--N 28°).
260      Londres [etc.] : |b Se vend ... chez Mrs. Nourse, Vaillant, Millar,
Rocque & Sayer [etc.], |c [1755]
300      1 map : |b hand col. ; |c 40 x 56 cm.
500      Wash coloring with some French forts circled in red.
500      Relief shown pictorially.
500      Appears in Palairet's Atlas méthodique. 1755.
500      Shows "Longitude occidentale de l'Île de Fer" and "de Londres."
500      Watermark includes fleur-de-lis.
500      Accompanied by: Description abrégée des possessions angloises et
françoises du continent septentrionale de l'Amérique. 62 p.
510 4    Karpinski, L.C. Printed maps of Michigan, |c LXIXa
510 4    LC Maps of North America, 1750-1789, |c 56
510 4    Phillips |c 3502
546      Legend in French and English.
590      NYPL lacks accompanying text.
590      NYPL copy has ms. notation in ink in upper right margin: 14.
651  0   North America |v Maps |v Early works to 1800.
700 1    Kitchin, Thomas, |d d. 1784.
700 1    Palairet, Jean, |d 1697-1774. |t Atlas méthodique.
799      Lawrence H. Slaughter Collection ; 550.t.
```

EXAMPLE 9

```
245 04  The state of New York, compiled from the most authentic information,
1796 / |c Martin sculpt.
255     Scale [ca. 1:1,325,000] |c (W 79°--W 72°/N 45°--N 41°).
260     N[ew] York : |b J. Reid, |c [1796]
300     1 map ; |c 39 x 47 cm.
500     Relief shown pictorially.
500     Differs from earlier state by corrected spelling in title.
500     From: The American atlas. New York : John Reid, [1796]. [No.] 10.
510 4   Wheat & Brun |c 372
510 4   Phillips |c 1216.
590     NYPL copy annotated with color in cartouche; also includes manuscript
notation.
651 0   New York (State) |v Maps |v Early works to 1800.
700 1   Martin, |e engraver.
700 1   Reid, John, |c publisher.
740 0   American atlas.
```

EXAMPLE 10

```
100 1   Thomson, John, |c geographer.
245 10  Northern provinces of the United States / |c drawn & engraved for
Thomson's New general atlas, 1817 ; Hewitt Sc. ...
255     Scale [ca. 1:2,250,000] |c (W 82°--W 66°/N 47°--N 38°).
260     [Edinburgh : |b Printed by George Ramsay and Company for John Thomson
and Company, etc., |c 1817]
300     1 map : |b hand col. ; |c 48 x 57 cm.
500     Relief shown by hachures.
500     From the author's New general atlas. 1817.
500     Includes ill.: The great falls of Niagara.
500     "No. 56."
510 4   Phillips, |c 731
651 0   Northeastern States |v Maps.
700 1   Hewitt, N. R.
700 1   Thomson, John, |c geographer. |t New general atlas.
799     Lawrence H. Slaughter Collection ; 528.
```

Cataloging Early Atlases:
A Reference Source

Lisa Romero
Nancy Romero

SUMMARY. The cataloging of early atlases presents a multitude of challenges to catalogers. This is true primarily because an atlas is a collection of maps in "book" form but also because the item is an early or "rare" item. This article attempts to provide the cataloger with the necessary guidance for cataloging early atlases by reviewing the relevant cataloging sources, discussing the issues relevant to early atlas cataloging, and providing examples of early atlas cataloging. The article is intended to serve as a "reference source" for those individuals who will be cataloging early atlases. *[Article copies available for a fee from The Haworth Document Delivery Service: 1-800-342-9678. E-mail address: getinfo@haworthpressinc.com <Website: http://www.haworthpressinc.com>]*

KEYWORDS. Early atlases, rare atlases, cataloging, rare book cataloging, cataloging of atlases

INTRODUCTION

The geographic atlas is probably the most unorthodox type of publication known to the cataloging world. With its irregular format and unusual con-

---

Lisa Romero is Acting Communications Librarian and Associate Professor, University of Illinois at Urbana-Champaign, 122 Gregory Hall, 810 South Wright, Urbana, IL 61801 (E-mail: L-ROMERO@UIUC.EDU).

Nancy Romero is Rare Book and Special Collections Cataloger and Assistant Professor, University of Illinois at Urbana-Champaign, 346 Main Library, 1408 West Gregory, Urbana, IL 61801 (E-mail: N-ROMERO@UIUC.EDU).

[Haworth co-indexing entry note]: "Cataloging Early Atlases: A Reference Source." Romero, Lisa, and Nancy Romero. Co-published simultaneously in *Cataloging & Classification Quarterly* (The Haworth Information Press, an imprint of The Haworth Press, Inc.) Vol. 27, No. 3/4, 1999, pp. 265-284; and: *Maps and Related Cartographic Materials: Cataloging, Classification, and Bibliographic Control* (ed: Paige G. Andrew, and Mary Lynette Larsgaard) The Haworth Information Press, an imprint of The Haworth Press, Inc., 1999, pp. 265-284. Single or multiple copies of this article are available for a fee from The Haworth Document Delivery Service [1-800-342-9678, 9:00 a.m. - 5:00 p.m. (EST). E-mail address: getinfo@haworth pressinc.com].

tents, the early atlas is–at least to its adherents–among the most interesting of the book world's denizens. About the year 1569 or 1570 Aegidius Hooftman complained to a friend named Radermacher about the inconvenience of having to roll and unroll large, cumbersome maps for use. He wondered whether something could be done about this problem. Radermacher suggested gathering several sheet maps and binding them together like a book. Hooftman agreed. Radermacher then called upon a friend, Abraham Ortel, map dealer and colorer, and commissioned him to assemble all the different sheet maps available in Holland, as well as those current in France and Italy. The result was the 1570 publication of 53 maps bearing the title *Theatrum Orbis Terrarum (The Theatre of the Whole World)*. The book was so popular that a second edition was published the same year. It was the opening of a new field of publishing. The mythical symbol of Atlas supporting the pillars of the sky, was suggested by Gerard Kremer (Mercator), a colleague in the map publishing business, to display on the title page. Hence, it was that the term "Atlas" became a generic title as well as symbol of a bound collection of maps.[1]

In preparing this article it seemed most logical to the authors to organize the information in the format in which a catalog record is constructed. To the authors this type of organization would be the most logical for the purpose for which it is intended–to serve as a "reference source" for those individuals who will be cataloging early atlases. All of the sources mentioned in this article may or may not be used by the cataloger in preparing copy, but it is important to know about them and how they would apply.

DEFINITIONS

Harrod's Librarians' Glossary defines an atlas as: "A volume of maps, with or without descriptive letterpress. It may be issued to supplement or accompany a text, or be published independently. Also, a volume of plates, engravings, etc., illustrating any subject; a large size of drawing paper measuring 26 1/2 x 34 inches; a large folio volume, resembling a volume of maps, sometimes called 'Atlas Folio'."[2]

The ALA Glossary of Library and Information Science defines rare book as: "A desirable book, sufficiently difficult to find that it seldom, or at least only occasionally, appears in the antiquarian trade. Among rare books are traditionally included such categories as incunables, American imprints before 1800, first editions of important literary and other texts, books in fine bindings, unique copies, books of interest for their associations; but the degrees of rarity are as infinite as the needs of the antiquarian trade, and the term is decreasingly used in libraries and other repositories, many of which prefer the terms special or research collection to rare book collection."[3]

Map Librarianship, 3rd ed.: "Definitions as to what constitutes a rare map vary from country to country, depending upon the history of that country. For the United States, any map published prior to 1900 (and especially of the western United States) should be considered to fall into this category–probably disintegrating as it descends if it was published during the nineteenth century or in newsprint. In Great Britain, 1825 is the cutoff date for rare maps."[4]

The term "rare" is relative pending on institutional guidelines/interpretation. However, the term "early" when applied to published atlases can be limited to time period. For the purposes of this article, the authors consider the term "early" to be limited to atlases issued prior to the year 1900.

SOURCES

In cataloging an early printed atlas the cataloger is faced with the issue of which cataloging sources to use. There are two important factors to consider: the volume is a "non-book" item (even though its physical format is book-like), and the item is considered rare, or early, by the cataloging institution. As a result, when cataloging early atlases the cataloger is not limited to using one cataloging source.

Anglo-American Cataloging Rules, Second Edition (AACR2): Chapter 3 of *AACR2* (Cataloging of Cartographic Materials) includes: atlases, globes, sheet maps, navigational charts, and aerial photographs. Upon the publication of this source the cartographic community expressed concerns in applying these rules to cartographic materials. Their concerns centered around two areas: lack of input on the development of the rules and objections to rule 21.1B2, which ruled out any entry under corporate body for cartographic materials. On the other hand, map librarians were pleased over all with the change in terminology from the use of the word "map" to "cartographic" and that it was given considerable treatment in *AACR2,* over that in *AACR.* In answer to map librarians' need for standardization of map cataloging rules, Canadian, British, and American members of the map community formed the Anglo-American Cataloguing Committee for Cartographic Materials (AACCCM).[5]

In 1982 *Cartographic Materials: A Manual of Interpretation for AACR2*[6] (Cartographic Materials) the guidelines for descriptive cataloging of cartographic material, was published. It was endorsed by the map community through the Map and Geography Round Table of the American Library Association (MAGERT) and other similar map librarianship organizations such as the Special Libraries Association's Geography and Map Division (SLAG&M). *Cartographic Materials,* produced by an international committee formed in 1979, illustrates, clarifies and expands upon the rules given in Chapter 3 of *AACR2.* It indicates the position of the Library of Congress, and the other country's national libraries, regarding the application of optional rules and provides guidance in the selection of access points for cartographic materials.[7] However dated it has become, it is the manual of choice for cataloging cartographic materials, though it does not provide adequate direction for the cataloging of early materials. The rules governing early printed materials are commonly implemented applying the bibliographic description of rare books. For example, when providing the descriptive data for those aspects of the piece as a cartographic item the cataloger applies the rule interpretations and options using *Cartographic Materials*. When providing data for those aspects which address the rarity of the item the cataloger uses the *Descriptive Cataloging of Rare Books* (DCRB).

Descriptive Cataloging of Rare Books (DCRB), published in 1991 by the Library of Congress for its own catalogers, is the successor to *Bibliographic Description of Rare Books* (BDRB). The *DCRB* is a manual of rules designed primarily for librarians, either within LC or outside, who catalog early printed materials. It is to be considered supplementary to AACR2, Chapter 2 and remains generally in harmony with the basic rules. In addition to *AACR2*, *DCRB* also tries, as much as possible, to conform to the *International Standard Bibliographic Description for Older Monographic Publications (Antiquarian)*. However, the fact that *DCRB* is a set of cataloging rules and *ISBD(A)* is a set of standards results in there being no competition between the two as the work of cataloging takes place in the rare book arena. It was prepared as a collaborative effort of the Association of College and Research Libraries, Rare Books and Manuscripts Section's (ACRL, RBMS) Bibliographic Standards Committee and the Library of Congress and is to be considered the general rules for the description of rare books, pamphlets and printed sheets. Although *DCRB* is predominately for the description of rare printed books there are portions of the "rules" that could have application in the cataloging of early printed atlases. For example, rule O.C1 regarding the "chief source of information (title page)" could readily be applied to early atlases since most atlas volumes, like printed books, have pages, unlike the early single sheet map.[9]

The Library of Congress, Geography and Map Division's *Map Cataloging Manual* provides guidelines covering all aspects of the cataloging record. It is an effort to assemble all information, practice, and policy regarding the cataloging of maps by the LC G&M Division. It includes information on encoding fixed fields, description, subject analysis and LC classification. However, it focuses on twentieth century maps, not other cartographic material and, in most cases, does not apply to pre-1900 maps.[8]

MAIN ENTRY

With cartographic materials published after the 19th century it is necessary for the cataloger to take into consideration corporate main entry since a good portion of cartographic materials issued after that time are usually prepared by a government agency such as the U.S. Geological Survey or a corporate body such as Rockford Map Publishers, Rand McNally and others. Rule 21.1B2 (Category f): "Category "f" was added with the 1982 revisions of *AACR2*. In this category, a cartographic work is entered under corporate body if the work emanated from a corporate body other than a body that is merely responsible for the publication or distribution of the materials. However, prior to the 1900s atlases were considered the sole creation of the cartographer. Therefore, the main entry for early atlases was, and in most cases still is, the name of the cartographer (see Appendix A).

TITLE AND STATEMENT OF RESPONSIBILITY AREA

Since the physical format of an early atlas is usually that of a book, the inclusion of a page which includes the normal information needed for cataloging is not unique. For instance, the 1778 *Atlas Ameriquain Septentrional* contains a page which provides a title, a place of publication, a publisher, and a date placed in the normal sequence of a "title page." In cataloging the item this information would provide the basis of the record. Likewise, Mercator's famous 1595 *Atlas sive Cosmographicae . . .* contains a beautifully illustrated page which includes the title, the author, and the place of publication. When cataloging this item the cataloger would no doubt use this page as the "chief source of information."

Title

According to *Cartographic Materials,* rules 1B-1G4, title proper information is recorded as instructed in *AACR2,* rule 1.1B, which merely instructs how to record the information. However, in *Cartographic Materials,* rule OB4, "Chief source of information for an atlas," the instruction there is to use the title page as the chief source of information for an atlas, or if there is no title page, the source from within the publication that can be substituted. The remainder of this rule provides alternate sources from which to obtain this information. An interesting footnote in this section states, "hereafter in the rules pertaining to atlases the term 'title page' is used to include any substitute." In the rule for scope in this section there is a disclaimer sentence which states that these rules "do not cover in detail the description of early or manuscript cartographic materials . . . ," which when considering the cataloging of early atlases could be considered not applicable. However, rule OB5 which is titled, "Chief source for early printed atlases," specifically states that if the early atlas has a title page, use it as the chief source of information.

If there is no title page, the rule then offers a list of alternatives in which material for the record can be used. Since rule OB5 does not define the term "early printed atlases" one could assume that the rules in this section may be applied in cataloging rare atlases. In rule 1A2 "Sources of information," the instruction is merely to take the needed information from the chief source of information for the material to which the item being described belongs. There is nothing in either of these instructions pertaining to the description of atlases. However, *Cartographic Materials* does provide consideration of atlases in its Appendix F: Geographical Atlases. In the introduction it states "Atlases share the characteristics of maps and books, having the content of maps and often the format of books." Further in this section it states that, "the item is treated as an atlas if it is principally a collection of maps issued in a form intended to be shelved." It would appear that any instructions in this section could be applied to any atlas, regardless of whether or not it would classify as "early."

Elsewhere in *Cartographic Materials* there are sections called "APPLICATION" which can be applied to cataloging early atlases. For example, in the rules for punctuation (OC1) there is an application that refers to "Early cartographic items" which gives instruction for recording punctuation which is characteristic of early printed items.

In using *AACR2,* rule 3.0A states that the rules in Chapter 3 cover cartographic materials of all kinds. This would include atlases. However, rule 3.0B1 refers the cataloger to rule 2.0B for sources of information for an atlas. Chapter 3 of *AACR2* is the basis for description while *Cartographic Materials* elaborates on *AACR2* and provides necessary background information on the application of the rules.

Descriptive Cataloging of Rare Books does not specifically mention in its paragraph on "Scope and Purpose" the use of rules for cataloging early atlases. However, due to the physical format of the early atlas there are rules that can be applied. For example, in *DCRB* rule O.C1, it states "The chief source of information for a publication other than a broadside or a single sheet . . . is the title page, or, if there is no title page, the source from within the publication that is used as a substitute for it." As one can see, this instruction is not too different from what is instructed in rule OB4 of *Cartographic Materials*. In *DCRB* Appendix E: DCRB Code for Records, paragraph four: Nonbook formats, the instructions here specify that the rules are not to be applied to "nonbook materials such as maps, music, and graphics . . ." It can be assumed from this statement that this disclaimer applies to single-sheet maps and not to atlases. In looking over the contents of *DCRB* one would see that the majority of the rules would not apply directly to early atlases.

STATEMENT OF RESPONSIBILITY

Recording the statement of responsibility for early atlases follows the same routine as that of the title proper. In the case of early printed atlases the cartographer is considered the author of the atlas. Even though there is a title page appearing at the beginning of the volume, it is not an uncommon occurrence for the cartographer/author of an early atlas not to appear on the title page in the traditional statement of responsibility area normally applied to books. For example, with the LeRouge *Atlas* (1778) the cartographer/author appears in a publication statement. However, accompanying wording clearly implies LeRouge's responsibility as "author."

In *Cartographic Materials* rule OB6, "Prescribed sources of information" states the chief source of information–which is the same instruction as that for title. Then rule OB7, which follows, instructs the cataloger to use the title page for atlases. The APPLICATION note to this rule states: "In all cases in which data for the title and statement of responsibility area . . . are taken from other than prescribed sources, make a note to indicate the source of the data."

Since the title proper and the statement of responsibility are so closely linked in all rule provisions their pattern of application is also similar. *AACR2* rule 3.1F refers the cataloger to Chapter 1, just as rule 3.1B did for recording the title proper. *DCRB* rule 1.G1 seems less restrictive when it comes to locating sources for the statement of responsibility: "Transcribe statements of responsibility appearing in the preliminaries, (cover, and any page preceding the title page) or in the colophon." The use of preliminaries and the colophon allows the use of areas not permitted to be used for describing modern books. In all instructions the use of a source other than the title page requires the data to be enclosed in brackets if used in the title and statement of responsibility area. Otherwise, the information can only be supplied in a note. All cataloging sources seem to be in agreement in this respect (see Appendix B).

MATHEMATICAL DATA AREA

Statement of Scale

Because an atlas is a bound (usually) collection of maps, sometimes the work of a single cartographer and sometimes not, the information regarding "scale" may be specific or generic. For an early atlas with maps at one scale only, *Cartographic Materials* rule 3B1a instructs the cataloger to give the scale as a representative fraction and precede the ratio by the word "scale." For an atlas with maps mainly at one or two scales, the cataloger should record one or both scales, and in the case of two scales the larger scale is given first. However, because the maps in early atlases oftentimes are not the result of a single cartographer and, as a result, whose scales may not be identical, *Cartographic Materials* rule 3B5 instructs the cataloger to consider the atlas a multipart item with three or more scales and to give the statement "Scales differ."

If the scale is not stated in representative fraction form on the item rule 3B1d instructs the cataloger to compute a representative fraction from a bar scale, graticule or grid and to record the scale in brackets preceded by "ca.". If the scale cannot be determined by the cataloger, rule 3B1e says to give the statement "Scale indeterminable," however, the Library of Congress follows a different practice and in such cases prefers to use the statement "Scale not given."[10]

Statement of Projection

Recording the information regarding projection for an early atlas follows similar guidelines as the information regarding scale. If all of the maps are drawn using the same projection, rule 3C1 of *Cartographic Materials* instructs the cataloger to record the projection using standard abbreviations and numerals in place of words. If the maps are drawn using two projections, both projections may be recorded connected by the word "and." If more than two projections are used, do not include a statement of projection. The cataloger should also add additional phrases connected with the projection statement, such as meridians and parallels, if they are found on the item.

Statement of Coordinates and Equinox

Cartographic Materials also instructs the cataloger that when recording the statement of coordinates and equinox for an early atlas, to record the coordinates for the maximum extent of the area covered by the atlas in the statement of coordinates, exclusive of reference maps. In addition, for early cartographic items where the coordinates are distorted or inaccurate, the cataloger should record the present-day coordinates for that area. The coordinates should be enclosed in parentheses and expressed in degrees, minutes, and seconds with each coordinate preceded by the appropriate designation, "W," "E," "N," or "S," in that order. Each longitude and latitude should be separated from its counterpart by a dash, while a diagonal slash is used to separate longitude from latitude.

PUBLICATION AREA

Rule 1.4 in *AACR2R* is followed when recording information for the name, place and date of publication and distribution, etc., information for cartographic materials.[11]

Place of Publication

Cartographic Materials rule 4C8 instructs the cataloger to record the place of publication as it is found on the item. In some cases, due to the age of the item it may be necessary for identification to add the modern name of the place, in brackets. The cataloger should add the full address or the sign of the publisher if it appears in the source and if it aids in identifying or dating the item. In the case where only the address or sign appears in the prescribed source of information, the name of the place of publication in English should be supplied, in brackets. If more than one place of publication appears on the item the first should always be recorded. The cataloger may record the others in the order in which they appear.

Publisher

The same guidelines for recording the details relating to the publisher should be followed when cataloging early atlases as when cataloging other cartographic materials, with additional attention to one important aspect of early printed materials, and that is the significance of the printer. The details relating to the publisher/printer should be recorded as they appear in the item (see Appendix B). The guidelines under *Cartographic Materials* rule 4G can be followed if the name of the publisher is unknown or if the information for the printer differs from the publisher. If the name of the publisher is unknown, give the place and name of the printer or manufacturer in parentheses. Or, if the printer of an early printed atlas is named separately in the item and the function as printer can be clearly distinguished from that of the publisher or bookseller, give the place of printing and the name of the printer, also in parentheses. In addition, rule 4E1 requires the addition of a term describing the function of the named body such as "distributor" or "producer," but only if the publication phrase does *not* include words that indicate the function performed or if the function of the publisher is not clear in the context of the publication statement.

Date of Publication

Cartographic Materials instructs the cataloger in rule 4F10 to give the date of publication or printing and to include the day and month as found in the item. The cataloger should change the Roman numerals to Arabic and add the date in the modern chronology if necessary. As is the case with many early printed items the date may be unknown. In this case, an approximate date should be given and bracketed, and one can also add a question mark to help indicate that the date is an approximation.

PHYSICAL DESCRIPTION AREA

Extent of Item

Cartographic Materials rule 5B1a includes a list of acceptable terms to be used as the specific material designation (SMD) when recording the extent of item. The appropriate SMD for atlases then, as given in the APPLICATION to this rule, would be "atlas." The extent of item statement for an atlas, therefore, should include the specific material designation and the pagination or number of volumes, or both.

Other Physical Details

According to *Cartographic Materials* rule 5C1 the other physical details to be recorded for atlases should include: the number of maps in the atlas, color, material, and mounting (see Appendix A). In addition, the APPLICATION for this rule specifies that if the atlas in hand is in manuscript form this is the place to record that information (using the term "ms." as indicated in the fourth example listed under "Describe coloured illustrations in an atlas . . . as such" following rule 5C3).

Dimensions

Cartographic Materials rule 5D2 indicates that when an atlas is bound the dimensions are taken from the height of the volume in centimeters, to the next whole centimeter, if unbound give the height of the item itself. As with most printed material, if the width of the material is less than half the height or if the width is greater than the height, give the width following the height preceded by a multiplication sign, e.g., 34 × 56 cm.

NOTE AREA

The application of notes for early atlases does not vary significantly in philosophy and application from any other material cataloged. However, due to the nature of early atlases there are certain notes that are specifically applicable. If one looks at the rules provided for notes in all three sources of cataloging rules (*AACR2R, Cartographic Materials,* and *DCRB*), a good many of the notes are applicable to atlases in general, but very few bring out the special characteristics of being "rare," or early. However, of the three sources *AACR2R* provides the most comprehensive coverage, whereas *Car-*

tographic Materials provides the least. In general, the rules for notes are comparable in each of the three cataloging sources as far as identifying terms are concerned.

For example, in Area 7 of *Cartographic Materials* the rules are numbered and listed in the same order as those in *AACR2R,* Chapter 3. In footnote 13 in Area 7 of *Cartographic Materials* it states "As the notes given from *AACR2* 3.7B1 to 3.7B21 are not sufficiently comprehensive for cataloging cartographic materials, a number of additional notes have been appropriated from other chapters of *AACR2* and inserted in their correct sequence according to the general outline given in *AACR2* 1.7B. One new note is included (relief features), and has been incorporated in 7B1." The implication here is that *AACR2R* can be the general guide for applying notes with only those specific rules from *Cartographic Materials* being applied as needed.

Note fields in *Descriptive Cataloging of Rare Books* are not too different from that in *AACR2R* and *Cartographic Materials.* Most of the rules in *DCRB's* Chapter 7 do not come right out and state that the notes are specifically for early atlases. As it implies, this source is for cataloging rare books. However, lest we forget, an atlas, in its bound physical format is a book–a book of maps, sometimes single-sheet maps. As with the other two cataloging sources, the *DCRB* note section pretty much follows the same pattern. Similarly, *DCRB* includes notes which by nature cannot be applied to early atlases and those which can be applied to early atlases simply by the type of information required by the note.

AACR2R Rule 3.7B1 "Nature and scope of the item," and rule 7.C1 in *DCRB* opens the door for application to early atlases. In the last sentence of rule 3.7B1 it states: "Also make a note on unusual or unexpected features of the item." If one considers rare as "unusual" then this note area could be applied when cataloging early atlases. *Cartographic Materials* does not contain a note for this area (see the notes in Appendix C).

A note for language of the item is covered in both *AACR2R,* rule 3.7B2 and *DCRB,* rule 7.C2. Since early atlases often have textual materials in more than one language which precedes the maps themselves, or is presented at the beginning of the volume, rules 3.7B2 and 7.C2 can be applied. Frequently, the textual portion which accompanies the maps will be in more than one language, whether or not the title and/or statement of responsibility makes this clear, thus this is an important note.

In light of the publishing style of early atlases it would not be uncommon for the same atlas title to exist with unique features. For example, one institution could hold a copy of the Le Rouge *Atlas* with a very plain binding with no unique features. Whereas another library could hold the same edition of the atlas rebound with a variant spine title or a title label on the front cover. These specific characteristics would need to be brought out in a "Source of title" note, if not on the title page, in the cataloging record for this copy. Since early atlases are published in the book format it would be safe to say that the majority have a title page from which to record title information.

However, due to the nature of early atlas publishing there is nothing to say that an early atlas could have been published without an "official title page," but perhaps with the title proper on the cover or spine. The rules in *AACR2R* dealing with title, 3.7B3 Source of title and 3.7B4 Title variations, are comparable to *DCRB*'s rules 7.C3 and 7.C4. Rules 3.7B3 and 7.C3 allow the cataloger to record in a note the source from which the title was taken should there not be a chief source of information for the atlas volume. In connection, rules 3.7B4 of *AACR2R* and 7.C4 of *DCRB* allow the cataloger the option of recording variant forms of the title, which in the case of early atlases can appear on the spine, cover or in an "at head of title" position.

In close connection with the rules in the above paragraph and controlled by some of the same factors, is *AACR2R* rule 3.7B5, "Parallel titles and other title information," *Cartographic Materials* rule 7B5, and *DCRB*'s rule 7.C5 The application of all three of these rules is controlled by the guideline that one can record in a note parallel title and other title information not recorded in the title and statement of responsibility area if it is considered to be important. It is left up to the judgement of the cataloger as to whether access to this information is of importance and where it should be placed.

Statement of responsibility, *AACR2R* 3.7B6 and *DCRB* 7.C6, can be applied to an early atlas in the case of a situation where the cartographer's name does not appear within the chief source of information but is obtained from a reliable external reference source. Even though an atlas can appear in a book-like format it can be issued without attribution to an "author," not unlike a modern book publication. With this note the cataloger is allowed to provide a connection to the cartographer by providing relevant external source information, e.g., by using a cartobibliography, and citing this source. In the case of early atlases this can provide a valuable link for publishing continuity.

The "Edition and history note" (*AACR2R* rule 3.7B7 and *DCRB* rule 7.C7) for early atlases allows the cataloger to compose a note about the work that cannot be provided elsewhere in the record. This note can be very important to the cataloger in that information can be supplied here that will then allow added entry access to information. For example, a note such as: "Maps 3 and 17 first appeared in Le Rouge's *Atlas Ameriquain Septentrional*" can provide valuable information to the patron since the publishing history and pattern of early atlases can be nebulous. This type of "connecting" note, therefore, is always included when necessary. The examples displayed under this rule in *AACR2R* are very good in illustrating the options of application for the cataloger to use (see Appendix C).

The note field for recording mathematical data (*AACR2R* rule 3.7B8) is only applied in a general way rather than providing specific information. In *AACR2R* the cataloger is instructed to consult rule 3.3B5 in the case of a title in which the scales vary, which would be the case of an early atlas where each map has its own scale. (In reality this is an unfortunate and incorrect use of the word "vary," which should read "differ" whenever it is applied to more than one map.) Notes which are supplied in this area include those that "Give other mathematical and cartographic data additional to or elaborating on that given in the mathematical data area" (*Cartographic Materials* rule 7B8) such as those for orientation if other than north, correctness of the given scale, or prime meridian when Greenwich is not used on the maps in the atlas.

DCRB's rule numbering departs here from that of *AACR2R* and *Cartographic Materials*. Rule 7.C8 does not deal with mathematical data but instead deals with publication details. It is this note rule in *DCRB* that allows the cataloger to record publication information not permitted in the publication area. With early atlases this note provides the cataloger the option of recording important imprint information that is taken from a source other than the title page–this other source can be from somewhere in the piece itself or from an external reference source, e.g., "Publisher statement on cancel slip. Original publisher statement reads: Sold by G. Walsh."

The "Publication, distribution, etc." note in all three cataloging sources (*AACR2R* rule 3.7B9, *Cartographic Materials* rule 7B9, and *DCRB*'s rule 7C8) do not state that they apply specifically to early atlases. However, looking at these three rules and the examples shown, the application to early atlases is not totally excluded either. In cataloging early atlases there are frequently choices that need to be made concerning publication information and how it should be recorded in the record. This note provides the cataloger the opportunity to provide additional publication information not found elsewhere in the record.

The "Physical description" note and the "Accompanying material" note both have limited applicability to early atlases. The statements given in these two note areas in all three cataloging sources are non-descriptive enough to be applied to early atlases in regard to physical description. Accompanying material being published with an early atlas is not a common situation, therefore use of this particular note may have limited applicability.

Except for the note rule for "Contents," the remaining kinds of notes in all three cataloging sources have little, or no, applicability to early atlases. The use of the contents note for early atlases as to whether access was desired rather than just a listing of contents, is left to the discretion of the cataloger. If the cataloger is aware that multiple copies of a particular atlas exist a contents note in the record would provide a means of comparing the individual maps a particular copy contains against other copies. Copies of early atlases have been known to be composed of different maps even though they were issued with the same title in the same year.

Application of the rules for notes from any of the three cataloging sources remains totally up to the needs and discretion of the cataloger. Even though it may not be readily apparent from the rules themselves, the examples under each rule provides the cataloger with an "open door" for adapting notes as they are needed.

SUBJECT HEADINGS

Subject application for early atlases is not that much different from that of modern atlases and single-sheet maps. Section H1865 of the *Subject Cataloging Manual: Subject Headings* states: "Map headings are assigned not only to maps and atlases, but also to works about maps and atlases, and to works with both textual and cartographic content. . . . "[12] Assigning subject headings to a work whose content consists mainly of maps follows a fairly "narrow road." For instance, if the atlas consists of maps of a particular geographic area, such as Spain, the subject heading: Spain–Maps would be the logical selection. However, if the work dealt with geological maps of Spain, then the subject heading would be: Geology–Spain–Maps. If this same atlas also had much textual content related to the geology of Spain then one would add as a second subject heading: Geology–Spain. The use of the first heading is to bring out subject content, while the addition of the second heading is to indicate the nature of the textual content of the work. The subdivision "maps" can be used under names of countries, cities, etc., and under individual corporate bodies, as well as with topical headings. In addition to the general subdivision "maps" there are also nine other subdivisions beginning with the term "maps" listed in the *Subject Cataloging Manual: Subject Headings,* in Section H1095. As in the basic subdivision "maps" these subdivisions are also textual content designators, e.g., the subdivision "Maps, Manuscript."

A new source for map cataloging, which appeared in 1991, the *Map Cataloging Manual*[13] (prepared by the Geography and Map Division, Library of Congress for their catalogers) has an excellent section on subject access for cartographic materials. Even though it does not touch on early atlases in particular, the instructions for application may still be applied to these works.

CLASSIFICATION

Since geographic area is the key point of interest for most cartographic users, most classification schemes use geography, or area, as its focal point. Materials with additional characteristics such as economics or political science would be given additional subject access points in the record, and in some classification schemes these additional subjects can also be indicated as a part of the class number system.

There are five classification schemes that may be used when classifying atlases. Three of the systems, Library of Congress, Dewey, and Universal Decimal can be used for all types of materials; two other systems, Boggs and Lewis and the American Geographic Society, are specifically designed for cartographic materials.

The Library of Congress Classification (LCC) system as a whole is designed primarily for books, the exception being Class G. But, the numerical notation in the Library of Congress Schedule-Class G is used first to classify by major geographical, political, or cultural units, then this notation can be subdivided according to smaller regions or local places. Alphanumerical cutters are then added to designate major topics. Following the class number geographic sub-area and/or subject cutter codes can be applied. Next, LCC adds the date of situation and concludes the call number with an author cutter. Because the scheme meets the requirements of large cartographic collections by arranging materials geographically first, and topically second, and is frequently updated, LCC is used in most map collections. However, the LCC system may also be thought of as having an Anglo-American bias.

Within the Dewey Decimal Classification (DDC) system cartographic materials are classified under the number "912." To this base number is added a number for subject or area. As with the Library of Congress system, DDC has an Anglo-American bias, specifically for the United States. But, perhaps its main drawback for map librarians is that it classes cartographic materials according to subject categories first, making location its secondary aspect.

The Universal Decimal Classification (UDC) system is based on DDC and is also oriented to books and not cartographic materials, but is less biased toward the United States. The notation "912" is used for cartographic materials and is subdivided by subject. But with the UDC system cartographic material may be classed first by area and then by subject or vice versa.

The Boggs and Lewis classification system is specifically devoted to maps, and was developed to meet the needs of the U.S. State Department's Map Library. The system, based on a numerical schedule for geographic areas and an alphabetic list for subjects, consists of four elements: area, subject, date for the situation portrayed on the map, and the initials of the author or publisher. While map librarians approved of the system because of its attention given to both subject and area and its freedom from North American bias, its major criticism is that it has never been updated since being issued in 1945, quite a limitation when classifying cartographic materials from ever-changing areas of the world.

Like Boggs and Lewis, the American Geographical Society classification system is devoted to cartographic materials. It uses a numeric notation to represent geographic area and an alphabetic notation for the subject, followed by a date. A limitation of the system is that its broad classification scheme is unable to accommodate materials with specialized subjects.[14]

Of the above classification schemes reviewed, LCC Class G is the most commonly used in map collections, especially ones that are larger in size such as those found in academic libraries.

CONCLUSION

When cataloging an early atlas the cataloger needs to remember three factors: the physical format of the item is like that of a book, it is usually bound and usually has a title page containing the necessary "cataloging information" from which to construct a record; the content, which consists of several individual items, the maps themselves–each an independent item but adding to the subject matter of the whole; and its "publishing" history which makes it rare, or early. In the case of each of these factors either a set of rules and/or manuals of rule interpretation need to be consulted and taken into consideration when an early atlas is cataloged. Unfortunately, at this time there is no single source to consult when cataloging an early atlas. (However, a new edition of *Cartographic Materials* is nearing completion, and it will contain much more guidance for describing early cartographic materials.)

On the more positive side, a recent development which is quite welcome to catalogers of cartographic materials is the change in coding for atlases in the "Type:" fixed field of OCLC's *Bibliographic Formats and Standards* document. In 1997 the code for this field was changed from that for a book, code "a" to that for cartographic material, code "e." This shows a change in thinking from considering the format of atlases as most important to realizing that the content of atlases is the prime concern for our users.

NOTES

1. Lloyd D. Brown, *Notes on the Care & Cataloging of Old Maps* (Windham, Conn.: Hawthorn House, 1941), 79-80.

2. Ray Prytherch, comp., *Harrod's Librarians' Glossary of Terms Used in Librarianship, Documentation and the Book Crafts, 6th ed.* (Brookfield, Vt.: Gower, 1987), 49.

3. Heartsill Young, ed., *The ALA Glossary of Library and Information Science* (Chicago, IL: American Library Association, 1983), 185.

4. Mary Lynette Larsgaard, *Map Librarianship: An Introduction, 3rd ed.* (Littleton, Colo.: Libraries Unlimited, 1998), 31.

5. Kathryn Womble and Mary Larsgaard, "The Map Cataloging Manual: Autobiography or Leadership Manual," *Meridian*, 1992, no.7: 34.

6. Hugo L. P. Stibbe, ed., Cartographic Materials: A Manual of Interpretation for AACR2 (Chicago, IL: American Library Association, 1982).

7. Jo Ann V. Rogers, *Nonprint Cataloging for Multimedia Collections, 2nd ed.* (Littleton, Colo.: Libraries Unlimited, 1987), 43.

8. Womble, "The Map Cataloging Manual," 33.

9. *Descriptive Cataloging of Rare Books, 2nd ed.* (Washington, D.C.: Cataloging Distribution Service, Library of Congress, 1991).

10. Carolyn O. Frost, *Media Access and Organization* (Littleton, Colo.: Libraries Unlimited, 1989), 40.

11. Rogers, *"Nonprint Cataloging for Multimedia Collections,"* 48.

12. Cataloging Policy and Support Office, Library of Congress, *Subject Cataloging Manual: Subject Headings, 5th ed.* (Washington, D.C.: Cataloging Distribution Service, Library of Congress, 1996).

13. Geography and Map Division, Library of Congress, *Map Cataloging Manual* (Washington, D.C.: Cataloging Distribution Service, Library of Congress, 1991). (Also available electronically as a part of *Cataloger's Desktop* software.)

14. Frost, *"Media Access and Organization,"* 48-50.

[**Warning**: The following appendices, of actual bibliographic records taken from OCLC, were done using different cataloging rules and standards at the times they were created. Do not use them as actual examples for cataloging purposes, use only for illustrative purposes. Instead, be sure to follow current rules and standards as found in *AACR2R* and other rare materials tools as outlined in this article.]

APPENDIX A

```
OCLC: 15097957        Rec stat:  n
Entered:  19870121    Replaced:  19950313    Used:  19960707
Type: e      ELvl: K      Srce: d      Relf:        Ctrl:         Lang: lat
BLvl: m      SpFm:        GPub: 0      Prme:        MRec:         Ctry: gw
CrTp: a      Indx: 0      Proj:        DtSt: s      Dates: 1595,
Desc: i
```

1 040   RQE |c RQE

2 090   G113 |b .M4

3 090 |b

4 049   UPMM

5 100 1   Mercator, Gerhard, |d 1512-1594.

6 245 10 Atlas sive Cosmographicae meditationes de fabrica mvndi ct fabricati figvra. |c Gerardo

Mercatore Rupelmundano, Illustrissimi Ducis Iulie Clivie & Motis &c. Cosmographo Autore...

7 260   Dvisbvrgi Clivorvm |c [1595]

APPENDIX B

```
OCLC: 6744056        Rec stat:  c
Entered:  19800923    Replaced:  19970929    Used:  19930326
Type: e        ELvl: 1        Srce: d       Relf:      Ctrl:          Lang: eng
BLvl: m        SpFm:          GPub:         Prme:      MRec:          Ctry: enk
CrTp: e        Indx: 1        Proj:         DtSt: s    Dates: 1635,
Desc: i
```

1 040   YHM |c YHM |d OCL

2 006   [aabc   001 0]

3 090   G1007 |b .H67 1635a

4 090 |b

5 049   UPMM

6 100 1   Mercator, Gerhard, |d 1512-1594.

7 245 10 Historia mundi, or, Mercator's atlas : |b containing his cosmographicall description of the fabricke and figure of the world / |c lately rectified in divers places, as also beautified and enlarged with new mappes and tables by the studious industry of Iudocus Hondy ; English by W. S. [i.e. Wye Saltonstall].

8 250   [1st ed.].

9 260   London : |b Printed by T. Cotes for Michael Sparke and Samuel Cartwright, |c 1635.

APPENDIX C

```
OCLC: 40341501        Rec stat:   n
Entered:  19981118     Replaced:  19981118     Used:   19981118
Type: e        ELvl: K       Srce: d       Relf:       Ctrl:          Lang: lat
BLvl: m        SpFm:         GPub:         Prme:       MRec:          Ctry: ne
CrTp: a        Indx: 0       Proj:         DtSt: s     Dates: 1616,
Desc:
```

1 040   NYP |c NYP

2 090 |b

3 049   UPMM

4 100 1   Mercator, Gerhard, |d 1512-1594.

5 245 10   Atlas sive Cosmographicae meditationes de fabrica mvndi et fabricati figvra.

|b Deuáo auctus.

6 250   Ed. 4.

7 260   Amsterodami, |b Sumptibus & typis aeneis H. Hondij, |c 1616.

8 300   374 p. |b col. maps |c 50 cm.

9 500   The 9th Latin edition was called the 5th edition by error. Cf. U.S. Library of Congress. A list of geographical atlases, I. p. 170.

10 500   The first part, originally published in 1585, is composed of three fascicles, each with a special t.p.: Galliae tabule geographicae; Belgii Inferioris geographicae tabulñe; Germaniae tabulñe geographicae.

11 500   The second part, originally published in 1589, has special t.p.: Italiae, Sclavoniae, et Graeciae tabulñe geographicñe.

12 500   The third part, under whose title the work is published, precedes the first and second part. Left unfinished by the author, it was completed and edited by his son, Rumold Mercator.

13 500   Printing on verso of maps.

14 650 0   Atlases |v Early works to 1800.

15 700 1   Mercator, Rumold, |d 1541-1600.

16 700 1   Hondius, Jodocus, |d 1563-1612.

Early Maps
With or In Printed Publications:
Description and Access

Dorothy F. Prescott

SUMMARY. This article identifies the types of publications containing maps and the need for access to maps contained in them. The emphasis is on older maps; the comments are apposite for cataloging of current maps. The needs of map users are discussed, identifying the points of access that are critical to successful map retrieval. Main entry for maps is discussed. Various categories of associated map and book items are identified, and suggestions made, with USMARC examples, as to how these maps might be treated by the cataloger. *[Article copies available for a fee from The Haworth Document Delivery Service: 1-800-342-9678. E-mail address: getinfo@haworthpressinc.com <Website: http://www.haworthpressinc.com>]*

Dorothy Prescott is a Director of Prescott Associates, specialising in map and boundary problems. She is an information specialist on maps and map libraries and a Commonwealth Approved Valuer of maps and cartographic literature. Her last appointment was as Map Curator for the National Library of Australia during the period 1979-83; prior to that she was Map Curator for the University of Melbourne Library. She is a joint author of *A Guide to Maps of Australia in Books Published 1780-1830* (1996) and *Frontiers of Asia and Southeast Asia* (1977).

Address correspondence to: Dorothy F. Prescott, BA, AALIA, MMSIA, 44 Lucas Street, East Brighton, 3187, Victoria, Australia.

The author would like to thank Pam Dunlop of the National Library for assistance and comments on the Marc 21 fields and Hal Cain of Ormond College Library, University of Melbourne for his assistance and comments on this paper.

[Haworth co-indexing entry note]: "Early Maps With or In Printed Publications: Description and Access." Prescott, Dorothy F. Co-published simultaneously in *Cataloging & Classification Quarterly* (The Haworth Information Press, an imprint of The Haworth Press, Inc.) Vol. 27, No. 3/4, 1999, pp. 285-301; and: *Maps and Related Cartographic Materials: Cataloging, Classification, and Bibliographic Control* (ed: Paige G. Andrew, and Mary Lynette Larsgaard) The Haworth Information Press, an imprint of The Haworth Press, Inc., 1999, pp. 285-301. Single or multiple copies of this article are available for a fee from The Haworth Document Delivery Service [1-800-342-9678, 9:00 a.m. - 5:00 p.m. (EST). E-mail address: getinfo@haworth pressinc.com].

KEYWORDS. Map cataloging, MARC 21, map users, map catalogers, linking entry fields

INTRODUCTION

What does one do if maps within an early printed volume are important enough to require cataloging? How does one treat maps appearing with associated textual material? How are such multi-format publications handled? Is there only one way of doing this or is there a choice of methods? While this article will focus on early maps, the techniques to provide access to these maps may be profitably used with maps of any date of publication or issuance.

Maps occur in many different kinds of publications, ranging from atlases and pilot books, through monographs of many kinds–including travel guides, voyages and travel accounts, diaries, religious societies' reports, scientific and historical accounts, and other geographical works such as gazetteers, encyclopedias, dictionaries and grammars, texts, including those for children–and serial publications–such as the reports of Parliament, government reports, yearbooks, directories, and geographical magazines–to the more ephemeral material such as tourist brochures and guide books. There is thus a considerable corpus of potential cartographic material to be retrieved. Among the categories that have been overlooked in bibliographic control have been Parliamentary reports, where indexing has concentrated mainly on the text rather than on the maps. Another category of uncontrolled matter is the pilot book, a scarce and elusive medium, but one of great importance for the discovery record.

Maps offer a type of information that the printed word cannot. This one fact alone is sufficient reason to justify the cataloging of maps. When the maps are contained with or in another publication, certain problems arise. We can categorize these problems into two principal groups, those for the potential map user and those faced by the librarian. These will be discussed in turn.

Let us take first the requirements of the map user. Maps contained in books are buried treasure as far as the map user is concerned. There is no way to tell by looking at the outside of a book whether it contains maps, and if these maps are not cataloged in some way or brought to the user's attention, they are inaccessible. They then represent from all viewpoints a wasted resource, information available but not controlled sufficiently to be retrieved from the library.

When we are talking about older materials this can be quite a significant wastage. In the case of a project recently completed[1] on Australian maps, this was very significant because the first maps of the newly colonized continent of Australia were contained only in books, such as the diaries and official accounts of the earliest administrators, naval and military personnel, clergy, judicial officers and well-to-do settlers. Add to this the fact that reports and maps sent back to Britain by the Governor of New South Wales were housed in the Colonial Secretary's Office as government documents, and thus were available only to government officials or by formal release by an official! Some of these maps saw the light of day as source material for the maps of John Arrowsmith, but a significant few appeared as appendices to reports of Parliament and were well and truly 'lost' in the volumes of written material emanating from Parliament. This information has remained buried except to the knowledgeable until the middle eighties of this century, for lack of indexing of the maps in the papers of Parliament.

Liberating these maps for the user is the task faced by the librarian. But how to do this and which is the best way–*that* is the burning question.

INFORMATION REQUIRED BY MAP USERS

Many catalogers fear maps. This is analogous to the situation anyone may face when going into hospital for an operation. In both situations it is a matter of confronting the unknown. For the cataloger therefore information is required on how maps are used, what are the key data that users need, why the techniques for compiling book catalogs are less than satisfactory for map users, and how these inadequacies may be addressed. Until the cataloger understands these requirements, s/he is not in a position to make an informed decision as to the best method for handling a particular cataloging problem.

ACCESS POINTS

Addressing firstly the question of how maps are used, we need to identify which are the most important access points. These are geographical area, subject content, title and author entry. Many users would add to this list map scale, particularly military users.[2]

Geographical Area

A user will always require the geographical area portrayed by the map and usually the time period portrayed by the map, for example, a map of Hudson Bay in 1860. In cataloging parlance this is referred to as the situation date. A user requiring a map of a city would be less than happy to be given an out-of-date map if s/he were a tourist. If the user's need is for a "time shot," then the request becomes very specific, down to a narrow range of only one or two years.

Another user may need to do a subject search to find out what maps are held on a particular subject. What postal-code maps does the library hold? What are the subject strengths of the collection? Does the collection have only modern maps or is there a significant representation of older maps, and of what areas?

Cataloging codes[3] devised for map collections provide access to maps by both the geographical area shown on the map and by the subject content. The geographical area is always qualified by the date of the situation shown on the map. These codes predate many of the international developments in cataloging, but nevertheless have the most sympathetic approach to maps, one that has the interests of the map user uppermost, principally because they are not restrained by any necessity to consider other formats.

Topic or Subject

With the appearance of the second edition of the *Anglo-American Cataloguing Rules*[4] in 1978, a serious attempt was made to provide fuller guidance for the cataloging of maps. The *Rules* were far more receptive to the needs of map catalogers and consequently were accepted by many who previously had rejected the earlier provisions for map cataloging. Turning from descriptive to subject cataloging, it is important for the map cataloger to note that the Library of Congress subject headings were designed in the period when the card catalog was the medium for holding information; it was also the period when the predominant format collected by libraries was the book. Consequently LCSH reflects the requirements of book users, and other formats have had to adapt to them to the disadvantage of the particular groups of users wanting to access these other types of formats.

For map users consulting catalogs built on AACR, major inadequacies that require remedying are subject access to maps and immediate access to situation date. A map user needs to be able to access a map from several approaches. There are users who do not know precisely what they want; they may even ask for maps that do not exist. The catalog should be able to guide such inquirers by offering an approach to the collection by geographical area and by thematic or topical content of the map. For example, a user should be able to find the same map under both these two headings: "Canada–Geology;" and "Geology–Canada." All subjects should be searchable in this fashion, and the hierarchical nature of LC headings which prevents direct entry for some subjects should be ignored in hardcopy catalogs; online catalogs that allow keyword searching, or that allow subject searching with the words in any order, in some ways solve this problem. The argument for providing this approach is that the user has only to look under the country or topic of interest to find a complete record of all the different types of maps available for that area or subject. What a boon! It immediately permits a user to evaluate the usefulness of a collection for that user's intended purposes.

The present Library of Congress approach to subject headings would require for many subjects a search of the entire catalog to determine this same information, an unrealistic expectation given the technology now available to catalogers. It ought to be possible to ascertain the extent and quality of any particular subject coverage of maps in the collection. With the advent of machine-readable cataloging and keyword searching, catalogers are able to include terms in their notes to eliminate this problem of inadequate subject access.

Situation Date

The situation date is an important aspect that qualifies the usefulness or lack thereof for the user of a particular map. In fact it is often critically important. It is normally used in conjunction with geographical area or subject; in classification codes such as those of the American Geographical Society's, Boggs and Lewis, and the British Defence Department, this date is considered vital. Date of situation may appear in headings as the last element after geographical area or subject, and it is very useful to have this date prominently displayed, especially in card catalogs, where scrutiny can be rapid. In Library of Congress cataloging[5] the situation date appears in the call number except for map series, where scale is substituted. Its omission weakens the usefulness of the records produced and assumptions or deductions have to be made by the would-be user from the date of publication if the map title does not provide this information. This takes time when working through a large file of records. Catalogers should always supply this information in a note if it is not provided elsewhere.

Author

The author approach for maps is assumed to include surveyor, hydrographer, cartographer, engraver, and sometimes printer and publisher. It is most likely to be needed when seeking maps made by a specific person, and is essential for retrieving early maps created in the period before corporate bodies became the dominant map-makers.

Title

Title is also a very important access point for older maps and is frequently needed as in many older maps there is no creator statement. Titles for maps within atlases are not supplied nearly enough as an access point in map catalogs; for example, prior to the advent of automated cataloging, many early map collections had classified catalogs, not dictionary catalogs, and thus had no title entries. Also, given that pre-1900 atlases are unfortunately often physically disassembled and the maps sold separately, lack of title entry for individual maps with or in other publications can create considerable problems.

MAIN ENTRY FOR MAPS

The last two items above introduce a discussion on main entry for maps. Many would question whether there is any necessity for this, to which the response might be perhaps we could make life a little easier for ourselves and perhaps we should rethink its purpose. The availability of keyword searching and multiple access points permits as many points of access as thought necessary and desirable; consequently not a great deal hangs on the debate except perhaps the points mentioned by Martin[6] and Dunlop. Martin calls for uniformity of heading for a work "so that whenever a work is referred to in a catalog it will always have the same access points regardless of what those works are called in particular items." Dunlop[7] argues against the irrelevance of main entry but opts for a simpler type, namely, title main entry in opposition to author main entry.

There is a great deal of common sense in this suggestion and although it is not new perhaps it is time to reconsider it yet again. As Dunlop points out, almost a quarter of cataloging queries by cataloging reviewers on the Australian Bibliographic Network are concerned with main entry. On these grounds alone, there would be a considerable positive economic impact were these problems removed.

Earlier map cataloging systems made title the main access point and all other approaches were considered to be added entries. Title is still a very important access point for maps, and consistently making it the main entry could save considerable time in deciding which should be the main entry when–as so often happens–a number of creators exist for any particular map.

THE PURPOSE OF BIBLIOGRAPHIC DESCRIPTION

Description is the substitute or surrogate for the real thing; therefore it should deliver information that the would-be user would seek if he had the book or map in hand. Map descriptions can serve as a useful aid in preserving maps, in that clients should be encouraged to peruse the catalog rather than browsing through the map drawers in a search for appropriate maps. As the life of a map is not very long in comparison with a book, map librarians and curators should consciously seek to keep map handling to a minimum. Until digital map images are included as part of the cataloging record as a matter of course, the notes must be the substitute. While descriptions for maps may be prepared to a variety of levels of detail depending on the nature of the listing or catalog in which they are to be included, in the interests of keeping browsing of maps by users to a minimum, fuller description is a good investment of the cataloger's time. Cataloging time is a significant cost in the operation of a library, so the sooner digital map images are incorporated as part of the record, the sooner the cataloging rate should improve.

ELEMENTS OF DESCRIPTION

The simplest listings are vendors' stock lists and various accession lists, where usually only minimal description is employed. In such lists, items appear with title, scale, publisher and date of publication. Sometimes, in the case of series maps, the number of sheets is added. Where only a few sheets are available, the individual sheet names and numbers may be given.[8] Such listings are usually acceptable for recently published material, but would not be acceptable for lists of out-of-print or rare materials.

The minimum description for a library catalog should contain: details of the map title and statement of responsibility; material- (or type of publication) specific details, such as map scale, publisher and date of publication; as necessary, edition or state (this latter for older maps only); extent of item (the number of pieces making up the whole); note(s)–for example, situation date of the map where no date is given in the title; subject where not included in the map title; and standard number. Some maps will require all these elements, others not. The aim should be to provide sufficient information for the user to be able to make an informed decision as to the appropriateness or otherwise of a title for his particular use. Cataloging codes give no latitude in this area as they need to maintain a standard to which all must work. Therefore as far as catalogers are concerned, they will need to include the above elements in each minimum description. For compilers of accession and stock lists, the shorter list given first would be adequate.

At the other extreme is the cartobibliography of early maps where a fuller description is called for. In such a description, include in addition to those elements in the previous list as appropriate the following: parallel titles and other title information; subsequent statements of responsibility; edition and statement of responsibility relating to edition; subsequent publisher statements; other physical details and dimensions of the item; title proper of series, statement of responsibility for series, the ISSN of series, the numbering within the series, title of sub series, ISSN of sub series, numbering within subseries; and extensive notes.

MAP AND BOOK COMPOSITE ITEMS

A decision with which many catalogers have problems is that concerning primacy, that is, which item can be considered to be the principal publication of a composite production. The intent of the publisher should be the first guide as to how to treat the material. If this is not obvious, the criterion should be whether one format is dependent upon the other to be effective. If this question can be answered in the affirmative for one of the formats, this is the secondary item which is dependent upon the other. An example of maps with accompanying texts is a geological map series which has individual text for each map sheet. The text would not exist without the primary item, which is the map. The third factor that may have some relevance is the mission statement of the library and its stated aims.

Catalogers face yet another problem in taking upon themselves the onus of making a value judgement as to what is significant and what is not. What is significant to one reader may be trivial to another and vice versa. It is never possible to encompass the totality of what uses the readership might have for the materials in hand. Therefore it is wiser to treat materials by the other criteria identified above.

There are many sorts of combinations of map and text. For the purposes of cataloging, the most useful categorization is possibly by format, by which is meant according to the method by which the items have been produced. Below are set out the most common associations encountered, which will be considered in turn following the listing.

Attached maps and text

- Book with map(s)
- Foldout map at end/within book
- Atlas maps
- Serial with maps included
- Multiple maps on one sheet in book

Detached maps and text

- Loose map in pocket of book
- Map series with pamphlets or books produced separately
- Maps in soft covers
- Loose maps and loose text
- Tipped in map and text

Attached Maps and Text

Book with Map, Foldout Map, Atlas Map

These types of formats are the ones most often encountered in cataloging older books and atlases. It is important from several viewpoints to be able to retrieve contemporary maps of a period being researched. They are primarily important as reflecting the state of knowledge at that time–for example, of the geography of a country, the place names in use, the existence of specific features, the limits of the communication network, the stages of discovery and exploration, etc.

Maps in such publications can best be cataloged by the use of the "In" mechanism, where the author and title of the book or atlas in which the map appears is cited, with short publication and date details, followed by the location of the map within the volume.

In all the following examples it is assumed that the main entry is author except for the serial examples; therefore the indicators given in the title field (245) reflect this situation.

MARC 21 Tag	Map Description
245 10	‡a New Holland & the adjacent islands agreeable to the latest ... |h [map]
300	‡a 1 map ; ‡c 19.0 x 24.0 cm.
773 0	‡7 p1am ‡a Kincaid, Alexander. ‡t A new geographical commercial and historical grammar ... ‡b 3rd ed. ‡d Edinburgh, 1799. ‡g Vol. II facing page 535.

The MARC 773 field is one of the linking fields specifically designed for treating such situations as above, but records must exist for each item, e.g., the map and the book. The field is infrequently used, mainly because of the lack of ability of library online-catalog software to utilize its capabilities. This field is to be used for a bibliographic item that physically contains the component part described by the target record (e.g., the map contained in the book). When the 773 field is used, there is a facility to produce a display constant which would automatically produce the required "In:". The use of fields such as this one (773) and similar linking fields such as 772 for supplements and 777 for "Issued with:" items is dependent on the ability of the library computer system to handle reciprocal linkages: "Very few library systems have been developed to provide these linkages and therefore the[se] fields have not been overly used."[9]

If the 773 field is not used, it is suggested that the same information be put into a general-note field (field 500); if this is done, in addition, an "In" note will have to be input by the cataloger. A policy should be established as to where this note should appear when there are other notes. Should the determining factor be prominence, the note could be either first or last in order.

Serial with Map Included

MARC 21 Tag	Map Description
245 12	‡a A chart of the Great Ocean or South Sea ... in the years 1785, 86, 87 and 88 ‡h [map] / ‡c Baker sc.
773 0	$7 nnas ‡t *Universal magazine of knowledge and pleasure*. ‡g Vol. 104 (March 1799), facing page 153.

In the above example a series has been chosen using the same 773 field. The print constant, "In:," is automatically produced and appears preceding the series title. In both these examples of the 773 field, the call number for the map is the same as that for the book and will appear in the 090 (local system call number, Library of Congress schedule G) field.

Multiple Maps on One Sheet

These are treated basically as the first example provided below. But in the recording of the map titles the cataloger has a choice of methods; either the item may be described as a unit, or a separate description may be made for each separately titled part (AACR2R 3.1G).[10] While the first method is appropriate for current maps, early maps with very few exceptions should be treated individually, as in the example below:

MARC 21 Tag	Map Description
245 10	‡a Plan du Comté de Cumberland (Nouvelle-Galles du Sud) d'après les cartes Anglaises ‡h [map]
300	‡a 3 maps on 1 sheet ; ‡c 16.8 x 10.9 cm. and 8.5 x 10.7 cm. and 8.3 x 10.5 cm., on sheet 35.0 x 26.8 cm.
773 0	‡7 p1am ‡a Freycinet, Louis de, 1779-1842. ‡t Atlas du voyage de découvertes aux terres Australes. ‡d Paris, 1811. ‡g No. 9.
501	‡a With: Plan du Comté de Cumberland. -- Plan des Îles Berthier.

Note that the USMARC 501 field is used for "With" items. This expression is reserved for describing two or more distinct maps contained within the same physical item, in this instance the page; it is not to be used for accompanying material. The collation statement records the existence of all the items, while the "With" note provides their details. As in the previous example, the call number is that for the atlas in which these maps appear.

Below is the entry for the second map on this sheet of three maps.

MARC 21 Tag	Map Description
245 10	‡a Plan des Îles Jerome ‡h [map] / ‡c par MM. H. Freycinet et Bernier an 1802.
300	‡a 3 maps on 1 sheet ; ‡c 8.3 x 10.5 cm. and 8.5 x 10.7 cm. and 16.8 x 10.9 cm., on sheet 35.0 x 26.8 cm.
773 0	|7 p1am Freycinet, Louis de. Atlas du voyage de découvertes aux terres Australes. Paris, 1811. ‡g No. 9.
501	‡a With: Plan des Îles Berthier. -- Plan du Comté de Cumberland.

The third plan is treated in a similar fashion to the first two. In this way each map is given the prominence due to historical items of this nature.

Detached Maps and Text

Detached maps and text come in a variety of forms. "Detached" has been given a very loose interpretation in this discussion, to include anything that might be removed from the primary item in which it is included. It also includes truly separate pieces, which were originally issued as separate items.

To refresh our memories, included in this group are:

- Loose map in pocket of book
- Map series with pamphlets or books produced separately
- Maps in soft covers
  - Loose maps and loose text
  - Tipped in maps and text

Libraries need policies to safeguard such materials, simply because loose items are in danger of being lost or removed legally or otherwise from the host.

Possible treatments are:

- Each part of the combination may be classified separately and may have a different call number according to the classification systems used for books and maps. This is applicable in libraries where differing systems are employed for maps and books.
- Each part may carry the same call number, but one part will be treated as the attachment and will have an addition to the call number to indicate this and to show where it is filed.

Of course, the book and map must be linked bibliographically in both instances.

The first situation is possible in libraries that use a different classification system for maps. A simple solution which avoids complicated cataloging is the use of a prefix to the call number–such as "AT" for attachment–which is added to the classification number of any item which has been physically separated from its host. It is useful in the situation where the different parts are housed in the same area but in different filing sequences. Using this technique, an "AT" filing sequence would be established for attachments. In a map section where the map is considered the primary document, the book would be treated as the attachment and placed on shelving located either in the closed or open stacks adjacent to the maps. It would be considered as an "Accompanied by:" item and briefly described in a note in the 500 field. The terminology, "Issued with:", has been avoided, as this phraseology is used in the 777 field.

Loose Map in Pocket of Book

The loose map in the pocket of a book is the most obvious candidate for being lost. These items are put in pockets for a variety of reasons but the maps are usually of some significance, so it is wiser to treat such maps as items to be individually cataloged and housed in the map files rather than leaving them folded up in a pocket, where they will certainly begin to deteriorate on the folds. The associated texts should be shelved close by, or at least in the same general area of the library. The following example illustrates a method for treating such material. The 773 field might be considered a possible choice for dealing with such items if it can be argued that a map in a pocket is considered to be a physically contained component part. But if the decision is to actually remove the item from the pocket and house it elsewhere, then it is probably more accurate to describe as an "accompanying" item and use the 500 field to express the relationship.

MARC 21   Map Description
  Tag

245 10   ‡a Map of the south eastern portion of Australia shewing the routes of the three

           expeditions, and the surveyed territory ‡h [map] / ‡c Maj. T.L. Mitchell del. ; engraved

           by J.W. Lowry.

300   ‡a 1 map : ‡b col. ; ‡c 54.0 x 72.5 cm., folded to 21.0 x 17 cm. + ‡e book.

500   ‡a Accompanied by: Three expeditions into the interior of Eastern Australia / by Major

           T.L. Mitchell. 2nd ed. London : T. & W. Boone, 1839. (2 v. : ill., maps ; 23 cm.). [Call

           no. for book]

The map in this instance was issued as a physically separate item, and a double-fronted hard cover was fashioned for the book into which the map was placed. Maps issued in this way disappear all too frequently and are more safely treated as loose maps and placed in the map files. The above example fits beautifully into the "Issued with" situation (MARC 777 field) which would be most appropriate for such situations, if your library system can support this field. The examples give both methods which are equally acceptable. In the case of the 777 field being used, a display constant, "Issued with:", can be generated which will precede the book citation in this field.

MARC 21   Map Description
  Tag

245 10 |a Map of the south eastern portion of Australia shewing the routes of the three

           expeditions, and the surveyed territory |h [map] / |c Maj. T.L. Mitchell, del. ;

           engraved by J.W. Lowry.

300 |a 1 map : |b col. ; |c 54.0 x 72.5 cm., folded to 21.0 x 17.0 cm. + |e book.

777 1 |7 p1am |a Mitchell, Thomas Livingstone, 1792-1885. |t Three expeditions into the

           interior of Eastern Australia / by Major T.L. Mitchell. |b 2nd ed. |d London : T. &

           W. Boone, 1839.

An example such as this could be categorized as a book with a map, the map being an outcome of the prior existence of the book. However, it is possible that the map existed in its own right as a separately published item before the account was written. The first example given above shows the map record with a mention in the physical-description area of the book and with a note giving bibliographical details for the associated book plus its call number in the 500 note field. In a library where the book was housed in a location other than the Map Collection it could be useful for readers to know that the map issued with the book was located elsewhere. The catalog should in this case also contain a similar record for the book with the map treated as the accompanying item in the note field and with the map call number. Readers have difficulties with combination records of this type and the addition of the accompanying item's call number in the note saves many frustrations.

As AACR2R has four options for describing accompanying materials, the treatment of the collation (MARC 300) is at the discretion of the cataloger; note that since we are dealing with early maps, we have taken the liberty of using the earlier term for physical description, "collation"! The above example fits the third option in rule 1.5E1 (AACR2R 1.5E1c), which is to describe the accompanying text briefly in a 500 field and to make a collation note. The above information is sufficient to permit identification of the host item. As cataloging time is money, the aim should be an adequacy of information with minimal effort.

The first option in AACR2R rule 5E1a is to record the details of the accompanying material in a separate entry. The second option (AACR2R 1.5E1b) provides for use of multilevel description. This is dependent upon institution policy; it seems to be normally used only by national cataloging agencies. The third option is to make an "Accompanied by" note (AACR2R 1.5E1c), which quite a number of cataloging agencies use. The USMARC field 500 could be used to accommodate this note, which in effect is the AACR2R equivalent of the USMARC 777 "Issued with" note.

The last option (AACR2R 1.5E1d) is to give the numbers and types of physical units as simple statements:

1 map : col. ; 54 x 72.5 cm. + 2 v.

An option for method four permitting slightly more detail in the physical description area is described in Chapter 3 (Cartographic Materials) of AACR2R.

Map Series with Pamphlet or Books Produced Separately

Loose Map and Loose Text

Map Tipped into Soft Cover with Separate Tipped in Text

All these items are usually presented as a separately published map and text. The map may be folded to slip along with the booklet into a cover fashioned to contain a pocket. Alternatively, the map and text may be tipped in lightly to the soft card cover, both thus unfortunately being easily detached at will.

The map is considered to be the primary publication when issued with explanatory notes. Likewise, an accompanying printed index of features on a map is subordinate to the map. Therefore they would not be treated as having equal weight as publications

Map series with explanatory notes are a common format in map publishing, and are usually produced by major government map agencies. They are extremely important items in almost all map collections. It is safer to put such material in a closed stack inaccessible to users, and to separate and file each part according to format.

The use of USMARC field 777 implies that map and booklet have equal status, and records must exist for both. As was previously mentioned, the booklet is dependent on the map and therefore series of this nature should only be treated as an "Accompanied by:" note used in the 500 field.

An entry for a sheet plus accompanying text in a typical series would include these fields among others as below:

MARC 21 Tag	Map Description
245 10	‡a Australia 1:250 000 geological series. ‡n Sheet SI/55-5, ‡p Booligal ‡h [map] / ‡c Geological Survey of New South Wales.
300	‡a 1 map : ‡b col. ; ‡c 45 x 56 cm., on sheet 78 x 115 cm. + |e 1 booklet (x, 94 p. : ill., maps ; 30 cm.)
500	‡a Accompanied by booklet: Booligal 1:250 000 geological sheet, SI/55-5 / by Roger G. Cameron. St. Leonards, N.S.W. : Geological Survey of New South Wales, Dept. of Mineral Resources, 1997. (Explanatory notes : ISSN 1321-828X).

Another field which is extremely useful for maps is the "parent record entry" or 772 field. This field contains information concerning the related parent record when the target item has a vertical relationship with the parent item. Examples which spring to mind are the map supplements issued by the National Geographic Society and the Annals of the Association of American Geographers. In both cases the map and text have an independent existence and both could stand alone and would not suffer by so doing. They are in fact classic examples of publications of equal weight. USMARC 772 may also be used, in records for individual sheets in map series, to hold information for the map series to which the sheet belongs. Parker (1990) describes this well in her article of 1990 in the *Information Bulletin* of the Western Association of Map Libraries, but as this article is concerned with maps in publications, it is only alluded to here as a further reference for those concerned with the cataloging of maps issued in parts.

MARC 21 Tag	Map Description
245 10	‡a South America ‡h [map] / ‡c produced by the Cartographic Division, National Geographic Society.
300	‡a 1 map : ‡b col. ; ‡c 74 x 55 cm.
772 00	‡7 nnas ‡t National Geographic. ‡g October 1972, Vol. 142, no. 4.

CONCLUSION

In conclusion, it is obvious that better access points are needed to meet the requirements of map users. These should be for situation date and subject access in particular. Failing legitimate avenues offered by cataloging codes and rules, the map cataloger should be aware of these needs and provide the vital information through the use in particular of the note fields. In seeking methods to present both map and associated text, utilizing an informal label such as the suffix "AT" in the call number for the map for "Accompanied by" items stored in the Map Collection is simple but effective. The book item of course carries the prefix "AT" followed by the map call number. Other entries that would assist users are added author entries in older map cataloging for all persons associated with the creation of the item.

The title is in many instances the first access point sought by users wanting older maps and is vital for comparative cartography in tracing states of a particular title. There would therefore be a significant value to map users in particular for main entry for maps to be title rather than author. In printed cartobibliographies the entries consisting of individual map descriptions forming the body of the work may be accessed from indexes. These indexes are the cataloging equivalent of tracing notes and in the case of the cartobibliography on Australian maps previously discussed (Prescott 1997), there are indexes to map titles, inset-map titles, map makers, geographical areas, and subjects, and a separate index to ships' tracks. All these fields should be supplied in cartobibliographies where fortunately some of the problems impeding catalogers are not encountered.

NOTES

1. T.M. Perry and Dorothy F. Prescott, *A Guide to Maps of Australia in Books Published 1780 to 1830: an Annotated Cartobibliography* (Canberra: National Library of Australia, 1996).

2. Great Britain, Dept. of Defence, General Staff, *Manual of map library classification and cataloguing* (Feltham: Ministry of Defence, 1978), 6-1, 18-1.

3. S.W. Boggs and Dorothy Lewis, *Classification and cataloging of maps and atlases* (New York: Special Libraries Association, 1945). American Geographical Society, *Manual for the classification and cataloguing of maps in the Society's collection* (New York: American Geographical Society, 1952).

4. *Anglo American Cataloguing Rules, 2nd ed.* (Chicago: American Library Association, 1978).

5. Library of Congress. Geography and Map Division, *[Cataloging manual] Subject analysis"* (Washington, D.C.: Geography and Map Division, 19–), 2E1, p.1. Typescript.

6. Giles Martin, "Main Entry: the Argument For A Defence of Main Entry in the OPAC," *Cataloguing Australia*, 22(1,2, 1996): 21-25.

7. Pam Dunlop, "Main entry: the Argument Against The Irrelevance of the Main Entry Concept," *Cataloguing Australia* 22(1,2, 1996): 26-33.

8. Bodleian Library, Map Section, Oxford, *Selected map and book accessions* No. 528(July 1998).

9. Personal communication from Pam Dunlop, Manager, Special Materials Formats, National Library of Australia.

10. *Anglo-American Cataloguing Rules, 2nd ed.* (Chicago: American Library Association, 1978).

REFERENCES

Parker, Velma. "Multilevel Cataloging/Description for Cartographic Materials." *Information Bulletin*, Western Association of Map Libraries 21(2, March 1990): 86-96.
Prescott, Dorothy F. "Maps of Australia in Books: A Cartobibliography." *The Globe* 45(1997): 19-31.

DIGITAL CARTOGRAPHIC MATERIALS

Metadata:
An Introduction

Jan Smits

SUMMARY. With the transition from cartographic materials to spatial information the nature and amount of access data for the library field is changing. Besides bibliographic data there exists now a range of metadata, each kind for specific purposes within specific user fields. To define their relation to each other they have been put into a diagram. Through the Resource Description Framework these should all be available through a common interface for Internet-searching. To prevent confusion spatial metadata is defined. Spatial metadata introduces new elements to descriptions with new application possibilities. *[Article copies available for a fee from The Haworth Document Delivery Service: 1-800-342-9678. E-mail address: getinfo@haworthpressinc.com <Website: http://www.haworthpressinc.com>]*

Jan Smits is Map Librarian at Koninklijke Bibliotheek,[1] National Library of The Netherlands, P.O. Box 90.407, NL-2509 LK Den Haag, The Netherlands (e-mail: jan.smits@konbib.nl). He is President, Groupe des Cartothecaires de LIBER (GdC, European Map Curators Group) and IFLA Representative for the ICA Commission on Standards for the Transfer of Spatial Data.

[WARNING: As this is a field in development users should always check current literature and Websites.]

[Haworth co-indexing entry note]: "Metadata: An Introduction." Smits, Jan. Co-published simultaneously in *Cataloging & Classification Quarterly* (The Haworth Information Press, an imprint of The Haworth Press, Inc.) Vol. 27, No. 3/4, 1999, pp. 303-319; and: *Maps and Related Cartographic Materials: Cataloging, Classification, and Bibliographic Control* (ed: Paige G. Andrew, and Mary Lynette Larsgaard) The Haworth Information Press, an imprint of The Haworth Press, Inc., 1999, pp. 303-319. Single or multiple copies of this article are available for a fee from The Haworth Document Delivery Service [1-800-342-9678, 9:00 a.m. - 5:00 p.m. (EST). E-mail address: getinfo@haworthpressinc.com].

KEYWORDS. Metadata, spatial metadata, Dublin Core, RDF, migration

FROM CARTOGRAPHIC INFORMATION TOWARDS SPATIAL INFORMATION

In this digital age where cartography and GIS become more integrated and where it is possible to integrate disparate sources of information, there seems for the cartographic professions to be a shift from purely cartographic visualization to the use of spatial data in all its manifestations. For map librarians this means that in future they not only will occupy themselves with the pure visualization of spatial data in the form of traditional cartographic documents–such as maps, atlases, globes, etc.–but also with the processes used to create cartographic visualization from spatial data. As spatial data are a new phenomenon for map librarians we first have to define what we are talking about:

> [Geo]spatial data are data describing phenomena directly or indirectly associated with a location and time relative to the surface of the Earth.[2]

This means not only the common data (which can be localised) we know of and which are traditionally represented in cartographic documents, but also data such as social statistics with statistical offices, zip-codes, telephone numbers, addresses, etc. Also included within spatial information are those documents concerned with history, religion, biology, economics, etc., which can be connected with a certain locality. As all these kinds of data are added, the domain with which we (shall) occupy ourselves has greatly grown. In line with this growth lies the growing belief that GIS will be one of the common tools for map librarians in future. *Map librarianship will come closer to 'geo-informatics,' that is, a discipline concerned with the modelling of spatial data and the processing techniques in spatial information systems. Modern developments in these fields show that data acquisition and processing are becoming more and more closely related.*[3] Public services and (supporting) processing will also become more closely related as we cannot assume that our customers will know the ins and outs of processing the spatial data they request. Though the focus of activities for map librarians may still be with visualisation, it could well be that we have to enlarge the field of activity, either ourselves or in co-operation with colleagues from other departments, to the whole field of GIS and similar technologies.

Because of the developments sketched above, there also will be a shift in the accessing of purely cartographic materials towards creating and/or using bibliographic data and spatial metadata. The rest of this article will mainly be concerned with the latter.

WHAT IS METADATA

In essence metadata is data about data. Taken literally this makes almost all data except primary sources metadata and that is a bit too much. Since documentation of Internet and digital sources became an item of research, defining metadata has led to confusion. To cut a long story short we shall restrict ourselves to the documentary field. Here metadata is in the first instance data, which helps us to locate and select sources of information. They are identical to bibliographic descriptions together with access data. In libraries for a long time this was the only use for metadata, except when analytical bibliographies are concerned which might also support evaluation processes.

In this electronic age, location and selection are only a few of the processes needed to come to a decision which source is best for the purpose information is needed for. In order to fulfil this function, certain metadata also must make it possible to evaluate and/or analyse a source before a decision is made.

"Metadata" remains hard to define. This sometimes is such an ambiguous term that the Task Force on Archiving of Digital Information avoids this term.[4] Fortunately, I can make use of other bodies who do this work for us, and thus have selected two definitions of mainstream documentary fields with which we occupy ourselves. The first is the library field, the second the digital spatial community.[5]

The first definition comes from the glossary of Biblink[6] "D1.1 Metadata Formats" and reads as follows:

> [Metadata is] information about a publication as opposed to the content of the publication; [it] includes not only [a] bibliographic description but also other relevant information such as its subject, price, conditions of use, etc.

The second is the working definition as adopted by the ICA (International Cartographic Association) Commission on Standards for the Transfer of Spatial Data at their Summer meeting, 1996, in Den Haag, The Netherlands, and reads:

> Metadata are data that describe the content, data definition and structural representation, extent (both geographic and temporal), spatial reference, quality, availability, status and administration of a geographic dataset.

It seems that between these definitions there is an apparent controversy with which we have been confronted before in our everyday practise. Or it might be a lasting controversy between keepers of DLOs (Document-Like Objects[7]) and keepers of images, though I hope not! Those who work with DLOs mainly focus on location and selection, while those who work with images focus more on evaluation and analysis. In the analog world, this is books and periodicals contra visual/cartographic materials.

Another contrast is that libraries do not see a main role for producers, while the spatial data community clearly reckons with the fact that producers will create the bulk of database metadata. The spatial data community also clearly aims at the transfer of the underlying spatial data.

But as we say in the Netherlands, the soup doesn't have to be eaten as hot as it is served, or in proper English, things are sure to simmer down. The latter seems to seep through the Biblink study, as it creates a continuum from metadata for locational purposes to metadata for analytical purposes.

To prevent confusion with other disciplines it may be wise always to refer to "*spatial metadata.*"

TYPES OF METADATA

This is illustrated in the diagram, which shows a typology of metadata for cartographic and spatial data, which I modeled after one of the diagrams used in the Biblink study. The color intensity shows to what extent map curators probably will produce and/or use metadata. The easiest is to follow the diagram in the quality levels to explain the differences.

Color intensities in Band One to Four signify use by map curators/librarians/spatial data specialists

Quality	Simple	⇒	⇒	Rich
Quality Level	Band I	Band II	Band III	Band IV
Diffusion	⇒	⇒	⇐	⇐
Availability	Internet	Internet	Internet/ Intranet	Internet/ Intranet
Purpose	Location	Selection	Evaluation	Analysis
Unit of information	Individual digital information object	Logical set of digital objects; no links between documents	Publication: links between whole and parts	Databases with links between whole and parts on all levels

			Generic standards used in information world	standards used in specialist subject domains
Standards[8]	Proprietary	emerging standards	Generic standards used in information world	standards used in specialist subject domains
Form of record[9]	**Proprietary simple records**	**Dublin Core**	**ISBD**	**FGDC,CEN, ISO, ANZLIC**
Format[10]	Unstructured	Attribute value pairs **(Dublin Core DTD)**[11]	Subfields qualifiers **(MARC)**,[12] **(USMARC)**,[13] **(UKMARC)**,[14] **(UNIMARC)**[15]	Highly structured mark up **(FGDC DTD**,[16] ANZMETA)[17]
Input	Robot generated	Robot plus manual input	Manually input	High level of manual input
Protocol	http with CGI form interface	directory service protocols (whois++ with query routing (Common Indexing Protocol)	Z39.50	Z39.50 (in future with collection navigation
User Community	Internet-surfers	Producers	Libraries/ Cartographic Information Centres	Producers/ Documentation Centres/ Clearinghouses

Modified after *BIBLINK-LB 4034, D1.1 Metadata Formats. 23 December 1996.*

Band I

These are simple records, which are used mainly by robot search engines such as NetFirst, AltaVista and Infoseek for full-text Internet indexing, and tend to be associated with directory service protocols. They are mainly used by the unaware netsurfers who start out to explore the available information in a random way.

Band II

These kinds of metadata are at the moment probably the most on research agendas of traditional metadata creators. On the instigation of OCLC Inc., workshops have been held since 1995 to try to find a modus in which these metadata can be formulated.

The first workshop held in 1995 in Dublin, Ohio, found consensus on a set of elements, called since then the *Dublin Core*. It is intended to be sufficiently rich to support useful fielded retrieval but simple enough not to require specialist expertise or extensive manual effort to create. The participants agreed on:

- A concrete syntax for the Dublin Core expressed as a Document Type Definition (DTD) in Standard Generalised Mark-up Language (SGML).
- A mapping of this syntax to existing HyperText Mark-up Language (HTML) tags to enable a consistent means for embedding author-generated description metadata in web documents.

The Dublin Core Elements[18]		
Scope	**Element**	**Description**
Content	Title	The name of the resource.
Content	Subject	The topic addressed by the resource.
Content	Description	A textual description of the content of the resource
Content	Source	Objects, either print or electronic, from which this object is derived, if applicable.
Content	Language	Language of the intellectual content.
Content	Relation	Relationship to other resources.
Content	Coverage	The spatial location and/or temporal duration characteristics of the resource.
Intellectual property	Creator	The person(s) or organization primarily responsible for creating the intellectual content of the resource.
Intellectual property	Publisher	The agent or agency responsible for making the object available in its current form.

Intellectual property	Contributor	The person(s), such as editors, transcribers, and illustrators who have made other significant intellectual contributions to the work.
Intellectual property	Rights	A rights management statement.
Instantiation	Date	The date associated with the creation or availability of the resource.
Instantiation	Type	The genre of the object, such as novel, poem, dictionary, etc.
Instantiation	Format	The physical manifestation of the object, such as PostScript file or Windows executable file.
Instantiation	Identifier	String or number used to uniquely identify the object.

We can use the fifteen Dublin Core Elements to create metadata with the maps we ourselves put on the net. Most HTML-editors have possibilities to create templates, which makes creation of this metadata easy. Dublin Core records can be used in the same way as traditional CIP-records, which in a later stage can be enriched to become full MARC-records.

One of the remaining problems is to find a transfer syntax, which makes it easy to embed the Dublin Core elements. This problem is treated in the paper, *A Proposed Convention for Embedding Metadata in HTML*, which might result in a "Dublin Core DTD."[19]

During the 5th Dublin Core-meeting in Helsinki, Finland, 1997, there was a growing consensus for the use of the *Coverage* element to support search for spatially referenced resources.[20]

At this moment the best sources for literature on the Dublin Core are *D-Lib Magazine* (ISSN: 1082-9873),[21] *Ariadne* (ISSN: 1361-3200)[22] and *The Dublin Core Metadata Element Set Home Page*.[23]

The Dublin Core is not the only system available in Band Two. Other resource description models are RFC 1807,[24] IAFA[25] (Internet Anonymous FTP Archive) and SOIF[26] (Summary of Object Interchange Formats) to name but a few, but for us the Dublin Core most probably is the nearest to our everyday practices, in large part because of the work done by the Library of Congress to map the Dublin Core metadata elements to MARC 21.[27] Probably other MARCs will follow soon.

Band III

In whatever way electronic publications and spatial databases will develop, our main access vehicle most probably will remain the ISBDs and the MARC formats. We would like to offer our users an access-continuum from manuscript and printed records to electronic databases, in order that they do not have to search several differently formatted databases.

During the 9th Conference of the Groupe des Cartothécaires de LIBER in Zürich, Switzerland (1994), this author presented a paper demonstrating that it is possible with ISBD and MARC to incorporate descriptions of dynamic electronic maps and databases in our current catalogue, making some proposals to adapt to both current and future practises.[28] One of the preconditions for MARCs, however, should be that they can be easily extended, especially when it concerns coded data (e.g., Unimarc tag 100-199). All MARCs are in the process of adapting their format to be able to incorporate data, verbal or encoded, which makes retrieval of electronic documents and data easier. They are also extending the formats to be better able to contain the bibliographic data.

There are currently available two crosswalks, the first created by the Geography and Map Division of the Library of Congress and the second by the Alexandria Digital Library–*Crosswalk: FGDC Content Standards for Digital Geospatial Metadata to USMARC*[29] and *Crosswalk: USMARC to FGDC Content Standards for Digital Geospatial Metadata.*[30] I expect that when the ISO and CEN standards are approved, other crosswalks will be created.

Other formats in this range are TEI[31] (Text Encoding Initiative) independent headers and EDI[32] (Electronic Data Interchange) messages.

Band IV

Digital spatial databases have been created from the late 1970s onwards, and have nowadays reached the phase where there is countrywide coverage (on municipal, provincial and state level) and through GIS they can be easily integrated with other databases. Usually they are a continuation of existing analogue processes, but for the fact that they are often in vector-format and that they are built up of layers of information which can be manipulated independently from each other or in concert with each other. To extend operability, many producers are digitizing their existing analogue data, usually in raster-format. Seeing the benefit of promotion and the need for a higher return on operating costs, producers have started to think of ways to access these data. As the economic stakes are higher then before, they have sought to create a description system, which incorporates not only data usually associated with ISBDs but also data, that could help their users to evaluate and analyse the fitness of use and quality of the digital spatial data offered.

Examples of these standards are ANZLIC[33] (Australia & New Zealand), FGDC[34] (USA), CEN[35] (Europe) and ISO[36] (global).

RESOURCE DESCRIPTION FRAMEWORK (RDF)
AND EXTENSIBLE MARKUP LANGUAGE (XML)

The Resource Description Framework (RDF) is an infrastructure that enables the encoding, exchange and reuse of structured metadata. RDF is an application of XML that imposes needed structural constraints to provide unambiguous methods of expressing semantics. RDF additionally provides a means for publishing both human-readable and machine-processable vocabularies designed to encourage the reuse and extension of metadata semantics among disparate information communities. The structural constraints RDF imposes, to support the consistent encoding and exchange of standardised metadata, provide for the interchangeability of separate packages of metadata defined by different resource description communities.[37]

In other words, RDF is designed to provide an infrastructure to support metadata across many web-based activities. It is a result of a number of metadata communities to bring together their needs to provide a robust and flexible architecture for supporting metadata on the Internet and WWW.

One obvious application for RDF is in the description of web pages. This is one of the basic functions of the Dublin Core initiative. The Dublin Core has been a major influence on the development of RDF. An important consideration in the development of the Dublin Core was to allow simple descriptions, but also to provide the ability to qualify descriptions in order to provide both domain specific elaboration and descriptive precision. The RDF Schema mechanism proposed provides a machine-understandable system for defining 'schemas' for descriptive vocabularies like the Dublin Core. It allows designers to specify classes of resource types, property types to convey descriptions of those classes, and constraints on the allowed combinations of classes, property types, and values.[38]

The same is true of other Document Type Definitions (DTD), like the schema for FGDC-metadata,[39] but maybe also for Internet and WWW-compatible MARC-records.

To be able to read the descriptions, a syntax is needed. The WorldWideWeb-consortium (W3C) for this purpose is developing eXtensible Markup Language (XML). The XML syntax is a subset of the international text-processing standard SGML (Standard Generalized Markup Language–SGML) specifically intended for use on the Web.[40] It promises to make the web smarter by including machine-readable information about the structure and content of Web pages. As a basic rule of XML, content and presentation are separated. XML-tags contain no information about how they should be displayed. Before a web-browser can display an XML-page, it will have to get the corresponding "stylesheet" first and format the page accordingly.[41]

NEW ELEMENTS

Most spatial metadata standards will have the same kind of structure as the image-map of the FGDC standard (first edition, 1994; second edition, 1998).[42] The amount of fields can vary, depending upon how groups of metadata are aggregated, and all fields are divided into elements. Fields 1, 4, and 6 can be easily translated to ISBDs, field 7 pertains to the standard itself, but the other fields are new to librarians and contain mainly technical information. Some of this new information is explained below.

Temporal Information

For a long time, three-dimensional properties of maps were the most important data for documentary systems. Only a few library-systems were able to incorporate searchable temporal characteristics. Digital spatial data-sets can be easily compared or overlayed in time-series to show situational changes than traditional maps, and therefore temporal characteristics become more important. Standards for spatial data provide more possibilities to include this aspect and make it searchable.

Data Quality Information

Good-quality information is essential to both the use and reuse of spatial information. This information is added as quality parameters to data to enable a better evaluation of the fitness-of-use of data by providing a model by which the information necessary to judge the quality of geographic information against its specification can be defined. Quality parameters are, e.g., accuracy, logical consistency, derivation, and quantitative measures of quality against a pre-established scale or relative to similar information.

Quality information concerning analog maps can usually be found in the notes area of ISBDs.

Feature and Attribute Information

This is information about the information content of the data set, including the entities or feature types, their attributes, and the domains from which attribute values may be assigned. A feature or entity type is a characteristic of a spatial object (point, line, polygon, etc.) by which this type can be distinguished or identified. An attribute is a non-spatial characteristic of a spatial object.

POTENTIAL OF METADATA

The major uses of metadata, according to the FGDC, are:

- to maintain an organisation's internal investment in geospatial data,[43]
- to provide information about an organisation's data holdings to data catalogues, clearinghouses, and brokerages,
- to provide information needed to process and interpret data to be received through a transfer from an external source.

These uses can be translated in the sense that metadata are part of a spatial-data model which helps to facilitate interoperability and data exchange. The model is completed with information-technology standards. Together they form the building blocks of national, supranational and global spatial data infrastructures.

I would add at least the following additional purposes:

- to provide information to assist in developing migration processes and to check whether these processes have functioned adequately,[44] and
- to be a source for bibliographic data.

Metadata can help in ensuring the level of integrity of the data after necessary manipulations to preserve them for future use, e.g., conversion, transformation, migration, etc. They help to answer the question whether the data have the same information value and quality as before the manipulations took place and at the same time whether unavoidable losses can be qualified. The same is true for the accompanying metadata itself.

Secondary benefits for users are:

- greater precision of results
- fielded search
- boolean support
- less information overload

Many electronic datasets are complemented with metadata in a postcoordinated instance. To fulfil the potential use of metadata, it is necessary that producers begin with creating metadata for a dataset from the first instance of

FIGURE 1

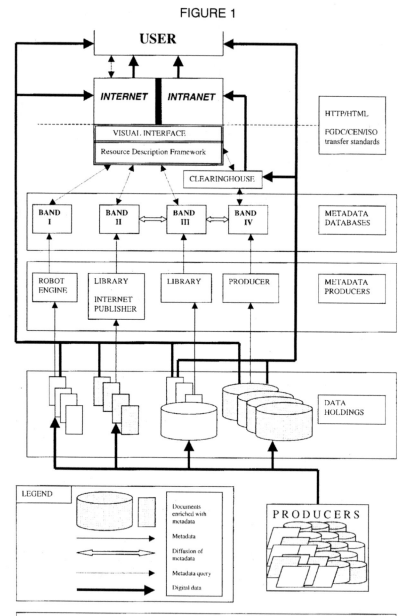

SIMPLE (META)DATA MODEL FOR ANALOGUE AND ELECTRONIC DOCUMENTS AND INFORMATION

creating the dataset. When they deliver these datasets to third parties the metadata should inherently be part of the dataset.

Should producers develop technologies to authenticate their metadata, this will provide legal documentation to protect their organisation if conflict arises over the use/misuse of their geospatial data. (See Figure 1.)

To create the shortest route for users, this model should have a single interface through which (indexes to) Dublin Core metadata, bibliographic data, and spatial metadata can be queried through graphical tools (maps), area-directories and subject-directories. In this way, researchers will have only one-stop-windows to survey all publicly and commercially available spatial data.

ENDNOTES

1. National Library of The Netherlands, Department for Cartographical Documentation, P.O. Box 90.407, NL-2509 LK Den Haag. E-mail: jan.smits@konbib.nl

2. McKee, Lance. *Building the GSDI*–Discussion paper for the September 1996 Emerging Global Spatial Data Infrastructure Conference. 1998.

3. Molenaar, Martien. *To see or not to see.* Inaugural address for the acceptance of the Chair in Geoinformatics and Spatial data Acquisition at the ITC. 1998.

4. In note 10 to its report *Preserving digital information* <http://lyra.rlg.org/ArchTF/tfadi.index.htm> [Accessed: May 10, 1998] the Task Force notes: *Metadata, which refers to information about information, is sometimes used as a generic term for systems of reference. We avoid use of the term in this report because it conveys a tone of jargon and because its use in the literature is varied and imprecise. In its report on Preserving Scientific Data on Our Physical Universe (1995), for example, the National Research Council uses the term to include any and all documentation that serves to define and describe a particular scientific database. Other uses of the phrase elsewhere in the literature are closer to the more limited sense of referential systems that we use here to mean systems of citation, description and classification. The preference for the word metadata in those other cases appears to flow from the felt need to emphasize the special referential features needed in the digital environment and to distinguish those special features from those of more traditional systems of citation, description and classification.*

5. The World Wide Web Consortium defines metadata as "*. . . machine understandable information about web resources or other things.*"
See for further information:
Berners-Lee, Tim. *Metadata architecture: documents, metadata, and links.* 1997. <http://www.w3.org/DesignIssues/Metadata.html> [Accessed: May 10, 1998]
Other definitions are:
Ancillary data characterising a data set (e.g., date of production, instrument type, etc.).
[Index of Remote Sensing Terms / Committee on Earth Observation Satellites]

Information describing a data set, including data user guide, descriptions of the data set in directories, catalogs, and inventories, and any additional information required to define the relationships among these.
[Glossary of Oceanography and the Related Geosciences with References / Steven K. Baum (Texas Center for Climate Studies, Texas A&M University)]
Information describing the characteristics of data; data or information about data; descriptive information about an organization's data, data activities, systems, and holdings.
[Glossary of modelling and simulation (M&S) terms / U.S. Department of Defense]

(1) Information about a data set which is provided by the data supplier or the generating algorithm and which provides a description of the content, format, and utility of the data set. Metadata provide criteria which may be used to select data for a particular scientific investigation.

(2) Information describing a data set, including data user guide, descriptions of the data set in directories, and inventories, and any additional information required to define the relationships among these. Source: ESADS, EPO, IWGDMGC. (Earth observation system data and information system glossary / EOSDIS (NASA)]
All definitions taken from: Global Change Master Directory List of Earth Science Acronyms and Glossaries. <http://gcmd.nasa.gov/acronyms.html> [Accessed: June 25,1998]

6. <http://hosted.ukoln.ac.uk:/biblink/wp1/d1.1/> [Accessed: May 10, 1998]
Project BIBLINK was launched on 1st April 1996 with funding from the European Commission's Telematics Applications Programme. It aims to establish a relationship between national bibliographic agencies and publishers of electronic material, in order to establish authoritative bibliographic information that will benefit both sectors. The concept crystallised from the work of an EU concerted action known as Co-BRA. This forum recognised that the significant growth in electronic publishing raised issues that needed to be addressed at an international level. Project BIBLINK will call upon the bibliographic expertise of the national libraries of Europe, working in conjunction with partners in the book industry, to examine the way in which electronic publications are described for catalogues and other listings.

7. Though the term is much in use on Internet sites concerning the documentation of Internet sources it is hard to find a definition thereof. After reading through some 30 Internet sources I would describe a DLO as electronic versions of traditional analogue text-based materials, titled and all. The following quotes may lead to a better understanding of the nature of DLOs:
Document-like objects which share the characteristic that they can, but need not be, adequately represented in a print format, examples being pure text, textual images, text with printable illustrations, and photographs which can be printed. Pockley, Simon. *Killing the duck to keep the quack.* 1997. <http://www.cinemedia.net/FOD/FOD0055.html> (Accessed: June 25, 1998]
Non-document-like objects, on the other hand, include such resources as virtual experiences, databases (including ones that generate document-like outputs), business graphics, CAD/CAM or geographic information generated from database values, and interactive applications which might have different content for each user. In the context of image discovery, these sources do not "contain" images as much as they

"generate" images. The images they generate may be described as fixed document-like objects, but the metadata required to describe them (the systems doing the generating) are distinct. Weibel, Stuart, and Eric Miller. *Image Description on the Internet. A Summary of the CNI/OCLC Image Metadata Workshop,* September 24-25, 1996, Dublin, Ohio. 1996. <http://www.dlib.org/dlib/january97/oclc/01weibel.html> [Accessed: June 25, 1998]

8. This field shows the kind of standard in use at the moment.

9. This is the way in which records are presented, e.g., an ISBD-description.

10. A format is a coding system for input, storage and processing of data by means of a computer.

A review of description formats for various kinds of metadata can be found in the document, *A review of metadata: a survey of current resource description formats* <http://www.ukoln.ac.uk/metadata/DESIRE/overview/rev_ti.htm>, which is deliverable RE 1004 of the DESIRE project (Development of a European Service for Information on Research and Education).

11. <http://www.oclc.org:5046/~weibel/html-meta.html> [Accessed: May 10, 1998]

12. <http://www.ukoln.ac.uk/metadata/DESIRE/overview/rev_14.htm> [Accessed: May 10, 1998]

13. <http://www.ukoln.ac.uk/metadata/DESIRE/overview/rev_15.htm> [Accessed: May 10, 1998]

14. <http://www.ukoln.ac.uk/metadata/DESIRE/overview/rev_16.htm> [Accessed: May 10, 1998]

15. <http://www.ukoln.ac.uk/metadata/DESIRE/overview/rev_17.htm> [Accessed: May 10, 1998]

16. <http://www.fgdc.gov/Clearinghouse/Reference/encoding797.html> [Accessed: May 10, 1998]

17. <http://www.environment.gov.au/database/metadata/anzmeta/anzmeta-1.1.dtd> [Accessed: May 10, 1998]

18. This table has been constructed from an amalgamation of the tables in: *Description of Dublin Core Elements* <http://purl.oclc.org/metadata/dublin_core_elements> [Accessed: June 25, 1998], and Dempsey, Lorcan and Stuart L. Weibel (1996). The Warwick metadata workshop: a framework for the deployment of resource description. <http://www.ukoln.ac.uk/dlib/dlib/july96/07weibel.html> [Accessed: June 25, 1998]

19. The Dublin Metadata workshop of March 1995 and the Warwick Metadata Workshop of April 1996 <http://mirrored.ukoln.ac.uk/lis-journals/dlib/dlib/dlib/july96/07weibel.html> [Accessed: May 10, 1998] were convened to promote the development of consensus concerning network resource description across a broad spectrum of stakeholders, including the computer science community, text mark-up, and librarians. The result of the first workshop–the Dublin Core Metadata Element Set–represents a simple resource description record that has the potential to provide a foundation for electronic bibliographic description that may improve structured access to information on the Internet and promote interoperability among disparate description models. Its major significance, however, lies not so much in the precise character of the elements themselves, but rather in the consensus that was achieved across the many disciplines represented at the workshop. The latest update on the

Dublin Core Metadata Element Set can be found under <http://purl.oclc.org/metadata/dublin_core_elements> (Accessed: May 10, 1998].

20. For discussions see: <http://alexandria.sdc.ucsb.edu/public-documents/metadata/dc_coverage.html> [Accessed: May 10, 1998]

21. <http://www.dlib.org/> [Accessed: May 10, 1998]

22. <http://www.ariadne.ac.uk/> (Accessed: May 10, 1998]

23. <http://purl.oclc.org/metadata/dublin_core> [Accessed: May 10, 1998]

24. <http://www.cis.ohio-state.edu/htbin/rfc/rfc1807.html> [Accessed: July 1, 1998]

This RFC [Internet Request for Comments] defines a format for bibliographic records describing technical reports. This format is used by the Cornell University Dienst protocol and the Stanford University SIFT system. The original RFC (RFC 1357) was written by D. Cohen, ISI, July 1992. This is a revision of RFC 1357. New fields include handle, other_access, keyword, and withdraw.

25. <http://www.man.ac.uk/MVC/SIMA/MMFFDB/IAFA-help/document.html> [Accessed: July 1, 1998]

Anonymous FTP Archives have been a popular method of making material available to the Internet user community for some time. In recent years they have been joined by other mechanisms such as Gopher and HTTP servers to provide archives of information and software on the Internet. This document specifies a range of indexing information that can be used to describe the contents and services provided by such archives This information can be used directly by the user community when visiting parts of the archive. Furthermore, automatic indexing tools can gather and index this information, thus making it easier for users to find and access it. <http://www.roads.lut.ac.uk/Reports/iafa-draft/iafa-draft.txt> (Accessed: July 1, 1998]. This is an Internet Draft not meant for citation.

26. <http://www.cube.net/harvest-docs/user-manual/node141.html> [Accessed: May 10, 1998]

27. <http://www.loc.gov/marc/dccross.html> [Accessed: May 10, 1998]

The document describes a crosswalk between the fifteen elements in the Dublin Core Element Set on the one hand and both MARC bibliographic data elements and GILS attributes on the other. The crosswalk may be used in conversion of metadata from some other syntax into MARC 21. For conversion of USMARC into Dublin Core, many fields would be mapped into a single Dublin Core element.

For further information on this problem, see:

Caplan, Priscilla, and Rebecca Guenther. Metadata for Internet resources: The Dublin Core Metadata Element Set and its mapping to USMARC. In: *Cataloging & Classification Quarterly 22(3/4,1996): 43-58.*

28. Smits, Jan (1994). "Describing Geomatic Data Sets with ISBD and UNIMARC: Problems and Possible Solutions." *ERLC, The LIBER Quarterly* 5(3,1994): 292-311. Also: <http://www.konbib.nl/kb/skd/liber/articles/lmeta-01.htm> [Accessed: May 10, 1998]

29. <http://alexandria.sdc.ucsb.edu/public-documents/metadata/fgdc2marc.html> [Accessed: June 26, 1998]

30. <http://alexandria.sdc.ucsb.edu/public-documents/metadata/marc2fgdc.html> [Accessed: May 10, 1998]

31. <http://www.uic.edu:80/orgs/tei/> [Accessed: May 10, 1998]

32. <http://midir.ucd.ie/~mbsft-3j/gvt.html> [Accessed: May 10, 1998]

33. <http://www.anzlic.org.au/metaelem.htm> [Accessed: June 25, 1998]

34. <http://geochange.er.usgs.gov/pub/tools/metadata/standard/metadata.html> [Accessed: June 25, 1998]

35. <http://forum.afnor.fr/afnor/WORK/AFNOR/GPN2/Z13C/PUBLIC/WEB/ENGLISH/prog.htm>. To view the pre-standard a password is needed.

36. To get access to the documents and the document catalog, you need user-id and password. Please contact your national standards agency to get this information.

37. Miller, Eric. *An Introduction to the Resource Description Framework.* In: D-Lib Magazine, May 1998. <http://www.dlib.org/dlib/may98/miller/05miller.html> [Accessed: July 31, 1998]

38. W3C. *Resource Description Framework (RDF) Schemas,* W3C Working Draft 9 April 1998. <http://www.w3.org/TR//WD-rdf-schema/> [Accessed: July 31, 1998]

39. See note 16.

40. W3C. *Extensible Markup Language (XML^{TM}).* 1998. <http://www.w3.org/XML/> [Accessed: July 31, 1998)

41. Werf, Titia van der. *Metadata and libraries: background and introduction.* 1998. <http://www.konbib.nl/persons/titia/publ/index.html> (Accessed: July 31, 1998]

42. <http://www.its.nbs.gov/nbs/meta/meta.htm> [Accessed: May 10, 1998]

43. This means also facilitating the organization and management of their data-holdings.

44. Though migrating is currently a technology beyond reach it will hopefully be possible in the not too long future.
See for further information:
Rothenberg, Jeff. *Ensuring the longevity of digital information.* 1998. <http://www.clir.org/programs/otheractiv/ensuring.pdf> [Accessed: January 7, 1999]. This is an expanded version of the article that appeared in the January 1995 *Scientific American* 272: 24-29.

Spatial Metadata:
An International Survey
on Clearinghouses and Infrastructures

Jan Smits

SUMMARY. Consistency and interoperability are objectives when creating standards for spatial metadata. Besides FGDC-metadata standards some other international standards are in use or will be in use soon. The use of these standards forms the basis for geospatial data infrastructures (GDI) and clearinghouses. Though most GDIs and clearinghouses are in the planning stage, the contours of regional and a global geospatial data infrastructure (GGDI) are slowly emerging. Maps should be part of the interfaces, which provide access to the GDIs and clearinghouses. *[Article copies available for a fee from The Haworth Document Delivery Service: 1-800-342-9678. E-mail address: getinfo@haworth pressinc.com <Website: http://www.haworthpressinc.com>]*

KEYWORDS. Spatial metadata, Geospatial Data Infrastructure, geospatial clearinghouses, GGDI, GSDI, Visual Geographical Interface

Jan Smits is Map Librarian at Koninklijke Bibliotheek,[1] National Library of The Netherlands, P.O. Box 90.407, NL-2509 LK Den Haag, The Netherlands (e-mail: jan.smits@konbib.nl). He is President, Groupe des Cartothecaires de LIBER (GdC, European Map Curators Group) and IFLA Representative for the ICA Commission on Standards for the Transfer of Spatial Data.

[WARNING: As this is a field in development users should always check current literature and websites.]

[Haworth co-indexing entry note]: "Spatial Metadata: An International Survey on Clearinghouses and Infrastructure." Smits, Jan. Co-published simultaneously in *Cataloging & Classification Quarterly* (The Haworth Information Press, an imprint of The Haworth Press, Inc.) Vol. 27, No. 3/4, 1999, pp. 321-342; and: *Maps and Related Cartographic Materials: Cataloging, Classification, and Bibliographic Control* (ed: Paige G. Andrew, and Mary Lynette Larsgaard) The Haworth Information Press, an imprint of The Haworth Press, Inc., 1999, pp. 321-342. Single or multiple copies of this article are available for a fee from The Haworth Document Delivery Service [1-800-342-9678, 9:00 a.m. - 5:00 p.m. (EST). E-mail address: getinfo@haworth pressinc.com].

321

CONSISTENCY AND INTEROPERABILITY

The basic objective of standardization in this field is to enable geographic information to be accessed by different users, applications and systems, and from different locations. This requires a standard way of defining and describing this information, a standard method for structuring and encoding it, and a standard way of accessing, transferring and updating it via geographic information processing and communication functions, independent of any particular computer system. Standardization enables consistent implementations across multiple applications and systems, but permits different implementation technologies to be used for storing data in computer systems.[2]

Developments within CEN/TC 287[3] (Comité Europèen de Normalisation/ Technical Committee) and ISO/TC 211[4] (International Organization for Standardization/Technical Committee) show integrated and consistently structured families of standards for information concerning objects or phenomena that are directly or indirectly associated with a location relative to the Earth, that is, geographic information. They promote at the same time international interoperability. After "strong standards," this is one of the best solutions for interoperability.[5] Furthermore, there is a need to develop an open-GIS environment which makes the processed data compatible with CEN and ISO standards.[6]

Creating standards unfortunately may lead to Babel-like confusion as technical terms deriving from many fields and cultures are involved and many new terms are introduced, not to mention the fact that language itself is often a barrier. To help reduce the possible confusion ISO/TC 211 has made available two Webpages of terms and definitions pertaining to respectively the standards developed by ISO/TC 211 and CEN/TC 287.[7] Another list is included in the DIGEST standard.[8] At the moment of writing most definitions have the status of working definitions and it is hoped that final definition lists stay available on the Internet, to promote international understanding of geographic information standards and better understanding of geographic information concepts among other disciplines. At the same time keywords in "dictionaries" used in standards (such as those in, e.g., subject classification) should be complemented with definitions to prevent confusion.

GEOSPATIAL DATA INFRASTRUCTURES (GDI)

Too many times it is thought that infrastructure used in this sense signifies the physical possibility of transporting information. But when we talk about data infrastructures we mean a conglomerate encompassing content (data), physical environment (hardware, telecommunication lines, etc.), software (to

enable creating data, descriptions thereof, and communication and transfer), standards, policies (political, economic and social drives) and people (to create, upkeep and use data).

The statutes of the PCGISIAP (see 2.1 below) give a rather good summary of the goals of GDI's:

a. institutional framework, which defines the policy, legislative and administrative arrangement for building, maintaining, accessing and applying standards and fundamental datasets
b. technical standards, which define the technical characteristics of fundamental datasets
c. fundamental datasets, which include the geodetic framework, topographic databases and cadastral databases
d. technological framework, which enables users to identify and access fundamental datasets forming the basis of a national or regional land administration, land rights and tenure, resource management and conservation and economic development, that support the organization and the analysis of a range of spatial and related information for a wide range of social, economic and environmental purposes.[9]

Though most of us will busy ourselves with only part of this conglomerate, it is helpful to know the larger context we are working in. At the front end much has already been done–numerous standards and (meta)datasets have been produced already and, though with effort, we are able to transfer the (meta)datasets. But in the area of policies and usage much promotion has yet to be done. Social awareness of the value of geospatial data is at the top of list.[10]

There are many perspectives from which GDIs are built–data-driven, technology-driven, institutional, market-driven, and application-driven. Two perspectives are predominant–data-driven[11] for the "top-down" approach and market-driven[12] for the "bottom-up" approach. Non-governmental/not-for-profit organizations (such as libraries) and the academic sector are, unfortunately, rather under-represented in these initiatives. This may be due to the fact that they are not well organised as a group to be stakeholders or that their function is not appreciated in economically value-added environments in which GDIs seem to develop. Increasingly these organizations need to be proactive in the role of defending the public right to certain kinds of information and democratising the rights of access.

As part of the proposed EGII (European Geographic Information Infrastructure) there is a project defined as ESMI (European Spatial Metadata Infrastructure). This is an initiative by several European public and private organisations to establish a framework for the distribution of geographic information by creating a universal metadata service. ESMI is part-funded as an implementation phase project of the INFO2000 Programme of the European Commission DG XIII-E.[13] This author has the uncomfortable feeling that the term, "universal metadata," restricts itself to metadata concerning digital datasets. It would be a loss of effort through the past twenty years and more if the hundreds of thousands of digital bibliographic records of hard-copy maps held in libraries and archives are not included one way or the other into this metadata service. Fortunately the Alexandria Digital Library does not discriminate against hard-copy materials as more than 300,000 bibliographic records have been downloaded into its metadata catalog.[14]

Since 1996, annual high-level conferences have been held to provide for the emergence of a Global Spatial Data Infrastructure (GSDI).[15] The conferences are still in the formative stages of defining GSDI and its components and coordinating regional/national/international initiatives. GSDI is closely linked with the Global Mapping project.[16] Except for global initiatives like the Global Change program and international political co-operation GSDI for libraries and researchers will be in first instance an enabling infrastructure to locate certain digital datasets, unless we can also integrate our electronic map catalogues and through these can make our hard-copy holdings available.

Because transfer and metadata standards are basic tools for GDIs, the next paragraph will describe some of the international efforts outside the United States currently underway in this field. For information on the Content Standard for Digital Geospatial Metadata (first edition, 1994; second edition, 1998), formulated by the United States Federal Geographic Data Committee, see the Committee's Website (http://www.fgdc.gov) and in these special issues on cartographic-materials cataloging the articles by Welch and Williams and by Larsgaard, which discuss how these fields are used in cataloging.

NON-US INTERNATIONAL METADATA STANDARDS

ANZLIC (Australia New Zealand Land Information Council)

In 1994, ANZLIC adopted a policy on the Transfer of Metadata. The Policy is intended to apply to the highest summary level of metadata (called "core metadata") used in directory systems at the jurisdiction or national level, and not necessarily to all metadata. The United States approach, developed by the previously mentioned U.S. Federal Geographic Data Committee (FGDC), specifies the structure and expected content of some 220 items (elements) which are intended to describe digital geospatial datasets adequately for all purposes. In comparison, the ANZLIC approach is deliberately less ambitious than what has been attempted in the US. Arguments advanced in support of the more modest objective rely on experience to date with the creation of high-level directories in Australia. While ANZLIC has not adopted the US approach, the Australia New-Zealand framework is, as far as possible, consistent with the guidelines on digital geospatial metadata produced by the U.S. FGDC and with the Australia-New Zealand Standard on Spatial Data Transfer AS/NZS 4270.[17]

CEN/TC 287

As many representatives of European countries are participating in the Working Groups of CEN/TC 287 and as signatories are obliged to implement CEN standards once they have become official, there are scarcely any European countries which are creating their own independent standards. Seeing that this standards issue started in the early 1990s and that they will probably be finished in 1999, this is a most efficient way of working seen from the European perspective. The document *CEN/TC 287 N 577*[18] gives an overview of the twenty participating and fourteen affiliated (mainly central and eastern European, with the exception of former USSR states) countries participating and the program of the five working groups.

This family of ENs (European Standard) is intended for use by suppliers of data and developers of systems and applications as well as users. It is based on existing information-systems standards and methodologies, which shall be used where appropriate, especially those relating to open systems interconnection. To enable databases and applications that are different in structure, form and content to interconnect and inter-operate, a distinction is made between internal and external applicability of this standard, with the major emphasis placed on the external aspects. Additional benefits may be gained from application of this standard internally to databases and applications. How the standards are being created can be found in the document *Overview.*[19] The standards will be finalised and voted upon probably in 1999.

ISO/TC 211

It is not surprising that initially this standard was based on ANZLIC, CEN/TC 287, FGDC, DIGEST and S-57. The metadata elements are divided into these sections:

- Identification Information
- Data Quality Information
- Spatial Reference Information
- Spatial Representation Information
- Feature and Attribute Information

- Distribution Information
- Metadata Reference Information

As no standard can meet all needs, the ISO metadata standard provides ways for users to extend their metadata and still ensure interoperability.[20] Together with information-technology standards, the standards of ISO/TC 211 should serve as a tool for the GSDI.[21]

Some metadata standards are incorporated in data-transfer standards, which are created by specialist bodies such as:

1. IHO: DX-90 (International Hydrographic Office Transfer Standard for Digital Hydrographic Data).

This standard is also known as S-57. An initiative under the European Communities called ECHO 926 is bringing the European countries together to develop a common, networked electronic-chart infrastructure. All of the countries have adopted IHO's S57 standard and are moving into full ENC (Electronic Navigation Chart) production of their territorial waters. A similar co-operative initiative–to create ENCs of the St. Lawrence Seaway system–is now underway between the United States and Canada.[22] Current information on the standard is available on the Website *IHO News.*[23]

2. NATO: DIGEST (Digital Geographic Information Exchange Standard)

The initial goal of the DGIWG (Digital Geographic Information Working Group), to establish a comprehensive exchange standard for DGI, is complete with the issuance of DIGEST's edition 1.1. Efforts are being made to identify and resolve differences between DIGEST and other developing exchange standards. Over the years, DGIWG has become a unique international forum for other organizations (such as NATO) to accomplish detailed technical work in less time than normally required for NATO standardization activities. It is anticipated that NATO or other groups may take advantage of DGIWG to develop and test other dataset types. The issuance of the Digital Chart of the World (DCW) is the first major dataset published in compliance with DIGEST. As users become familiar with the data and DIGEST, they will certainly propose further improvements.[24] The full text of the DIGEST standard Edition 2.0 June 1997[25] is available on the WWW.

For other standards see *Spatial Database transfer standards 2*[26] and the European Communities' *Open Interchange Initiative's* Webpage on geographic data exchange standards.[27]

INFRASTRUCTURES AND CLEARINGHOUSES

Clearinghouses can be central (meta)data warehouses, but currently they are mainly brokers for metadata, though it is also possible that they function only as a common interface. More common in future will be that a clearing-house is part of a distributed network where each agency stores its own data and metadata.

Clearinghouses are becoming part of national and international geospatial data infrastructures (GDI). The latter are networks of databases, linked by common policies, standards and protocols to ensure compatibility. GDI not only provides improved access to geospatial data, but also helps investors to apply their capital more efficiently. This is shown in a report by the Economic Studies and Strategies Unit of Price Waterhouse on the economic benefits arising from the acquisition and maintenance of Australia's land and geographic information. It was estimated that for the period 1989-1994 approximately $1 billion has been spent in Australia on investment in geospatial data. This investment produced benefits within the economy in the order of $4.5 billion. The study also found that this investment has saved users approximately $5 billion.[28]

Today, the information traditionally held on maps, as well as other forms of geographic information, can be stored in digital form, enabling it to be handled by computers. Geographic information systems (GIS) enable different kinds of digital geographic information to be linked, so users can extract and analyse geographic information to support political, economic and scientific decision-processes. It is not surprising therefore that especially GIS-bodies and associations are involved with the creation of geospatial data infrastructures. This domination will show itself clearly within the next paragraphs.

Through EUROGI[29] (European Umbrella Organisation for Geographical Information), there is an overview available of Spatial Data Infrastructure Initiatives (SDI).[30] Most of the following SDIs are in the formative stage and will produce a viable infrastructure only after the year 2000, so information sometimes is focussed on the organising body.

1. Africa

1.1 AFRICAGIS

This is a programme under the aegis of UNITAR (United Nations Institute for Training and Research), mainly focussed on GIS-literacy and the building of an environmental spatial data infrastructure.[31]

1.2 South Africa: NSIF, National Spatial Information Framework[32]

Since 1998 the Southern Africa Metadata Clearinghouse Gateway[33] is operating using the Z39.50 retrieval tool, and containing metadata from several state departments. The FGDC standard is used, but migration to the ISO (TC211) standard is planned to take place before the end of 2000. Since June 15, 1998 a metadata audit takes place in the governmental and private sectors which in due time will lead to more metadatasets being available.

2. Asia and Pacific

2.1 PCGIAP (Permanent Committee on GIS Infrastructure for Asia and the Pacific)

The aims of the Committee are to maximise the economic, social and environmental benefits of geographic information in accordance with Agenda 21 by providing a forum for nations from Asia and the Pacific to:

a. co-operate in the development of a regional geographic information infrastructure; and
b. contribute to the development of the global geographic information infrastructure.[34]

The PCGIAP's vision for the Asia-Pacific Spatial Data Infrastructure (APSDI) is of a network of databases, located throughout the region, that together provide the fundamental data needed by the region in achieving its economic, social, human resources development and environmental objectives. At the moment only a policy-paper is available.[35]

2.2 Japan: GIDD, Geographical Information Directory Database[36]

Almost all GIDD Webpages, unfortunately for English-only readers, are available only in Japanese. Reading without Unicode[37] is impossible. More information is available from the Japanese Geographical Survey Institute.[38] Since 1995 there is a spatial infrastructure in preparation, sponsored by the National Spatial Data Infrastructure Promoting Association.[39]

2.3 Malaysia: NaLIS, National Infrastructure for Land Information System[40]

The whole Website is bilingual and contains a metadataserver[41] with a clickable map. *The NaLISCO Siri 1.0Kod NaLIS bagi Pencaman Lot Tanah* standard is used for descriptions and seems to be an independent standard. Though the site also promises the possibility to view data registration, it is only open to government department/agencies.

2.4 Australia: ASDD, Australian Spatial Data Directories

The ASDD is a national initiative supported by all governments under the auspices of the Australia New Zealand Land Information Council (ANZLIC). The ASDD aims to improve access to Australian spatial data for industry, government, education and the general community by providing a mechanism for its effective documentation, advertisement and distribution. The directory will link government and commercial nodes in each State/Territory and spatial data agencies within the Commonwealth Government.[42] The Z39.50 protocol is used to query each of the data directories that are linked to the ASDD. A single search is broadcast to each of the nodes, which then processes the query separately and returns a set of results, which are mapped to Z39.50-compliant attributes according to the ANZLIC spatial metadata standard. The Z39.50 gateway interface to the ASDD is based on a Java servlet called the MetaStar Gateway, which was developed by Bluangel.[43]

3. Europe

When not stated otherwise, these clearinghouses use (pre-)norms of CEN/TC 287 for their metadata. Information concerning countries marked with (*) does not derive from EUROGI.

3.1 MEGRIN: GDDD, Geographical Data Description Directory

MEGRIN (Multi-purpose European Ground Related Information Network) makes geographic information from the national mapping agencies of thirty-six European countries available to users of pan-European geographic data. In 1993 it launched GDDD in order to gather information on European digital geographic data, implementing the draft CEN/TC 287 metadata standard in a PC-based relation database system (MSAccess). From the GDDD homepage[44] three different paths lead to the dataset summaries–Country, Type of product (e.g., scanned maps, DEM, etc.), Suppliers list. The classification keywords are linked to a "dictionary" to allow users to look-up keyword definitions. In May 1996 the GDDD contained descriptions of 185 datasets provided by thirty-five European national mapping agencies. The now centralized GDDD will be replaced in future by a distributed database.

3.2 Centre for Earth Observation, Directorate General JRC, European Commission: EWSE–An Information Exchange for Earth Observation[45]

The EWSE is basically an on-line dynamic database. There are several views of this database offered via a WWW interface and several novel features. One such novel feature is a free-text search facility that allows the EWSE, and other, databases to be queried by typing in a query that may then, optionally, be expanded by the system to include (user selectable) synonyms and related words from an Earth Observation (EO) thesaurus. Another view of the database is by way of geographical location. Users may select an area of the Earth at different resolutions and list entries for that area. Alternatively one might select a country and look for products, services, or users in that country. Registration and use are free of charge.

3.3 EUROGI

Most European initiatives are co-ordinated by EUROGI. Through the European Commission, DG XIII has made 1 million ECU available to start BE-GIN (Building the European Geo-information Interoperable Network).

3.3.1 Belgium*: SPIDI, Spatial Information Directory[46]

Only a site for the Province of Flanders is available, in which some 200 datasets have been described.

3.3.2 Denmark: Infodatabase on Geodata (KMS)[47]

Ultimo December 1996, the project group has completed a provisional description of approximately 180 data collections. The metadata are available on the Internet free of charge, but all texts, except for info, are in Danish. Electronic supply forms for metadata are available in several formats.[48]

3.3.3 Finland: Paikkatietohakemisto[49]

The clearinghouse is already in full operation; the Website is only in Finnish. A policy-paper is available.[50] As part of the drive for dissemination of spatial data the National Land Survey of Finland provides free of charge (registration by e-mail) the Citizen's MapSite,[51] with maps up to the scale 1:40,000. There is also a for-fee Professional MapSite available, which includes topographic map datasets covering the whole of Finland in 2-m pixel size (1:8 000) and in 5-m pixel size (1:20 000), datasets (1:200 000 and 1:100 000) derived from a 1:250 000 database, and a coordinate search function.

3.3.4 France: AFIGEO

The French Société de l'Information has produced a socio-economic policy-paper, which describes the state of French cartography and cadastral survey, how to organise new initiatives and how to disseminate the spatial products.[52] A preliminary catalogue of digital datasets is available.[53]

3.3.5 Germany: DDGI, Deutschen Dachverbandes für Geoinformation[54]

There are some topical and state spatial infrastructures, but not yet a nation-wide infrastructure. To provide for a future clearinghouse the German

Government has created IMAGI (Interdepartemental Committee for Geographic Information) on June 17, 1998, tasking it with standardisation and dissemination of geographic information.[55] The Bundesamt fur Kartographie und Geodäsie (BKG, Federal Office for Cartography and Geodesy) is maintaining the ATKIS Meta Information System[56] for topographic data of the sixteen mapping agencies of the German states. Each state survey office generates its own metadata and transfers it to the geodata centre of the BKG. The site contains a clickable map.

3.3.6 Hungary: HUNAGI, Hungarian Association for Geo-Information[57]

HUNAGI seems to be in the same formative phase as AFRICAGIS and presently aimed at GIS-literacy. Close cooperation with EUROGI should ensure the eventual creation of a national spatial infrastructure. The META-TÉR project–the development of a unified spatial metadata service for the Hungarian public administration–is among the primary tasks of the Prime Minister's Department.

3.3.7 Ireland: IRLOGI, Irish Organisation for Geographic Information (Eagraíocht um Eolais Gheografaigh na hÉireann): Irish meta-data base[58]

The Irish Geospatial Information Directory (GeoID) will promote co-ordination and awareness within the Irish GI marketplace through the development of a metadatabase of available geographical information resources. IRLOGI, the Irish Organisation for Geographical Information, is also involved in the initiative. IRLOGI will be co-funding the development of the Geospatial Information Directory. There is a selection of links to existing metadata resources on the Internet as well as information on the different standards to which they comply.

3.3.8 Italy: AM/FM ITALY, Automated Mapping Facilities Management Geographic Information Systems[59]

Only a policy document in Italian, on the coordination and dissemination of geographic information, is available.[60]

3.3.9 Netherlands: Nationaal Clearinghouse Geo-Informatie (NCGI)[61]

The NCGI contains presently metadata descriptions of some 1,400 datasets from 14 providers. The descriptions give metadata-elements as well as a map-sample. Though the Topographical Survey (TDN) and the Central Office for Statistics (CBS) are participating in the NCGI, no datasets have yet been

included. At the moment experiments are in process to include descriptions of
the database of the Dutch Union Map Catalogue (CCK[62]). Should these experiments be successful, then some 70,000 descriptions of analogue material
could be included.

3.3.10 Norway*: Statens Kartverk

NGIS Katalogtjenesten[63] is a catalogue of datasets available from the
national mapping agency Statens kartverk. The site also has a clickable map.
Other organisations have their own metadata services.

3.3.11 Portugal: CNIG, Centro Nacional de Informação Geográfica[64]

A clearinghouse is available, which provides access to geographic information produced by national public agencies within environmental, agricultural, economic, scientific and cultural domains. Users can also find the latest
innovations and organisations associated with geographic information.[65]

3.3.12 Slovenia*: Geoinformation Centre[66]

The Geoinformation Centre of the Ministry of Environment and Physical
Planning has a national catalogue of geospatial data sources in which over
400 datasets are described from various sectors and ministries. The metadatabase is available only in Slovenian.

3.3.13 Spain: AESiG, Asociación Española de Sistemas de Información Geográfica[67]

It seems that AESIG is still organising the Spanish GIS field, and this
author cannot deduce from their info whether they are working already at a
spatial infrastructure.

3.3.14 Switzerland: SOGI/OGIS, Schweizerische Organisation für Geo-Information/Organisation Suisse pour l'Information[68]

Though one of the aims of SOGI/OGIS is to facilitate exchange of geographic information between its members there is no initiative yet for a
spatial infrastructure.

3.3.15 United Kingdom: AGI, Association for Geographic Information

Though there are some programmes like the National Land Information
Service (NLIS[69]) for England and Wales, and Scotland's Land Information

Service (ScotLIS[70]) in their formative stages, only recently has the programme for a UK National Geospatial Data Framework come into a more productive phase where it seems to become a reality.[71]

4. North America

4.1. Canada: CGDI, Canadian Geospatial Data Infrastructure[72]

Through CEOnet (Canadian Earth Observation Network) information on Canadian and international geomatics[73] and Earth observation organisations, products and services is available.[74] The database contains descriptions of several thousands of datasets according to the FGDC *Content Standards for Digital Geospatial Metadata*. Another important feature is the heading "services" which contains descriptions of various geomatic products and their suppliers. Another feature of the CGDI is the *National Atlas of Canada*.[75] The National Atlas of Canada is responsible for the development and maintenance of an authoritative synthesis of the geography of Canada. Products, in both digital and conventional form, include base maps, and thematic maps that reflect the social, economic, environmental and cultural fabric of Canada. Customized map services are also provided.

4.2. USA: NGDC, National Geospatial Data Clearinghouse

See the Website of FGDC, http://www.fgdc.gov.

EXPLANATION OF DIAGRAM

The (Appendix) diagram shows the possible geospatial (meta)dataflows in national, continental and global infrastructures. National clearinghouses are sketched horizontally and vertically. Horizontal are national drives (e.g., in the United Kingdom: OS (Ordnance Survey), AGI (Association for Geographic Information) and BL (British Library)). The other two horizontal flows are a Dutch and a French example of clearinghouses. Vertical are supranational drives (MEGRIN, EUROGI, LIBER) which aim at national clearinghouses and ESDI (European Spatial Data Infrastructure). In order not to confuse the diagram horizontal arrows are not drawn. As the diagram is schematic, not all players in the field are shown. The bar with standards [CEN/ISO–DUBLIN CORE/ISBD (MARC)] is shown to ensure interoperability between metadata and data transfer processing.

On top of the National (Meta)data Clearinghouses are placed Visual Geographical Interfaces. The interface should be medium to small scale topographical and thematic maps with underlying geographic grid to which are linked the indexes which are created by (meta)data bases. In this way centroid or bounding box geographical co-ordinate queries, as well as topographical thematical queries, can be made on the underlying databases.

In the Netherlands, we try to interest producers of spatial data in providing free-of-charge medium- and small-scale geographical data for the electronic third edition of the *Scientific Atlas of the Netherlands*. When this atlas could be used as the Visual Geographical Interface it can serve multiple purposes. The maps can provide general geographic information to the public at large and especially for educational purposes. When research is finished an on-line mapping application (OMA) will be available, which will make the maps interactive. But they can also be used as instruments for a visual geographical Internet-interface within the National Clearinghouse Geo-Information. We plan currently to provide three ways of querying this interface:

1. When a map of The Netherlands is shown queries can be made for land-covering datasets.
2. When multiple datasets for different scale-levels are available searchers can zoom in up to the largest scale-level and then make a query on the area available on screen.
3. On any map available bounding-box co-ordinates can be drawn and the underlying metadata databases can be queried for the requested area. Hopefully then also queries such as point-in-polygon, area, distance and buffer zone, and path will be also available.[76]

When this interface is realized, the general public and geographical-education programmes will have valuable geographical information at their disposal. At the same time, producers will have a better advertising opportunity for their more valuable data, which generally must be paid for.[77] The overall purpose is to create an environment in which economically viable data and freely available data can work in unison and profit from each other's existence within the same framework.

For the function of the Resource Description Framework see in this same volume another article by this author, "Metadata: An Introduction."

On top of the continental and GSDI-interface small scale topographical maps should make geographical queries easier.

The emphasis on visual geographical interfaces does not mean, however, that no topographical and thematical dictionaries should be made available for queries on all of the underlying databases.

The Visual Graphic Interface diagram (see Appendix) is an idealised visualisation of a rather chaotic structure as the GSDI will probably grow organically with many unguided cross-links between Websites. Not as symmetrical as a cobweb, but hopefully structured enough to find our way to the information we need. The diagram is meant to foster discussion.

NOTES

1. National Library of The Netherlands, Department for Cartographical Documentation, P.O. Box 90.407, NL-2509 LK Den Haag. E-mail: jan.smits@konbib.nl

2. CEN/TC 287 N 510 (1996). *Report of the Comment Resolution Meeting of prCEN 12160 (Geographic information–Data Description–Geometry)*, Paris, 1996-06-03/07.

3. CEN (Commission Europeènnes de Normalisation)/TC 287 Geographic Information creates the following family of standards:

ENV 12009, Geographic Information–Reference Model
ENV 12160, Geographic Information–Data Description–Spatial Schema
prEN 12656, Geographic Information–Data Description–Quality
prEN 12657, Geographic Information–Data Description–Metadata
prEN 12658, Geographic Information–Data Description–Transfer
CR 12660, Geographic Information–Query and Update–Spatial Aspects
ENV 12661, Geographic Information–Referencing–Geographic Identifiers
prEN 12762, Geographic Information–Referencing–Direct Position
CR 287002, Geographic Information–Overview
CR 287003, Geographic Information–Definitions
prENV 287006, Rules for Application Schemas
<http://forum.afnor.fr/afnor/WORK/AFNOR/GPN2/Z13C/PUBLIC/WEB/ENGLISH/pren.htm> [Accessed: October 13, 1998]. The full texts may only be downloaded with authorization for Z13C.

4. The projects under development with [SO (International Standards Organization)/TC 211 Geographic Information/Geomatics are:

15046-1: Geographic Information–Part 1: Reference Model
15046-2: Geographic Information–Part 2: Overview
15046-3: Geographic Information–Part 3: Conceptual Schema Language
15046-4: Geographic Information–Part 4: Terminology
15046-5: Geographic Information–Part 5: Conformance and Testing
15046-6: Geographic Information–Part 6: Profiles
15046-7: Geographic Information–Part 7: Spatial Subschema
15046-8: Geographic Information–Part 8: Temporal Subschema
15046-9: Geographic Information–Part 9: Rules for Application Schema
15046-10: Geographic Information–Part 10: Cataloguing
15046-11: Geographic Information–Part 11: Geodetic Reference Systems
15046-12: Geographic Information–Part 12: Indirect Reference Systems
15046-13: Geographic Information–Part 13: Quality
15046-14: Geographic Information–Part 14: Quality Evaluation Procedures
15046-15: Geographic Information–Part 15: Metadata
15046-16: Geographic Information–Part 16: Positioning Services
15046-17: Geographic Information–Part 17: Portrayal of Geographic Information
15046-18: Geographic Information–Part 18: Encoding
15046-19: Geographic Information–Part 19: Services
15046-20: Geographic Information–Part 20: Spatial Operators
15854: Geographic Information–Functional Standards
16569: Geographic Information–Imagery and Gridded Data
URL: <http://www.statkart.no/isotc211/pow.htm> [Accessed: October 12, 1998]

5. Paepcke, Andreas et al. "Interoperability for Digital Libraries Worldwide." *Communications of the Association for Computing Machines* 41(4,1998): 33-43. *An annotated bibliography of interoperability literature.* Available from <http://www-diglib.stanford.edu/diglib/pub/interopbib.htm> [Accessed: October 9, 1998]

6. See the White Paper on OGIS (Open Geographical Information Systems) and ISO/TC 211, which is available on the web sites of OGC <http://www.opengis.org> [Accessed: October 12, 1998], ISO/TC211 <http://www.statkart.no/isotc211/> [Accessed: October 12, 1998], and the Institute for Geoinformatics at the University of Muenster <http://ifgi.uni-muenster.de/> (Accessed: October 12, 1998].

7. <http://www.statkart.no/isotc211/terms/terms211.html> [Accessed: October 12, 1998]
<http://www.statkart.no/isotc211/terms/terms287.html> (Accessed: October 12, 1998]

8. DGIWG (Digital Geographic Information Working Group). *DIGEST Part 1: General description.* 1997. <http://www.j2geo.ndhq.dnd.ca/digest/html/GP10.HTM> [Accessed: October 12, 1998]

9. PCGISIAP (Permanent Committee on GIS Infrastructure for Asia & the Pacific). *Statutes–details.* 1997. <http://www.permcom.apgis.gov.au/pcstat.htm#a2> (Accessed: October 12,1998]

10. *"Looking at the collection, provision and use of geographic information in Europe as "economic activity" generated within a range of commercial, industrial and governmental sectors, it is possible to justify estimated figures of the order of 10 billion ECU. Several hundred thousand people, with high skill levels, are employed."* European Commission. Directorate General XIII. *Towards a European Policy Framework for Geographic Information: a working document.* 1996. <http://www2.echo.lu/gi/en/gi2000/gi2000dd.html> [Accessed: October 14, 1998] In a draft for this document it was said: *". . . about 0.1% of GNP per annum, corresponding to some 6 billion ECU, is already spent on topographical data alone, a sizeable market place by any standards"* (author's emphasis).

11. *"This perspective reflects the values and aspirations of those involved in the creation, maintenance and dissemination of selected geospatial "foundation data-sets" designed to be shared by a wide variety of users on a jurisdiction-wide basis. Much of the early impetus/or both spatial data distribution and the sharing of spatial data between organisations has been management-driven, with a few visionaries seeing the economic and operational benefits of early mission-driven, multipurpose mapping, charting and land records programs."* From: Coleman, David J. and John McLaughlin. *Defining global geospatial data infrastructure (GGDI): components, stakeholders and interfaces.* 1997. <http://www.eurogi.org/gsdi/ggdiwp1.html> [Accessed: November 1, 1998)

12. *"Advocates of this perspective emphasise the importance of market demand and customer satisfaction in defining and prioritizing products, product coverage, on-going maintenance and access to products and services. They are more concerned with nearer-term return on investment and are less burdened by public-sector concertos over jurisdiction-wide coverage or mission-driven program support unless there is a satisfactory guaranteed income stream associated with the venture."* From: Coleman, David J. and John McLaughlin. *Defining global geospatial data in-*

frastructure (GGDI): components, stakeholders and interfaces. 1997.
<http://www.eurogi.org/gsdi/ggdiwp1.html> [Accessed: November 1, 1998)

13. GEODAN. *ESMI, European Spatial Metadata Infrastructure.* 1998.
<http://www.geodan.nl/museum/index.htm> [Accessed: October 15, 1998)

14. Alexandria Digital Library (ADL). *Collection metadata for the ADL catalog.* 1998.
<http://www.alexandria.ucsb.edu/~gjanee/collection-metadata/adl_catalog/metadata.html> [Accessed: November 4, 1998]

15. EUROGI (European Umbrella Organisation for Geographical Information). *Global Spatial Data Infrastructure (GSDI).* 1998.
<http://www.eurogi.org/gsdi/#documents> [Accessed: November 4,1998]
There is some confusion about naming the global infrastructure. Global conferences concerning this subject use the name GGDI and on the EUROGI-websites it is named GSDI. The difference is in using either *"geospatial"* or simply *"spatial"* as defining the data concerned.

16. ISCGM. *Global mapping homepage.* 1998.
<http://www1.gsi-mc.go.jp/iscgm-sec/index.html> [Accessed: November 4, 1998]
The Global Mapping Concept calls for every nation and all concerned organizations to work together to develop and provide easy and open access to global geographic information at a scale of 1:1,000,000 or with a ground resolution of about one kilometer.

17. ANZLIC. *Core Metadata Elements for Land and Geographic Directories in Australia and New Zealand.* 1998.
<http://www.anzlic.org.au/metaelem.htm> (Accessed: October 13, 1998].
Through this address the complete document is available by FTP.

18. CEN/TC 287 Secretariat. *CEN/TC 287 Secretariat's report.* 1998.
<http://forum.afnor.fr/afnor/WORK/AFNOR/GPN2/Z13C/PUBLIC/DOC/287n577.doc>
[Accessed: October 13,1998]

19. Comité Europeèn de Normalisation (CEN). Technical Committee (TC) 287. Working Group 1..*[Draft European Standard (EN)].* 1998.
<http://forum.afnor.fr/afnor/WORK/AFNOR/GPN2/Z13C/PUBLIC/DOC/287n540.doc>
[Accessed: October 13, 1998]

20. Danko, David M. *Perspectives in the development of ISO metadata standards.* 1998.
<http://www.fgdc.gov/publications/documents/metadata/nimapaper.html> [Accessed: October 12, 1998]

21. Resolution 3 of the conference participants of the 1997 Conference on Global Spatial Data Infrastructures.
<http://www.eurogi.org/gsdi/gsdi97r.html> (Accessed: October 12, 1998]

22. TDC. *Electronic navigation on the St. Lawrence: a reality.* 1997.
<http://www.tc.gc.ca/TDC/news/ssafe.htm> [Accessed: October 13, 1998]

23. <http://www.iho.shom.fr/news.html> (Accessed: November 4, 1998]

24. Digital Geographic Information Working Group (DGIWG).
<http://www.tmpo.nima.mil/guides/dtf/dgiwg.html> [Accessed: October 13, 1998]

25. DGIWG. *The Digital Geographic Information Exchange Standard (DIGEST).* Edition 2.0. 1997.
<http://www.j2geo.ndhq.dnd.ca/digest/html/digest.htm> [Accessed: October 13, 1998]

26. Moellering, Harold (ed.). *Spatial data transfer standard 2: characteristics for assessing standards and full descriptions of the national and international standards in the world.* New York: Pergamon, 1997. On behalf of the ICA Commission on Standards for the Transfer of Spatial Data.

27. European Commission. *Geographic Data Exchange Standards.* 1998. <http://www2.echo.lu/oii/en/gis.html> [Accessed: October 12, 1998]

28. CSDC. *Commonwealth Position Paper on the Australian Spatial Data Infrastructure (ASDI).* 1998.
<http://www.auslig.gov.au/pipc/csdc/sdi4.htm> [Accessed: October 12, 1998]

29. EUROGI is an association of national or pan-European geographical information (GI) associations. It aims to promote, stimulate, encourage and support the development and use of geographic information and technology at the European level and to represent the common interest of the geographic information community in Europe.
<http://www.eurogi.org/> (Accessed: October 15, 1998]

30. EUROGI. *Global spatial data infrastructure.* 1998.
<http://www.eurogi.org/gsdi/> [Accessed: October 15, 1998]

31. *"Ce programme est un espace multipartenaires et un cadre de travail privilégié dans le domaine des Systèmes d'Information Intégrés sur l'Environnement en Afrique, tant sur le plan thématique (besoin d'approfondir des réflexions méthodologiques, d'avoir des normes standards, des bases de données harmonisées, d'avoir des moyens de communication efficaces . . .) que sur le plan institutionnel (plateforme d'échange et de concertation entre centres techniques et universitaires du sud et du nord et partenaires financiers bi et multilatéraux)."*
<http://antana.orstom.mg/africagis/>]Accessed: November 10, 1998]

32. NSIF. 1998.
<http://168.172.43.22/:> (Accessed: November 16, 1998]

33. NSIF. *Southern Africa Metadata Clearinghouse Gateway.* 1998.
<http://168.172.43.22/gateway.html> (Accessed: November 16, 1998]

34. Permanent Committee on GIS Infrastructure for Asia and the Pacific. *Statutes–Details.* 1997.
<http://www.permcom.apgis.gov.au/pcstat.htm#a2> (Accessed: October 12, 1998]

35. PCGIAP. *PCGIAP publication number 1.* 1998.
<http://www.permcom.apgis.gov.au/tech_paprs/apsdi_pub.htm> [Accessed: November 4, 1998]

36. <http://giddsurv.gsi-mc.go.jp/gidd/howtouse.asp> [Accessed: November 10, 1998]

37. UNICODE. 1998.
<http://www.unicode.org/> [Accessed: November 10, 1998]

38. Japan. Geographical Survey Institute. 1997.
<http://www.gsi-mc.go.jp/> [Accessed: November 10, 1998]

39. NSDIPA. 1998.

<http://ux01.so-net.or.jp/~nsdipa/public_htmlEhtml/indexE.html> [Accessed: November 16, 1998]

40. Malaysia. Ministry of Land and Cooperative Development. *NaLIS*. 1998. <http://www.nalis.gov.my/> [Accessed: November 16, 1998]

41. Malaysia. Ministry of Land and Cooperative Development. *NaLIS gateway server.* 1998. <http://www.nalis.gov.my/laluan/lalu03.htm> (Accessed: November 16, 1998]

42. Australian Surveying and Land Information Group (AUSLIG). *Australian Spatial Data Directory.* 1998. <http://www.auslig.gov.au/pipc/asdi/asdd.htm> [Accessed: October 12, 1998]

43. ANZLIC. *Technical Documentation.* 1998. <http://www.environment.gov.au/database/metadata/asdd/tech.html> [Accessed: October 12, 1998]

44. MEGRIN. *GDDD-Geographical Data Description Directory.* 1998. <http://www.ign.fr/megrin/gddd/gddd.html> [Accessed: October 15, 1998]

45. European Commission. *EWSE.* 1996. <http://ewse.ceo.org/> [Accessed: November 12, 1998]

46. GIS Vlaanderen. *SPIDI.* 1998. <http://193.58.158.196/metadata/> (Accessed: November 12, 1998]

47. Denmark. Kort and Matrikelstyrelsen. *Infodatabase om geodata–A Danish Metadata Catalogue of Spatial Data.* 1998. <http://www.geodata-info.dk/ig-about.htm> (Accessed: October 12, 1998]

48. <http://www.geodata-info.dk/ig-forms.htm> [Accessed: November 10, 1998]

49. Finland. National Survey. *Paikkatietohakemisto.* 1998. <http://www.nls.fi/ptk/aineistot/> (Accessed: October 12, 1998]. Available only in Finnish.

50. Consultative Committee for the Shared Use of Geographic Information–PYRY. *National Geographic Information Infrastructure of Finland–starting point and future objectives for the information society.* 1998. <http://www.nls.fi/ptk/infrastructure/index.html> (Accessed: November 10, 1998]

51. Finland. National Survey. *Mapsite.* 1998. <http://www.kartta.nls.fi/karttapaikka/eng/info/kansalaisen.html> [Accessed: November 10, 1998]

52. Societé de l'Information. *L'information géographique française dans La Société de l'Information.* État des lieux et propositions d'action. Version 3.0. 1998. <http://www.cnig.fr/livre_bl.html> [Accessed: October 12, 1998]

53. Conseil National de l'Information Géographique. *Catalogue des sources d'informations géographiques numériques.* 1996. <http://www.cnig.fr/catalogue.0.fr.html> [Accessed: November 12, 1998]

54. DDGI. *Homepage des Deutschen Dachverbandes für Geoinformation.* 1998. <http://www.ddgi.de/> (Accessed: October 12, 1998]. Available only in German.

55. DDGI. *Bundesregierung räumt Geoinformation hohe Priorität ein.* Einrichtung des ständigen Interministeriellen Ausschusses für Geoinformation 'IMAGI' auf Initiative des DDGI. 1998. <http://www.ddgi.de/> [Accessed: November 10, 1998]

56. ATKIS. *Meta Information System.* 1998.
<http://www.atkis.de/meta/meta_start_eng.htm> [Accessed: November 12, 1998]
57. <http://geo.cslm.hu:80/hunagi/> [Accessed: October 12, 1998]
58. Irish Organisation for Geographic Information. *Irish meta-data base.* 1998.
<http://www.irlogi.ie/gidirectory.html> [Accessed: October 12, 1998]
59. <http://195.110.158.11 l/> [Accessed: October 12, 1998]. Available only in Italian.
60. AM/FM Italia. *Coordinamento dei sistemi informativi geografici di interesse generale.* 1998.
<http://195.110.158.111/dati/coordinamento.html> (Accessed: November 10, 1998]
61. NCGI. *Welkom bij het National Clearinghouse Geo-Informatie.* 1998.
<http://www.geoplaza.nl/ncgi/plsql/splash> [Accessed: October 12, 1998]. Only available in Dutch.
62. Velden, G. J. K. M. van der et al. *CCK: Making cartographic materials accessible.* 1990.
<http://www.konbib.nl/kb/skd/liber/articles/cck.htm> [Accessed: November 10, 1998]
63. NGIS. *Statens Kartverk.* 1998.
<http://katalog.statkart.no/katalog/> [Accessed: November 12, 1998]. Available only in Norwegian.
64. Portugal. Centro Nacional de Informação Geográfica. *Welcome.* 1998.
<http://snig.cnig.pt/snig/english/index.html> [Accessed: October 12, 1998].
65. Portugal. Centro Nacional de Informação Geográfica. *Digital Geographic Information.* 1998.
<http://snig.cnig.pt/snig/english/infoing.html> (Accessed: November 12, 1998]
66. Slovenia. Ministry of Environment and Physical Planning. *Geoinformation Centre.* 1998.
<http://www.sigov.si:81/giceng/index.html> [Accessed: November 12, 1998]
67. AESIG. 1998.
<http://mercator.org/aesig/> (Accessed: October 12, 1998]. Available only in Spanish.
68. <http://www.sogi.ch/> [Accessed: October 12, 1998]. Available in German and French.
69. NLIS. *National Land Information Service.* 1998.
<http://www.nlis.org.uk/> (Accessed: November 11, 1998]
70. Harvey, Richard, and Bruce Gittings. *ScotLIS on the Web.* 1997.
<http://www.geo.ed.ac.uk/~scotlis/slis.html> [Accessed: November 11, 1998]
71. Nanson, Bryan, and David Rhind. *Establishing the Off National Geospatial Data Framework.* 1998.
<http://www.ngdf.ore.uk/whitepapers/ngdfcan.htm> [Accessed: October 10, 1998]
72. *Canadian Geospatial Data Infrastructure.* 1998.
<http://cgdi.gc.ca/> [Accessed: November 11, 1998]
73. The Canadians define "geomatics" as "*the scientific and technical domain concerned with methods, procedures and technologies associated with computer systems for the collection, manipulation, display and dissemination of geographically referenced data*" (see preceding footnote for Website). This may be the field with which map curators in future will occupy themselves, thereby broadening their field of work to include production and use. See further: Parker, Velma (ed.). *Geomatic*

data sets: cataloguing rules. Ottawa, Canada: Canadian General Standards Board, 1994.

74. Natural Resources Canada. *Welcome to CEONet.* 1997.
<http://ceonet.ccrs.nrcan.gc.ca/cs/en/> [Accessed: November 12, 1998]

75. Natural Resources Canada. *National atlas of Canada.* 1998.
<http://www-nais.CCRS.NRCan.GC.CA/english/home-english.html> [Accessed: November 11, 1998]

76. Larson, Ray R. *Geographic information retrieval and spatial browsing.* 1996.
<http://sherlock.sims.berkeley.edu/geo_ir/PART1.html> [Accessed: October 14, 1998]

77. Smits, Jan. "Developments in spatial data management: Dutch impressions." *Proceedings [of the] 35th Annual Symposium [of] the British Cartographic Society [and] Map Curators' Group Workshop.* Keele: University of Keele, 1998, pp. 10-19. As MS PowerPoint presentation: URL: <http://www.konbib.nl/persons/jan-smits/bcs/bcspres.PPT> [Accessed: October 14, 1998]

APPENDIX

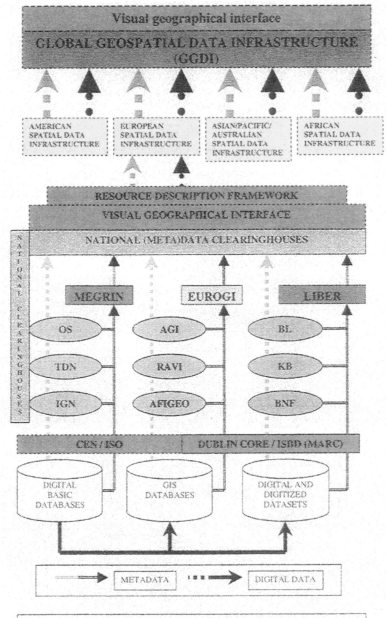

Idealized simple production flowmodel for spatial (meta)data infrastructures

Cataloguing
Digital Cartographic Materials

Grace D. Welch
Frank Williams

SUMMARY. Cartographic materials in digital format are now a reality in modern map libraries. Libraries have been reluctant to catalogue this type of material because of the lack of cataloguing rules and the highly technical nature of the information. This paper provides a status report on cataloguing rules for electronic cartographic materials, with particular emphasis on the new fields that have been created in USMARC to accommodate the special requirements for this material. For each part of the bibliographic description, both what current cataloguing rules allow and what is being recommended as part of the revision to *Cartographic Materials: A Manual of Interpretation for AACR2* is presented. The paper also looks at what is required to get started and identifies resource tools. *[Article copies available for a fee from The Haworth Document Delivery Service: 1-800-342-9678. E-mail address: getinfo@haworthpressinc.com <Website: http://www.haworthpressinc.com>]*

KEYWORDS. Cataloging, geospatial data, digital cartographic materials, electronic resources, cartographic materials

Grace D. Welch, BA, MLS, is Map Librarian, Map Library, Library Network, University of Ottawa, Ottawa, Ontario K1N 9A5, Canada and the Association of Canadian Map Libraries and Archives representative on the Anglo-American Committee on Cataloguing Cartographic Materials (E-mail: gwelch@uottawa.ca).

Frank Williams, BA, MMus, MLIS, is Map Cataloguer, Map Library, Library Network, University of Ottawa, Ottawa, Ontario K1N 9A5, Canada (E-mail: fwilliam@uottawa.ca).

[Haworth co-indexing entry note]: "Cataloguing Digital Cartographic Materials." Welch, Grace D., and Frank Williams. Co-published simultaneously in *Cataloging & Classification Quarterly* (The Haworth Information Press, an imprint of The Haworth Press, Inc.) Vol. 27, No. 3/4, 1999, pp. 343-362; and: *Maps and Related Cartographic Materials: Cataloging, Classification, and Bibliographic Control* (ed: Paige G. Andrew, and Mary Lynette Larsgaard) The Haworth Information Press, an imprint of The Haworth Press, Inc., 1999, pp. 343-362. Single or multiple copies of this article are available for a fee from The Haworth Document Delivery Service [1-800-342-9678, 9:00 a.m. - 5:00 p.m. (EST). E-mail address: getinfo@haworth pressinc.com].

INTRODUCTION

Cataloging cartographic materials is not without its challenges. When cataloguers first began to describe maps, they were confronted with trying to describe material that did not fit the model of book cataloging. They persevered though, and mastered the intricacies of determining title, extent, recording coordinates and projections, etc., codifying the results in a standard set of rules for describing printed maps in *AACR2* and *Cartographic Materials: A Manual of Interpretation for AACR2.*[1] We now face a new challenge: how to describe digital cartographic materials.

Today, digital cartographic materials are a reality in most map collections. They arrive on CD-ROM, on diskette, or as files downloaded from the Internet or even attached to email messages, in the form of electronic atlases, digitized maps, and vector and raster data sets to be manipulated using geographic information systems (GIS). We talk about maps in digital form, electronic cartographic resources, geomatic data, geospatial data, digital cartographic resources, GIS data, digital spatial data and the list goes on. These electronic data sets are installed on workstations (we hope!), listed on library web pages, advertised to users, but are they catalogued?

Purpose of Paper

As responsible map librarians and map cataloguers, we cannot continue to ignore the need for cataloguing geospatial data. This paper is intended to provide a status report on cataloguing rules for digital cartographic materials, with particular emphasis on the new fields that have been created to accommodate the special requirements for this material. For each of the cataloguing areas, both what current cataloguing rules allow and what is being recommended as a change will be presented. The paper will also look at what is required to get started, identify resource tools, and hopefully encourage map cataloguers to create bibliographic descriptions for this newest type of cartographic information.

GETTING STARTED

Definitions

Let us begin with some basic definitions. Digital cartographic resources generally divide into two types: electronic atlases and geospatial data. Electronic atlases are just that, the electronic counterpart to the traditional paper atlases in our collections; they allow the user to select and display information for a particular geographic area or on particular themes. Although these atlases have become more sophisticated, most still offer only a limited capability for altering the display. Examples include Microsoft *Encarta Virtual Globe,* PC Globe *Maps 'N Facts,* and *StreetAtlas USA.*

Geospatial data is information about the shape and location of objects on the Earth's surface which can be manipulated in desktop mapping or GIS programs such as ArcView, ArcInfo, MapInfo, or Intergraph. This data can be either vector, raster or tabular. Vector data represents geographic features (entities) as x,y coordinates. Features are described as points, lines and polygons. As an example, for a map of all of Canada, cities may be represented as points, rivers and roads would be examples of lines and the provinces would be polygons. Examples of common vector formats are Arc/Info Export (E00 files), MapInfo MID/MIF files, and DXF. Raster data is a cell- or pixel-based method of representing the Earth's features, with each cell or pixel having a value. Satellite images are raster data, as are maps or images created via scanning. Examples of raster formats are TIFF, GIF, JPEG, and BMP.

Tabular data (attributes) describe the geographic features and are linked to the features by means of a unique identifier. Attributes of a lake, for instance, might include the name, depth, and area. Vector data always includes an attribute table linked to the geographic features they describe.

Background Documentation

The rules for cataloguing geospatial data are still in the evolutionary stage. AACR2R Chapter 9 exists for computer data but its focus seems to be primarily on social-science numeric-data sets and software. The current cataloguing rules do not address the requirements for describing spatial data in its digital form. The bible for map cataloguers, *Cartographic Materials,* is now being revised by the Anglo-American Cataloguing Committee for Cartographic Materials (AACCCM) with publication expected in the year 2000. In addition to changes to bring the manual up-to-date with AACR2 revisions, extensive sections are being added for describing geospatial data. Some of the new rules for geospatial data have been derived from *Geomatic Data Sets Cataloguing Rules*[2] and work undertaken by the "Ad-Hoc Working Group on Cataloguing Digital Materials" which met in Santa Barbara in November 1995.[3]

Until the revision to *Cartographic Materials* is published, cataloguers can refer to the articles in this volume such as Larsgaard's description of cataloguing CD-ROMs, in conjunction with Chapters 3 and 9 of AACR2. Of particular importance are the new provisional fields added to USMARC (tags 342, 343, 352, 514, 551, and 786) to accommodate geospatial data. Some existing fields were also modified: 034, 037, 255, 355 and 787.[4] Detailed documentation on how to complete these new fields has been distributed to libraries as part of the MARC updates. Before creating bibliographic records using these new fields, it is important to verify with the library's systems staff to ensure that the fields have been validated in the library's cataloguing system.

Hardware and Software

Too often, adequate equipment for cataloguers is overlooked. In order to work with geospatial data, there is a certain minimum requirement for both hardware and software. These requirements may exceed the definition of the standard cataloguer's workstation, as created by the Library's Systems or Cataloguing Department, and may require discussion and negotiation to ensure that the appropriate equipment and software are supplied.

Let's start with hardware: the cataloguer's workstation should be equipped with a CD-ROM drive; CD-ROMs are now the preferred distribution medium for digital cartographic resources, especially with the increased availability of writable CD-ROMs. There are no hard-and-fast guidelines for the amount of RAM (Random-Access-Memory) required, but as a general rule of thumb, at least 32 MB (megabytes) of RAM should be considered a minimum. The workstation's hard drive or drive allocation on a local-area network should be sufficiently large to allow for the loading and storage of graphic files and related software; it is surprising how quickly a hard drive can fill up.

Moving to software, a few basic programs are a prerequisite. Because of the large size of graphic files, geospatial data is often distributed as an "executable" compressed file (the file has the extension .exe after the name) or in a compressed or zipped format (the filename ends in .zip or .tar). The file(s) have to be copied to the cataloguer's hard drive or network drive and then opened or unzipped before the contents can be described or to find the technical information required to create a description. Very often, clicking on one executable or zip file will create several new files as in the example below. The diskette for GSC (Geological Survey of Canada) Open File 3475, *Surficial Geology, Digital Map, Low Quebec 31G/13*[5] contained two files:

31G13.txt	15KB	Text	A readme file giving information about the title, author and origin of the data as well as definitions for the attribute data
31G13E00.exe	876KB	Application	An executable file

This second file, when opened, created three new ArcInfo Export files:

S31g13.e00	4,904KB	E00 File	Contains polygon data

| Ss131g13.e00 | 118KB | E00 File | Contains line data |
| Ssp31g13.e00 | 7KB | E00 File | Contains point data |

When the E00 files are imported and displayed in ArcView or ArcInfo, a surficial geology map corresponding to the area covered by Canadian topographic map sheet 31G/13 at a scale of 1:50,000 is displayed.

The distributor may supply a program with the data that will unzip the file, but this is happening less frequently now that WINZIP, a Windows-based program for compressing and decompressing files is widely available. This inexpensive software program should be installed on the cataloguer's workstation.

Web access via a browser such as Netscape or Explorer is another prerequisite. If one is fortunate, the documentation provided with the data set will be sufficient to create a bibliographic description but all too often it is necessary to consult the data provider's web site for critical information not included with the digital data. There also seems to be a recent trend toward providing accompanying documentation in HTML format which can only be read by an Internet browser.

Another commonly used method for delivering documentation is Adobe Acrobat files, recognized by the ".PDF" extension after the filename. The Adobe Acrobat reader can be downloaded from the Internet and installed locally.

It is not necessary to have a desktop GIS system such as ArcView or MapInfo in order to catalogue digital cartographic data, but there are instances (though infrequent) when it is helpful to have access to a GIS to visualize the data set or verify information. For those cataloguers of cartographic materials who are not administratively part of a map library, arrangements should be made with a GIS specialist in the map collection to load and display those data sets requiring additional exploration. The GIS specialist may also be of assistance in interpreting some of the more technical descriptions accompanying the geospatial data.

Preliminary Decisions

When should a computer file be considered a cartographic item or treated as an electronic resource and described according to Chapter 9 of AACR2? The Library of Congress has offered some guidance on this subject in the form of a Web document[6] which can assist cataloguers in the selection of the USMARC Type of Record code, type "e" for cartographic items or type "m" for computer files (Leader/06 position, USMARC).

This is a fundamental decision facing the cataloguer and one which may take some analysis to determine if the item to be catalogued is in fact cartographic and should be coded as "e" in Type of Record. Digital data to be treated as cartographic items include images of actual maps, remote-sensing images, electronic atlases, and data which can be processed by a GIS to produce images, such as ArcWorld or Digital Chart of the World. Software to manipulate geospatial data such as ArcView, MapInfo, IDRISI, or AutoCAD would be catalogued in type "m." Also included in type "m" are subject/topical databases which may have a geographic interface to the data such as the "GeoNames" CD-ROM.

Is the data set published? In the case of paper maps there are clear-cut definitions about whether an item has been published or whether it should be treated as a manuscript. In the area of electronic resources the rules for published versus unpublished do not apply. For the purposes of cataloguing, if a data set is publicly available, be it on CD-ROM, diskette or for downloading from a remote site, it should be considered as published and described as such.

SOURCES OF INFORMATION

Metadata

Before talking about the Chief Source of Information, the concept of metadata needs to be introduced. Metadata, simply defined, is data about data. Metadata, which is usually created by the producer of a geospatial data set, describes the content, extent (geographic and temporal), quality, availability and other characteristics of a data set. In most cases, metadata is the comparable equivalent of the title page and verso of the title page where cataloguers traditionally find most of the information for creating a bibliographic description. Metadata may be supplied with the digital data as an accompanying printed document, or a file included on the CD-ROM or diskette named "metadata" or "readme," or it may sometimes be necessary to refer to the metadata file on the data producer's web site.

In the United States, the Federal Geographic Data Committee (FGDC) has defined a metadata standard for data producers, *Content Standard for Digital Spatial Metadata,*[7] which outlines all of the elements to be included in a metadata description and includes a very useful glossary. U.S. federal-data producers are required by Executive Order 12906 (April 11, 1994) to produce metadata for all of their geospatial data and most contributions to State spatial data clearinghouses follow the FGDC standard.

In Canada, a national metadata standard has also been published: *Directory Information Describing Digital Geo-Referenced Data Sets.*[8] However, the developing Canadian Geo-Spatial Data Infrastructure (CGDI) has adopted the U.S. *Content Standard* as the metadata standard for contributions to the

CGDI. Both standards are expected to be replaced by the ISO (International Organization for Standardization) standard for metadata when it is completed.

Metadata descriptions produced according to the FGDC standard can be extremely detailed and may seem intimidating when approached for the first time. For a more detailed discussion of metadata, see the Smits' articles in this volume. There are also several excellent publications and fact-sheets available on the Web which help to demystify these descriptions.[9]

It should be emphasized that there is still wide variation in the quality and quantity of metadata that may be provided with a digital data set, particularly if it is an older data set or the data has been obtained from a non-government source. Unfortunately, there are still all too many cases where a data set does not have any metadata!

In a perfect world we would be able to simply take the metadata file and parse it directly into a catalogue description in our online system. A second best alternative is metadata in electronic format or from a Web site from which the cataloguer can cut and paste the relevant information required for the bibliographic description into their cataloguing interface. Cataloguers are aided in this task by two crosswalks prepared by Elizabeth Mangan, Geography and Map Division, Library of Congress.[10,11] These crosswalks identify which FGDC metadata field corresponds to which USMARC field, and vice-versa.

Chief Source of Information

The chief source of information defines the principle source of information for preparation of the bibliographic description. For electronic resources the chief source is the title screen or in the absence of a title screen, the cataloguer is to use other internally presented sources such as metadata, readme files, documentation files or header files. As mentioned earlier, some action by the cataloguer, such as unzipping a file, may be required before these internal sources can be identified.

In the absence of internal sources, the cataloguer may have to use information from accompanying documents (such as a metadata document provided by the producer or creator of the data), the physical carrier (such as a label on a CD-ROM or diskette) or information printed on a container issued by the publisher, distributor. Although some CD-ROMs come with a cover or an explanatory text that can be used for description, this is not always the case, especially for data provided on "writable CD-ROMs."

When there is variation in completeness of information in the sources mentioned above, the source with the most complete information should be treated as the chief source of information. In most cases, this will be the metadata description.

CREATING THE DESCRIPTION

Fixed fields for geospatial data are described in the Larsgaard article which follows and thus are not repeated here. As previously mentioned, the most important fixed field is "Type of Record" (leader 06 position in US-MARC) which determines whether an electronic resource will be treated as a cartographic item and catalogued according to Chapter 3 in AACR2 or considered a computer file and described according to Chapter 9.

Title and GMD

As with printed cartographic materials, finding the title of a geospatial data set can be problematic; a CD-ROM, diskette or data set downloaded from the Internet may have a short incomplete title or only a cryptic filename rather than a descriptive title. The cataloguer may have to refer to readme files or accompanying documentation to find the most complete title. When a title does not include an indication of the geographic coverage of the data, this information will need to be supplied by the cataloguer in square brackets, as other title information. As with all types of computer files, the source of the title proper should always be recorded in a note.

The General Material Designation (GMD), subfield h in the 245 field, specifies the type of material being catalogued and in a multi-media catalogue is an important element of the description. However, for cartographic materials in digital format, what GMD should be used? As the rules exist now, the GMD should be selected from Chapter 9. There is no indication for the researcher that the data is cartographic in nature, the recurrent "content versus carrier" problem in AACR2. To overcome this problem, the AACCCM is recommending to the Joint Steering Committee for the Revision of AACR (JSC) that multiple GMDs be permitted similar to those allowed for tactile and large print items (rule 1.1C1 in AACR2R). The first GMD would specify the nature of the item, i.e., cartographic material, qualified by the second GMD for the physical carrier, as per the example below:

[cartographic material (electronic resource)]

The Committee (AACCCM) suggests the term, "cartographic material," as a GMD in place of the term, "map," which is now in use in North America, as the former is more generic. It is recommending the qualifier, "electronic resource," in lieu of, "computer file," in anticipation of the incorporation of new ISBD (ER) specifications into AACR.

Mathematical Data Area

AACCCM is preparing a submission to the respective rule-making bodies of each country suggesting that this section be renamed "Mathematical and

Other Material-Specific Details Area" to take into account the several new areas added for describing geospatial data. Provisional fields have already been created in USMARC and new areas will be added to *Cartographic Materials*:

3E File Characteristics (MARC 256)

3F Digital Graphic Representation Area (MARC 352)

3G Geospatial Reference Data Area (MARC 342 and 343)

Before discussing these new areas, let's look at the traditional elements in the mathematical data area, beginning with scale (area 3A, $a in field 255, a mandatory subfield in MARC 21). The importance of recording the scale of a digital data set has generated considerable discussion. Does it have relevance in a GIS environment, when you can zoom in or out to any "scale"? AACCCM is recommending to JSC that there be a new phrase, "Scale not applicable," for use with digital data. Until this phrase is officially sanctioned, cataloguers must choose between four possibilities, "Scale indeterminable," "Scale not given," "Scale differs," or "Scale varies," none of which seem appropriate. As an interim measure, the Library of Congress has chosen to use the phrase, "Scale not given."

There is also the question of "input scale" when a digital cartographic item has been digitized from a paper map; the elements included in the electronic version have been selected on the basis of the scale. Scale then affects both the content of the item and the extent to which the data can be used for other purposes. For example, shoreline data digitized from a 1:1,000,000 map will be too generalized if a researcher is doing an analysis of the harbour of Halifax, Nova Scotia. Recording input scale would be appropriate, for instance, when describing the Digital Chart of the World which has been derived primarily from the 1:1,000,000 Operational Navigational Charts or for scanned paper maps such as the 1:24,000 Digital Raster Graphics:

255 ## $aInput scale 1:1,000,000

255 ## $aInput scale 1:24,000

The use of "input scale" for digital data will likely be an optional element for geospatial data digitized from paper maps leaving individual libraries to decide whether they wish to record input scale or not. The safest approach at this time is to record input scale in a note.

As the rules now stand, recording of the projection statement is called for in field 255 ($b) and the new field 342 ($a). From a practical point of view, the cataloguer should only have to enter the projection once. AACCCM is recommending that the field 255 be used for paper maps and field 342 for electronic cartographic resources where additional subfields permit the cataloguer to add detailed information such as the longitude of central meridians and latitude of projection centre, false easting and false northing, etc. Cataloguers of digital files will have to decide whether they will input the information twice or use only field 342; 255 $b is not a mandatory field and records will not be rejected should this subfield be absent.

Before further discussion about field 342, let us complete the field 255 information. The geographic extent (coordinates) of a geospatial data set are recorded here as subfield c. This information can usually be derived from the bounding coordinates supplied in the metadata description (in section 1, Identification Information, when the supplier has followed the FGDC standard). These coordinates are presented in decimal degrees with the locations south of the Equator and west of Greenwich preceded by a minus sign. In recognition of the special requirements for digital data, AACCCM is recommending that the cataloguer have the option of recording the coordinates in this field as either degrees, minutes and seconds, or as decimal degrees (or both). In determining cataloguing policy with respect to the method selected for transcribing geographic extent, users of the library catalogue must be considered: will the library's clients understand decimal degrees or are they more comfortable reading degrees, minutes and seconds?

When recording coordinates in decimal degrees, the plus sign is assumed for latitudes north of the equator and longitudes east of Greenwich but the minus sign should be included for locations south of the equator and west of Greenwich. If metadata have not been supplied with a digital cartographic item that has been created from a paper map, local policies will dictate whether the paper map should be consulted to determine the geographic extent of the data set.

Coordinates are also recorded in the fixed fields (034). A request will need to be made to MARBI to have new fields added to allow for the input of decimal degrees; fortunately, decimal degrees can be expressed in field 255 $c without the need to make a request to MARBI for new subfields. Until the new subfields have been implemented, coordinates in decimal degrees will require conversion to degrees, minutes and seconds. We hope that a conversion facility to automate this routine will soon be available on the Web.

Additional subfields have also been added to field 255 should the cataloguer wish to record what are referred to as "G-ring polygon coordinate pairs" which is another method of describing geographic extent for geospatial data. It is expected, though, that inclusion of this information will likely be limited to specialized catalogues.

File characteristics (field 256) are a required element for digital geospatial data, defining the type of file (e.g., computer file, computer images, computer data and programs), the number of files, and file sizes. These last two elements need only be recorded if they are readily available and not too numerous for inclusion. In practical terms, users find that total file size information is more relevant than a long list of individual file sizes, as total size determines whether the data can be copied easily onto a diskette. Although the existing rule from Chapter 9 does not explicitly state that total file size should be included, or give examples with total file size, there is nothing to prevent the cataloguer from including this information in this field. Some experienced cataloguers even consider that the total file size is more analogous to the number of pages in a book or atlas and therefore more appropriately recorded in the extent of a cartographic item (field 300). In terms of where field 256 should display in the ILS, it is suggested that it appear after 255, 342 and 343 but before 352.

We now arrive at the most daunting part of cataloguing digital cartographic materials–completing the areas for Digital Graphic Representation (Field 352) and Geospatial Reference Data (Field 342 and 343). GIS has a very technical, specialized vocabulary and many of the terms will be new to map cataloguers. Fortunately, most of the elements in these areas need only be supplied if they are readily available in the metadata. Until the revised *Cartographic Materials* is published, the articles in this volume and the provisional fields for USMARC for geospatial data provide guidance on what elements should be included in these new areas. As provisional fields, punctuation and print constants have not been defined. The examples given below reflect the punctuation recommended by AACCCM in its draft revision to *Cartographic Materials*. Until print constants have been defined, the cataloguer will need to supply explanatory text in these fields to make the elements comprehensible to the researcher. In terms of "what goes in where" in these new fields, cataloguers are referred to the crosswalks between USMARC and the FGDC Standard, mentioned earlier in this article.

The amount of information to be included in these fields will be determined by the availability and completeness of the metadata and local cataloguing policies. For instance, rather than create long, detailed bibliographic records for geospatial data, a library may choose to mount metadata files on their web page and then supply the Internet address, or point directly to the metadata description on the data supplier's Web site. The Web address can be recorded in the generic 500 note. Or, the cataloguer may choose to use field 856 which contains information required to locate an electronic item and, via the second indicator relationship (value 2), permits links to related documents such as metadata. Prior to using this note, the cataloguer should verify how field 856 is displayed in the public catalogue record.

What exactly is in field 342? This field is used to record information about the horizontal and vertical coordinate systems of a data set (projection, grid, datum) and may be repeated when multiple elements need to be identified. The first indicator value delineates whether the information is for a horizontal (0) or vertical (1) reference system; the second indicator value codes the specific geospatial reference method. For example, for a data set that has been saved in decimal degrees (lat/long) and based on the North American Datum 1927, two 342 fields would be required:

> 342 00 $a Geographic system: coordinates ;$c latitude resolution: 0.0004 ;$d longitude resolution: 0.0004 ;$b unit of measure: decimal degrees

> 342 05 $a North American Datum of 1927

If readily available, the cataloguer may provide additional information about the datum, such as illustrated in the following example:

> 342 05 $a North American Datum of 1927 :$q ellipsoid: Clarke 1866 $r (semi-major axis: 6378206.4 ;$s denominator of flattening ratio: 294.98)

Each 342 field should be separated by a full stop and space; each set of projection or ellipsoid parameters are enclosed in parentheses and multiple parameters/statements are separated with a semi-colon.

When the provisional fields for geospatial data were created, not all of the information from Spatial Reference Information (section 4 of the FGDC standard) could be incorporated into field 342. Thus field 343, Planar Coordinate Data, was created to provide users with additional measures of the accuracy of the data set:

> 343 ## $a Planar coordinate encoding method: coordinate pair ;$c abscissa resolution: 1;$d ordinate resolution: 1;$b planar distance units: meters

In completing fields 342 and 343, the cataloguer needs to be aware of how units of measurement are handled for those subfields that express measurement:

- the units of measure should be added for the false easting or northing, if the value is anything other than 0. The unit of measurement can be recorded either in the subfield:

> $i false easting: 1 000 000 m,

or as part of the explanatory text:

$i false easting in meters: 1 000 000.

- certain measurement units are never given as they are assumed to be well understood by the user community, such as those units for the semi-major axis and the flattening ratio for ellipsoids.
- the unit of measurement given in the field is assumed to be the unit for all of the values in the field. For example, in field 343, abscissa and ordinate resolution are expressed in the planar distance units, so the cataloguer does not need to repeat the unit of measurement, e.g., "abscissa resolution in meters: 1.0." This is also true in field 342 when the horizontal reference system is "geographic" (saved in decimal degrees). The latitude and longitude resolution are given in the same units as those recorded in 342 $b.

If there is doubt about whether to record a unit of measurement after a value in one of the subfields and it is supplied in the metadata (e.g., meters or feet), the cataloguer should add the unit of measurement. A note of caution when completing these two fields: the information should never be supplied by the cataloguer unless it is taken directly from source material.

Where should fields 342 and 343 display in the public catalogue? As noted in the Larsgaard article, negotiation with systems staff may be required to ensure that these fields appear immediately following field 255, as the three fields contain related information. At the same time, decisions about which subfields should be displayed in the public catalogue should be reviewed. Many integrated library systems (ILS) have two public displays, a brief and a full record. For example, for field 342, it is appropriate to limit the display on the brief record to subfield a, which gives the name of the horizontal reference system (projection or datum, for example).

One of the first questions users ask about a geospatial data is whether it is vector or raster. This information, derived from Spatial Data Organization (section 3 in the FGDC metadata), is recorded in field 352 $a as the direct reference method. Additional details may be added in this field such as the row and column count for raster data or the indication of the level of completeness of the topology for a vector data set. This field logically follows the computer file information (field 256) as it relates to how the data is stored. As noted earlier, the USMARC provisional fields do not include punctuation. The three examples below illustrate the proposed AACCCM punctuation: precede the object type by a colon; enclose each statement on the number of objects in parentheses; and, if both point/object count and VPF level are given, precede the VPF level by a semicolon.

352 ## $a Vector :$b object type: network chain

352 ## $a Raster :$b object type: pixel$c (5000 x 5000)

352 ## $a Vector :$b object type: points, lines and polygons ;$g VPF topology level: 3

Unfortunately, information about the file format name (usually the second question we are asked) is relegated to field 037, Source of Acquisitions, a field which does not display in the ILS. In 037, subfield g, the cataloguer can record the data transfer format (from section 6.4.2.1.1 of the FGDC metadata standard). The file format name such as Arc/Info Export (.e00), MapInfo MID/MIF, etc., tells the user whether the data set can be manipulated directly by a specific GIS software and needs to be available in eye-readable form in the catalogue record. At this time, this critical piece of information can be recorded in the notes area in 538 field, System requirements. The lack of display of the file format information in a prominent location in the catalogue record will be addressed by the AACCCM, likely with a recommendation to create a new subfield in field 352 that will display immediately after the $a, the direct reference method.

Detailed Examples of Mathematical Data and Other Physical Details[13]

Geological Map of Canada

Scale not given (W 141°00'–W 52°37'/N 83°10'–N 41°41'). – Projection: Lambert conformal conic (standard parallels: 49 00 N, 77 00 N ; longitude of central meridian: 95 00 W ; latitude of projection origin: 49 00 N ; false easting: 0 ; false northing : 0). – Horizontal datum: North American Datum 1927 : ellipsoid name : Clarke 1866. – Planar coordinate encoding method: coordinate pair ; planar distance units: meters. – Computer data (426,082,304 bytes). – Vector.

Circum-Arctic Map of Permafrost and Ground-Ice Conditions

Scale not given (W 180°–E 180°/N 90°–N 25°). – Projection: Lambert azimuthal equal area (longitude of central meridian: 180 ; latitude of projection centre: 90 ; false easting: 0 ; false northing: 0). – Horizontal datum: none ; ellipsoid name: Clarke 1866 (semi-major axis: 6378206.4 ; flattening ratio: 295.98). – Planar coordinate encoding method: coordinate pair ; abscissa resolution: 1 ; ordinate resolution: 1 ; planar distance units: meters. – Computer data (9 files ; total 26.6 MB). – Vector : point (13671) : string (20171) : GT-polygon composed of chains (13672).

As the above examples demonstrate, descriptions for geospatial data can become quite lengthy and each library will need to assess their clientele and their cataloguing workload to identify how much detail will be supplied in the mathematical data area. In defining this minimum set of elements there are a few critical elements that should always be included, if available:

Geospatial Reference Data (fields 342 and 343)

- Horizontal coordinate system, e.g., whether the data has been saved in decimal degrees (geographic) or saved as a projected file (map projection, grid system such as UTM or a local planar system).
- Resolution: defines the smallest object or feature that is discernable in the data; as defined by the FGDC,[12] "it is the minimum difference between two independently measured or computed values which can be distinguished by the measurement or analytical method being considered or used." For raster data, resolution is usually expressed as pixel size. For vector data, if the data set is coded geographic in field 342, with a second indicator value of "0," latitude and longitude resolution will be included in the metadata description. Resolution should not be confused with accuracy, which is a measure of the degree of correctness of a measurement. For example, accuracy might be recorded as + or − 10 metres, which means that any location on the map is correct within 10 metres of its precise location on the ground. This information, sometimes referred to as positional accuracy, should be recorded in a note (field 514, see below).
- Horizontal datum: defines the reference system used for defining the coordinate points (e.g., NAD27, NAD83, etc.).

Digital Graphic Representation (field 352)

- direct reference method: specifies whether the data is raster or vector data (352, $a). A note should be included in field 538 to give the file format name.

If there is any doubt about whether some of the technical information should be included in this area, it can always be added to the description as a free-text note.

The detailed description for the *Circum-Arctic Map of Permafrost and Ground-Ice Conditions* given above could be shortened to:

Scale not given (W 180°–E 180°/N 90°–N 25°). – Projection: Lambert azimuthal equal area. – Horizontal datum: none; ellipsoid name: Clarke 1866. – Computer data (9 files ; total 26.6 MB). – Vector.

Physical Description

Under the existing rules for physical description (section 5, *Cartographic Materials*), just the physical carrier is identified, e.g.: 1 computer disk : col. ; 3 1/2 in. To be more consistent and explicit about the item being described, AACCCM is considering a proposal that the description include the specific material designation and the physical carrier, as per the examples below:

1 map on 1 computer disk : col. ; 3 1/2 in.

2 maps on 1 computer optical disc : col. ; 4 3/4 in.

23 remote-sensing images on 2 computer optical discs : col. ; 4 3/4 in.

The choice of an appropriate specific material designation from the list of nine possibilities in AACR2 Chapter 3 can be problematic for digital data. If the cartographic item is a digital form of a printed map such as the surficial geology map of Low, Quebec, or a soil map of Ottawa-Carleton, the expression "1 map on 1 computer disk" or "1 map on 1 computer laser optical disc" is entirely appropriate. But what term describes the Digital Chart of the World (DCW) or ArcWorld, which provide the capability of producing an infinite number of maps? These products are essentially databases containing both tabular data and geographic features. Should the physical description have an entry such as: "   maps on 4 computer optical discs" or "1 atlas on 4 computer optical discs" or should a new SMD be created? AACCCM is now considering recommending that the phrase "cartographic database" be added to the list of recommended SMDs.

The issue of color with respect to geospatial data is always debated–does it have any applicability as an element of the description when color is assigned by the GIS or image processing software? In MapInfo, for example, all layers are initially displayed without color until the user selects an appropriate color. From a practical point of view, color should be assumed for most GIS data and recorded in other physical details.

For spatial-data sets that have been downloaded from the Internet, or ftp'ed from a remote server and saved on a hard drive, extent and other physical details such as color can be recorded in field 300. But there are no physical dimensions, in the traditional sense, to measure such matters as the size of the diskette or CD-ROM. At the same time, the size of the file occupying space on a computer drive is a manifestation of a physical dimension. The issue of whether file size should be included in field 300 $a is currently under discussion.

The last element of field 300 is accompanying material ($e) which can be used for printed metadata, user manuals, guides, or even print-outs of readme files. The cataloguer can decide whether the abbreviated entry in 300 $e is appropriate, or a separate entry or note may be preferred as a means of providing more detail to the user.

Notes

Notes are especially important for geospatial data as they allow the addition of information about the data which may be of critical importance for the user in deciding about the appropriateness of the data set. As with other types of media, the decision about what notes to include is subject to local policies except where required by certain rules, e.g., source of title proper. Each institution may wish to define a minimum set of notes for the cataloguing of their digital cartographic resources. However, as with the selection of elements to include in the mathematical data and other physical details area, there are certain notes that are more important than others, including new note fields 514 and 551, which were specifically created for geospatial data.

Field 538, a mandatory note for computer files, specifies systems requirements and should display as the first note. Many producers provide precise minimum system requirements with their digital products but it is often the case when a file has been received on diskette or downloaded from the Internet that system requirements are lacking. In preference to the phrase "System requirements unknown," the user is better served by a note in field 538 which gives system requirements as "GIS processing software capable of processing vector data" or "image-processing software." The version number of the software required to manipulate the file is another important element and should be supplied when known.

The format of the data, as mentioned earlier, is one of the most important elements for inclusion in this field, whether the file is in, e.g., ArcInfo Interchange (E00), MapInfo MID, Autocad (DXF), to name but a few of the common formats. Very often the file format only appears as a cryptic abbreviation such as BIL, DXF, TIFF. If the cataloguer is in doubt whether a particular abbreviation is a file format name, the list in section 6.4.2.1.1 of the FGDC standard is particularly helpful. When the data sets exists in another format, a note can be added in field 530, e.g., "Issued also in MapInfo format."

Two new notes were created specifically for geospatial data. In Field 514, Data Quality note, information about the accuracy and completeness of the data (section 2 in the FGDC standard) is recorded. Field 551, Entity and Attribute Information was added to MARC 21 to allow the cataloguer to record either detailed descriptions about the data or give only overview or summary of the entities/attributes. The level of detail provided on the entity/attribute information will again be based on local cataloguing policies.

It is likely that there will also be notes related to the mathematical data area for information that could not be expressed in fields 342, 343 and 352. If a data set has been acquired via a remote source and installed locally, a note will be needed to indicate that the digital data is available on a local computer. Our library, for example, uses field 535 to indicate library specific details.

In Canada, the Restrictions on Access note, field 506, is often applicable for geospatial data to explain restrictions on use such as "For teaching and research purposes only," "Use limited to faculty, students and staff of University XXX," etc.

CLASSIFICATION AND SUBJECT ANALYSIS

What call number should be assigned to digital cartographic items? If this type of material is treated as a form subdivision in the LC G schedule (e.g., .A4 pictomaps, .A9 transparencies, slides), there would be cases where the cataloguer would not be able to assign a subject cutter. Alternatively, the cataloguer could classify the geospatial data by map or atlas number without regard to form, and use a format indicator such as one uses for folio, x-folio, to indicate the location of the CD-ROM or diskette. Each library will have to make a decision with regard to which option is locally preferred.

As is often the case with rapidly developing fields, the choice of valid subject headings in the Library of Congress Subject Headings (LCSH) has not yet kept pace with the need for new subject terms for digital cartographic materials. Currently, the Library of Congress is using Field 653 for form, e.g., "Maps–Digital," or more specifically, "Maps–Digital–Raster" or "Maps–Digital–Vector" (if known), until the heading is formally approved as a form/genre index term. Use of the form division "Digital" or "Digital–Vector" or "Digital–Raster" can also be used after other specific materials designations, e.g., "Remote-sensing images–Digital–Raster."

The use of subdivision, "Databases," as the final element of the subject heading can be an important term to differentiate paper maps from their digital counterpart. This term should only be applied to data which includes both geographic (entity) information and attribute (tabular) data. For example, digital map sheets from the Ontario Digital Topographic Data Base were assigned the heading "Ontario–Maps, Topographic–Databases."

CONCLUSIONS

It is evident from this article that the rules for cataloguing digital spatial data are still evolving. As this article was being written, a sub-committee of AACCCM was reviewing the new rules for electronic cartographic resources in order to identify possible changes or additions to AACR2 for submission to the JSC.

As map cataloguers we are at a crossroads. Should we catalogue our cartographic material in digital form now or should we wait until *Cartographic Materials* is published later in 1999 or possibly 2000? Libraries will have to weigh working with evolving cataloguing rules against the need to make their users aware, through their public catalogue, of the valuable electronic resources they are now acquiring in the map collections.

There is no easy answer: cataloguing digital cartographic materials is going to be challenging, but over time, creating bibliographic descriptions from metadata descriptions will become easier because decisions will be recorded and policies and procedures documented. With the widespread availability of the Internet and electronic mail, cataloguers can be almost instantly informed of decisions made by national bodies on some of the outstanding issues identified in this article. In addition, more experienced map cataloguers are usually willing to provide assistance to their colleagues in other institutions. Map cataloguers have always accepted challenges–now it is time to take on the task of cataloguing the newest form of material to arrive in our collections.

NOTES AND REFERENCES

1. *Cartographic Materials: A Manual of Interpretation for AACR2.* Chicago: American Library Association, 1982.

2. Parker, Velma (ed.). *Geomatic Data Sets Cataloguing Rules.* Ottawa: Canadian General Standards Board and the Canadian Library Association, 1994.

3. The results of this meeting, together with information from a workshop given by Velma Parker at the 1996 Association of Canadian Map Libraries and Archives' annual conference were published in: ACMLA (Association of Canadian Map Libraries and Archives) *Bulletin* 97 (Fall 1996): 1-6.

4. For a list of the contents of these new tags see: Library of Congress, Network Development and MARC Standards Office. *USMARC Concise Format for Bibliographic Data.* November 12, 1997.
<http://www.loc.gov/marc/bibliographic/>

5. Geological Survey of Canada. *Surficial Geology, Digital Map: Low, Quebec (31G/13)* [computer data]. Input scale: 1:50,000. Ottawa: GSC, 1997.

6. Library of Congress. *Guidelines for Distinguishing Cartographic Materials on Computer File Carriers from Other Materials on Computer File Carriers.* Washington, D.C.: Library of Congress, January 1998.
<http://lcweb.loc.gov/marc/cfmap.html>

7. United States. Federal Geographic Data Committee. *Content Standard for Digital Spatial Metadata (revised June 1998).* Washington, D.C.: Federal Geographic Data Committee, 1998. This document may be downloaded from:
<http://www.fgdc.gov/metadata/contstan.html>

8. *Directory Information Describing Digital Geo-Referenced Data Sets.* Ottawa: Canadian General Standards Board, 1995.

9. Although primarily aimed at data producers, the following Web documents are useful:

- USGS maintains a *Frequently-asked questions on FGDC metadata,* October 28, 1998, available on the web:
  <http://geology.usgs.gov/tools/metadata/tools/doc/faq.html>.
- The National States Geographic Information Council has published a *Metadata Primer A "How To" Guide on Metadata Implementation,* June 10, 1998, available on the web: <http://rat.lic.wisc.edu/metadata/metahome.htm>.
- The U.S. Bureau of Land Management has a GIS tutorial, May 1995: <http://www.blm.gov/gis/meta/barney/tut_met1.html>.

10. Mangan, Elizabeth. *Crosswalk: FGDC Content Standards for Digital Geospatial Metadata to USMARC.* Washington, D.C.: Library of Congress, January 24, 1997.
<http://alexandria.sdc.ucsb.edu/publicdocuments/metadata/fgdc2marc.html>

11. Mangan, Elizabeth. *Crosswalk: USMARC to FGDC Content Standards for Digital Geospatial Metadata.* Washington, D.C.: Library of Congress, September 3, 1996.
<http://alexandria.sdc.ucsb.edu/public-documents/metadata/marc2fgdc.html>

12. United States. Federal Geographic Data Committee, 66.

13. Examples derived from:

- Geological Survey of Canada. *Geological Map of Canada, Map D1860A = Carte géologique du Canada, carte D1860A* [computer data]. Version 1.0. Ottawa: GSC, 1997.
- J. Brown et al. *Circum-arctic Map of Permafrost and Ground-Ice Conditions* [computer data]. Version 1.0. Reston, Va.: USGS, 1997.

Cataloging Cartographic Materials on CD-ROMs

Mary Lynette Larsgaard

SUMMARY. CD-ROMs are the most frequently seen varieties of digital geospatial data in map libraries. This article is an overview of how to catalog these CD-ROMs. It is to be used in concert with the Welch and Williams article which immediately precedes it. *[Article copies available for a fee from The Haworth Document Delivery Service: 1-800-342-9678. E-mail address: getinfo@haworthpressinc.com <Website: http://www.haworthpressinc.com>]*

KEYWORDS. Digital geospatial data, CD-ROMs, computer laser optical discs

INTRODUCTION

In the early 1990s, map libraries began receiving a few CD-ROMs, the modest harbingers of the plastic flood to come out of the computer world's answer to Frisbees. As the cost of the CD drives and of CD-mastering plummeted and as CDs took over in the Federal fields of digital information from the most current dinosaur–the 9-track tape–of the digital realm, CDs in ever-

Mary Lynette Larsgaard has since 1988 been Assistant Head of the Map and Imagery Laboratory, Davidson Library, University of California, Santa Barbara. She has a BA in geology, an MA in library science, and an MA in geography. She is the author of *Map Librarianship, An Introduction* (Third edition, 1998, Libraries Unlimited).

[Haworth co-indexing entry note]: "Cataloging Catrographic Materials on CD-ROMs." Larsgaard, Mary Lynette. Co-published simultaneously in *Cataloging & Classification Quarterly* (The Haworth Information Press, an imprint of The Haworth Press, Inc.) Vol. 27, No. 3/4, 1999, pp. 363-374; and: *Maps and Related Cartographic Materials: Cataloging, Classification, and Bibliographic Control* (ed: Paige G. Andrew, and Mary Lynette Larsgaard) The Haworth Information Press, an imprint of The Haworth Press, Inc., 1999, pp. 363-374. Single or multiple copies of this article are available for a fee from The Haworth Document Delivery Service [1-800-342-9678, 9:00 a.m. - 5:00 p.m. (EST). E-mail address: getinfo@haworth pressinc.com].

increasing numbers poured through the doors of the technical-processing areas in charge of by now phlegmatic "map" catalogers–what's yet another format to a person accustomed to air photos, globes, raised-relief models, books, microfiche, and, oh, yes, flat maps. It is rare for any map library, and especially those that are U.S. depositories, to have any less than a couple of hundred CDs–which means that the dread day of cataloging these latest denizens of the map collection sooner or later arrives.

The object of this article is to provide general guidance on the least painful way to get one's CD collection cataloged, focusing on the fields that are different from those prevalent in cataloging hard-copy maps. For detailed information concerning the relatively new geospatial digital-data fields that were provisionally approved in USMARC in 1995, the reader is referred to the article preceding this (by Welch and Williams), which deals with cataloging geospatial digital data, emphasizing the new USMARC fields for digital geospatial data. Recognizing that each ILS (integrated library system) software has its own methods of displaying the fields in the leader and in 001 through fields 008, the form used in certain cases is taken directly from USMARC and in others is as presented in OCLC. OCLC is very frequently used for map cataloging, and therefore its form of record is relatively widely recognized and understood. See the "Crosswalk Table for MARC 21/OCLC/ RLIN Fixed-Field Tags" in the previous double issue for details.

TYPE OF RECORD

In January of 1998, several offices of the Library of Congress (Cataloging Policy and Support Office, Network Development and MARC Standards Office, the Geography and Map Division, and the Special Materials Cataloging Division) issued guidelines for determining when to catalog materials on the so called map format and when to catalog them on the computer-file (formerly machine-readable data file) format, information which is carried in what OCLC terms the "Type" field in the fixed field area; it is the Leader/06 position in USMARC.[1] This was a follow-up to MARBI (Machine-Readable Bibliographic committee) Proposal 95-9, which redefined code "e" from "printed map" to "cartographic material," making it possible for catalogers to use "e" (and thus the map format) not only for printed maps but for all forms of cartographic materials in either hardcopy (as long as non-manuscript) or digital form.[2] Note that this value may be different in some integrated-library-system (ILS) software; for example, NOTIS uses "p" for the map format.

There are some forms of material that–even if they have a cartographic aspect or subject matter–are to be cataloged as computer files: computer games (e.g., "Where In The World Is Carmen Sandiego"); application software including GIS (geographic information systems) software (e.g., ArcInfo); multimedia works which do not have a single significant aspect or which have a non-cartographic significant aspect; materials that are primarily text in nature; and databases with a subject focus that have a geographic interface to the data (e.g., "International Station Meteorological Climate Summary").

It is this last category that is most likely to give the cataloger pause, since upon calling up the information on the computer, the cataloger may assume that any data presented cartographically are therefore cartographic in nature. It is made even more problematic by the point that GIS software results in databases that for as many geographic points as possible give textual information, generally in tabular form (for example, the "Info" side of ArcInfo). The best way to deal with this is to consider this last category of materials to be the inside-out version of a GIS database; that is, it is a system where all the information is collected generally in tabular form and then may be viewed by clicking on an area on a map display with the map display considered only as a display and secondary to the numeric or textual data that is presented. A quick way to determine this is to look at the file extensions; if the bulk are non-image file types, then the item is in all likelihood not considered to be cartographic material. The cataloger will need to look at the files anyway, in order to be sure that the files are indeed digital in nature and not analog. As it happens, a few CDs do have analog instead of digital files on them. Analog files are generally in a video format, and thus will not have the file extensions–e.g., .tif; .gif; etc.–that map librarians are accustomed to seeing.

It is in any case essential for the CD cataloger to have a computer with a CD drive, and–for those cases where there is a reasonable doubt as to the contents of the CD–software (for those cases when the CD does not contain software) to enable the cataloger to view enough to decide into what category the contents of the CD fall.

USMARC Fields

007, 008 and 006

Coded fields–especially when they are non-mnemonic–are easily the most irritating part of cataloging; and those catalogers working with a system that does not have pull-down menus for them are most painfully aware of this. Unfortunately, cartographic-materials catalogers have or quickly gain considerable familiarity at least with 007. It is a mandatory field for full-level cataloging, and is the field in which information relating to the primary nature of the item–its intellectual content and physical form–is placed.[3] At this point the next determination the cataloger must make is whether the digital data is a map, remote-sensing image, or globe, since each has a different set of codes for 007. While one recognizes that globes seem to be by definition three-dimensional:

> A model of a celestial body depicted on the surface of a sphere (US-MARC 007/globe)

yet it is possible to have a digital video of a spinning globe, and in this case one would need to use the 007 values for globe rather than the values for map or remote-sensing image. There has been some discussion concerning the use of 007 for remote-sensing images, given that in it there are no positions for physical medium, type of reproduction, production/reproduction details, or polarity. Thus in order to transmit this information in the coded fields, one would need to use also the 007 for map, which does include each of these fields (respectively positions 4, 5, 6, and 7). It is a bit confusing that the 007 for map still contains, in position 1, "r," for remote-sensing image, which it would seem should be removed since the 007 field for remote-sensing image now exists.

Be that as it may, it is this cataloger's interpretation that the values for an electronic atlas on CD, composed primarily or solely of maps displayed in color, would be, using the 007 for maps:

> 007 |a a [GMD: map] |b d [SMD: atlas] |c [do not use this subfield] |d c [multicolor] |e e [physical medium: synthetic] |f n [type of reproduction: not applicable] |g z [production/reproduction details: other] |h n [positive/negative aspect: not applicable]

It would be better if the position for production/reproduction details that is intended to indicate the photographic technique used to produce the item has the value "n," for "not applicable," rather than leaving the cataloger to decide between the Scylla of "u [unknown]" and the Charybdis of "z [other]." In contrast, the 007 for computer files may be filled in as:

> 007 |a c [computer file] |b o [optical disc] |c [do not use this subfield] |d c [multicolor] |e g [4 3/4 in.] |f [blank: no sound]

If the atlas were in the main or solely composed of satellite images from non-camera sensors, then the 007 field would be:

> 007 |a r [remote-sensing image] |b [blank–no type of SMD appropriate] |c c [altitude of sensor: spaceborne] |d c [attitude of sensor: vertical] |e 0 [percent cloud cover between 0 and 90%] |f f [unmanned spacecraft] |g b [platform-use category: surface-observing] h dv [combination of spectral characteristics]

Next we move on to the 008 and the 006 fields. 008 is intended for coding the primary characteristics of the material and 006 for coding any secondary characteristics of the material. Thus, if one had a serial cartographic item appearing on CD, the 008 would be intended for coding its cartographic aspect, and two 006s for coding the fact that it is a computer file and in serial form.

Values for the 008 field are often given in the mnemonic beginning lines of a record. Each 008 field begins with the same 00-17 and ends with the same 35-39 positions; these have to do with dates, language, when modified, and cataloging source (e.g., in OCLC, Entered, DtSt, Dates, Lang, MRec, Srce). The remaining fields are Relief (in OCLC, Relf), Projection (Proj), Type of cartographic material (CrTp), Government publication (GPub), Index (Indx), and Special format characteristics (SpFm). Mercifully, there is no separate 008 for remote-sensing images or globes (one hastens to add, "At least not yet," in order not to tempt the gods to chuckle, look ruminative, and then cause the fields to be added).

For the 006 field, the values for a commercially produced electronic atlas on CD would be:

> 006 |00 m [computer file] |05 g [general–target audience] |09 c [representational–pictorial or graphic information] |11 [blank; not a government publication]

For whatever reason, the use of the 006 (computer files) for cartographic items on CD is honored more in the breach than in the observance.

VARIABLE FIELDS EXCEPT SUBJECT HEADINGS

Gratefully moving from the fixed fields, we are able to move directly in the variable fields to that of general material designation (GMD) in the title (245|h). Currently what AACR2R states is that two options are possible–"cartographic material" for the British and for Australia, Canada, and the U.S. the word, "map." The Anglo-American Cataloguing Committee for Cartographic Materials (AACCCM) is strongly considering recommending that this be changed to "cartographic material" being appropriate for use by all–given that maps form only a part, not all, of cartographic items–and that one be able to acknowledge both the intellectual content and the physical carrier in the GMD, thusly:

> 245 |h [cartographic material (electronic resource)]

It seems likely that "electronic resource," which appears in the new ISBD(ER), will be accepted by the Joint Steering Committee (JSC) as a substitute for "computer file."[4]

Deciding what constitutes the chief source of information means looking both in Chapter 3 and Chapter 9 of AACR2R, specifically in 3.0B and 9.0B. Taking the two in combination, it appears that anything the cataloger can find on the CD is an acceptable source of information. The chief source for cartographic materials is the item itself (depending upon how one interprets this, either the files on the CD or the CD including the files) and any container or case (e.g., the jewel case), while for computer files it is the title screen; since one begins with Chapter 3, that seems to be the chapter to follow for the definition of chief source

The next field that is different from standard cartographic-material cataloging is that of mathematical data, specifically the sections on scale and on projection. Currently, the phrases that are accepted are, "Scale indeterminable," "Scale not given," "Scales differ" (for many different items in a collection, with more than two scales) and "Scale varies" (for any one item where scale varies within that one item). The problem here is that there is some discussion as to what meaning the concept of scale has when one is dealing with digital data, given that a digital cartographic item may be displayed at many different magnifications and reductions. The opinion amongst the U.S. digital-geospatial-data community is that the phrase should be something on the order of, "Scale not applicable," and that resolution (e.g., size of a pixel in meters for remote-sensing images, which in effect gives the size of the smallest object that may be discerned) is of importance. This latter is presented in a field that we shall deal with later on. In Canada, and probably in other countries, the belief is that the phrase to be used–as appropriate–is "Input scale . . . ," e.g., "Input scale 1:24,000" if digital maps are scans of paper maps at 1:24,000 scale. Neither of these two phrases–"Scale not applicable;" "Input scale . . ."–is approved yet; so the cataloger is well advised with trying to match the situation with the four permissible phrases given in the second sentence of this paragraph, and in having a note on input scale.

Next we come to projection, which now has two possible locations within USMARC. With the many new fields needed for projection (e.g., longitude of central point; etc.), there was insufficient space in USMARC field 255 to shoehorn in all of the new information. Not only that–there was not enough space in the 2xx fields to shoehorn in said information, and thus 342 and 343 were created. The implication is that catalogers will need to approach their systems staff and try to find ways for 342 and 343 (or at least the more important subfields of each) to appear right after 255. The next implication is that either the cataloger will have to enter projection twice–both in 255|b and in 342 01 |a–or that a rule change will need to be requested for AACR2R, in which projection is moved from its current location to immediately after coordinates (255|c), and that a parallel request will need to be made to

MARBI (Machine-Readable Bibliographic Information committee) to have USMARC field 255|b be declared obsolete.

For more detail on field 342, see the article, by Welch and Williams on cataloging geospatial data in digital form, that immediately precedes this article. Speaking generally, this field is used to give information on horizontal and vertical coordinate systems. Its most frequent use will be to indicate projection, grid, and geodetic model. This is still a provisional field and its final form has not been determined, but it seems likely that a display constant will be required for each of the different geospatial reference methods that it is used to indicate. For example, for projection, the first indicator would be 0 (horizontal coordinate system) and the second would be 1 (map projection), so what the library patron would most usefully see would be, e.g.:

Projection: Albers equal area.

Again, for detail on field 343, see the article that immediately precedes this one. This has information on planar coordinate data (e.g., abscissa and ordinate resolution), which in comparison with projection and geodetic datum is relatively seldom provided by the data producer.

Where field 352–Digital Graphic Representation–is to display is another one of those decisions that is not yet made. Given its importance to the user, immediately following 256 (Computer File Characteristics) seems appropriate. This is a field that the cataloger should make every effort to determine a correct value for, since its first component indicates whether the data is stored as raster, vector, or point; for example:

352 |a Raster ; |b pixel.

Also, somewhere in the record, we need the native file format, or file extension; that is, is the file a TIFF file (.tif), JPEG (.jpg), etc.? Here again, this is a question invariably asked by any user of digital data, and it should appear in the body of the record rather than in a note. Perhaps it could be a field in 352, following very soon after |a, Direct reference method; for example:

352 |a Raster ; |b pixel. |h File extension: .tif.

For the moment, the only place it can fit is in 538 (System Details Note), with the portion of that note that deals with software.

Field 256 is another field where there are changes in the works, once again because of the new ISBD(ER). What is currently given in AACR2R that is of most use to the cartographic-item cataloger are, most often, the phrases, "Computer data," and "Computer data and programs"; this cataloger has also seen "Computer documents and maps," but there is nothing in

AACR2R to justify that phrase. ISBD(ER) has in its Appendix C (pp. 90-91; see end note 4) a good many other possibilities, which include phrases such as, "Electronic representational data," with "Electronic map data" being listed as a subcategory under this. "Electronic image data" is yet another option, which unfortunately has the implication–but one suspects not the intent–of including remote-sensing images. Probably it would be more sensible to have something on the order of, "Electronic cartographic-materials data," but then this becomes just a repetition of the GMD and there seems little point–and over time much extra work–in that course of action.

Field 256 also includes information on number of files, number of records, and number of bytes. Counting files and records is analogous to counting plates, and unless the information is specifically given on the item or its accompanying material, it is not necessary to supply it. Given that file size is one of the first questions that any user of digital data asks, it is important to supply total file size if at all possible. While file size is analogous to number of pages or sheets and therefore seems more appropriate logically in the physical description, at the same time placing it here does get it up toward the front of the record.

Another field that has changes in the works, and for the same reasons– ISBD(ER) and the work of the AACCCM–is that of physical description (300). What has been done for digital cartographic items cataloged on the computer-files format is to use, ". . . computer optical discs," a phrase that it is doubtful many library users comprehend. ISBD(ER) proposes to make this a bit more like what the users understand–". . . computer optical discs (CD-ROM)." This will still leave the problem of the physical description looking to many users as if the physical carrier of the data, not the actual information on the carrier, is in color; e.g.:

300|a 1 computer optical disc : |b col. ; |c 4 3/4 cm.

But moving beyond that, for cartographic items cataloged on the map format, there is no specific guidance in AACR2R. The pattern in Chapter 3 of the latter is to use the appropriate specific material designation (SMD) in 300|a along with physical carrier as appropriate, then noting if color in 300|b, and then giving dimensions–either of the cartographic item, its physical carrier, both, or just the physical carrier if it would be difficult or impossible to supply the measurement of the cartographic item; for example:

300|a 1 map on 9 sheets : |b col. ; |c 239 x 200 cm., sheets 30 x 20 cm.

The following seem to be most in parallel with what is currently in Chapter 3 and in Chapter 9:

300|a 184 remote-sensing images on 10 computer laser optical discs : |b col. ; |c discs 4 3/4 in.

Here is another possibility, which is attractive in that it moves the size of the file to physical description:

300|a 184 remote-sensing images (ca. 5 gigabytes) on 10 computer laser optical discs : |b col. ; |c ca. 30 megabytes, discs 4 3/4 in.

This brings up a problem almost immediately. Whether a digital file is presented in gray-scale (which would be considered to be black and white) or in color is a mixture of a capability of the software and of the user's requirements. What we mean in this case is that provided the user has software with this capability, the image can be viewed in color.

The subfield for accompanying material, 300|e, seems most appropriately used only for print texts and guides that accompany the CD, although it would be possible to make a case for the "readme" files that thankfully are included on most CDs. Certainly these files could be indicated in this subfield if they are printed out; or their presence on the CD may be indicated in a note.

Within the notes fields, the ones of special interest in cataloging geospatial digital data are 514 (Data quality), 516 (Type of computer file or data), 530 (Additional physical form available), 538, and 551 (Entity and attribute information). For discussion of fields 514 and 551, see the article preceding this one. Just as a general statement, the most frequently used of the subfields for these two notes seems likely to be:

a. 514 |a Attribute accuracy report, |g, Horizontal position accuracy value in meters, and |j, Vertical positional accuracy value in meters;
b. 551 |o, Entity and attribute overview.

While the order of notes is again a matter of discussion, keep in mind that–provided your ILS allows it–538 (Systems details) should come right after nature-and-scope notes, such as 516. Note 516–which this author has not seen used in records for geospatial-data CDs–seems to be a slightly different version of 352, as per the following example from the USMARC homepage (http://www.loc.gov/marc/bibliographic/ecbdnot1.html#mrcb516):

516 |a Numeric (Spatial data: Point)

although perhaps the following would be more appropriate:

516 |a Representational (Spatial data: Point)

Both indicators are blank in this example; the first indicator blanked means that the display constant, "Type of file:" may be generated (a value of 8 would mean that no display constant is generated), and the second indicator is undefined.

Note 530 comes just before 520 (Summary); it is used to give information on the hardcopy version of the digital item, e.g.:

530 |a Electronic version of: Times atlas of the world.

Its indicators are both undefined.

SUBJECT HEADINGS

Here again, there is a gulf between what the cataloger would like to do in order to provide maximum access to the items, and what the rules–in this case, the *Library of Congress Subject Headings* (LCSH)–permit; and once again, work is proceeding within the map-cataloging community to change matters.[5] The core of the situation is that there are several different aspects of the material that one wishes to present to the user, and preferably all in one subject heading:

a. intellectual content:

    i. geographic area

    ii. theme or topic (e.g., geology)

b. physical carrier

    i. digital

        a. raster, vector, point

    ii. type of digital carrier, e.g., CD-ROM, magnetic tape, computer hard drive, etc.

Physical-carrier information does appear in other fields (e.g., 300; 352), so it may not be necessary to have subject headings that incorporate it; this depends upon the abilities of the ILS in one's library. LCSH takes care of intellectual content in a satisfactory manner. Where we have troubles is with the physical carrier. An Ad Hoc Task Force on Form/Genre, organized by the Geography and Map Division of the Library of Congress, is recommending that headings such as, "Maps–Digital–Raster" be added to LCSH. Until that, libraries are thrown back on the use of 653 (Uncontrolled subject heading), or 655 (Index term–genre/form), specifying local form/genre headings (|2 local), in order to indicate that the item is digital and the carrier is a CD-ROM.

The exception to this would be for GISs–the data, not the software to run them–which are indeed databases and for which one may then therefore use the subdivision, "Databases." At about the same time as the substantial narrowing of the situations for which the computer-files format of USMARC may be used, the use of "Databases" was also much restricted. Prior to May of 1996, this subdivision was used for any computer file except software; now it is restricted to databases; see, in the LC *Subject Cataloging Manual* "Databases H 1520."[6] This excludes its use for remote-sensing images in digital form, which are not databases, although they may be part of a database, such as a GIS. Should software appear on the CD, then a subject heading that has "Software" as a final subdivision must be present in the record.

This leads nicely into what is an easy error to fall into while cataloging CDs, and that is the use of subject headings meant to be applied to works ABOUT a topic, rather than to items that ARE that topic. Classic examples of this are: Computer mapping; CD-ROMs; Digital Mapping; Maps, Statistical. These are intended for use in bibliographic records for works ABOUT each of these topics, and are not to be used as, in effect, free-floating subdivisions such as, "Maps."

CONCLUSION

Although we have come a long way since our beginning to catalog CDs–as evidenced in earlier articles such as Kollen and Baldwin 1993[7]–this is still very much in process, given the work both of the AACCCM and of ISBD(ER). The former would like to have a second edition of *Cartographic Materials: A Manual of Interpretation for AACR2*[8] published no later than the year 2000, and preferably in late 1999; the Task Group on Harmonization of ISBD(ER) of the American Library Association's Committee on Cataloging:Description and Access (CC:DA) will have a report in by mid-1999 as to recommendations for change to AACR2R. Until then, the intrepid cataloger of digital geospatial data on CD-ROM is advised to hew to the first principle of cataloging–giving an accurate, brief description of the item in hand–keep advised of changes in the rules, and use the manual, Cataloging Internet Resources, as an assist.[9]

ENDNOTES

1. *Guidelines for Distinguishing Cartographic Materials on Computer File Carriers from Other Materials on Computer File Carriers.* Washington, D.C.: Library of Congress, 1998. Available over the Web at: http://lcweb.loc.gov/marc/cfmap/htm

2. *Encoding of Digital Maps in the USMARC Bibliographic Format.* Washington, D.C.: Network Development and MARC Standards Office, 1995. (MARBI Proposal no. 95-9) MARBI proposals are available over the Web at: http://lcweb.loc.gov/marc/

3. *USMARC Concise Bibliographic*. Washington, D.C.: Library of Congress, [1969?-] Available over Web at: http://lcweb.loc.gov/marc/

4. *International Standard Bibliographic Description (Electronic Resources), ISBD(ER)*. Wiesbaden: Saur, 1998.

5. Library of Congress. Subject Cataloging Division. *Library of Congress Subject Headings*. Washington, D.C.: Library of Congress, 1986-

6. Library of Congress. Cataloging Policy and Support Office. *Subject Cataloging Manual*. 5th ed. Washington, D.C.: Library of Congress, Cataloging Distribution Service, 1996-

7. Kollen, Chris, and Charlene Baldwin. "Automation and Map Librarianship: Three Issues." Special Libraries Association Geography and Map Division *Bulletin* 173(1993):24-38.

8. *Cartographic Materials: A Manual of Interpretation for AACR2*. Chicago: American Library Association, 1982.

9. Olson, Nancy B. *Cataloging Internet Resources: A Manual and Practical Guide*. 2d ed. Dublin, OH: OCLC Inc., 1997.

CLASSIFICATION AND SUBJECT ACCESS OF CARTOGRAPHIC MATERIALS

Navigating the G Schedule

Susan Moore

SUMMARY. The author explores the development and use of the Library of Congress Classification G schedule. The organization of the schedule is discussed, as is the application of the schedule to the creation of class numbers. Some weaknesses of the schedule are also discussed. *[Article copies available for a fee from The Haworth Document Delivery Service: 1-800-342-9678. E-mail address: getinfo@haworthpressinc.com <Website: http://www.haworthpressinc.com>]*

KEYWORDS. Cartographic materials, maps, atlases, classification, Library of Congress, G schedule

Susan Moore is Assistant Professor, Cataloging Department, Rod Library, University of Northern Iowa, Cedar Falls, IA 50613.

The author gratefully acknowledges the assistance of Joan Loslo and Beth Clausen in the preparation of this article. She also thanks the editors for their time and perseverance.

[Haworth co-indexing entry note]: "Navigating the G Schedule." Moore, Susan. Co-published simultaneously in *Cataloging & Classification Quarterly* (The Haworth Information Press, an imprint of The Haworth Press, Inc.) Vol. 27, No. 3/4, 1999, pp. 375-384; and: *Maps and Related Cartographic Materials: Cataloging, Classification, and Bibliographic Control* (ed: Paige G. Andrew, and Mary Lynette Larsgaard) The Haworth Information Press, an imprint of The Haworth Press, Inc., 1999, pp. 375-384. Single or multiple copies of this article are available for a fee from The Haworth Document Delivery Service [1-800-342-9678, 9:00 a.m. - 5:00 p.m. (EST). E-mail address: getinfo@haworthpressinc.com].

INTRODUCTION

There have been a number of different classification systems developed for use in map librarianship over the years. The one commonality most of these systems share is that maps are classified first by area covered, and then by subject, followed often by scale or date. Currently, the classification probably most widely used that follows this theoretical basis is the Library of Congress' *Library of Congress Classification, Class G: Geography, Maps, Anthropology, Recreation,* most often referred to as the "G schedule." The G schedule was first published in 1910, but this first edition did not include a classification schedule for atlases or maps (Chan 1990, p. 166). The second edition of the schedule, published in 1928, included a provisional scheme for atlases. The third edition, published in 1954, included a provisional scheme for maps that had been drafted in 1945. The fourth edition–as noted earlier, *Library of Congress Classification, Class G: Geography, Maps, Anthropology, Recreation*–was released in 1976, and included major revisions to both the atlas and maps portion. The Library of Congress has been converting the schedules into the USMARC Format for Classification Data and should have the complete G schedule in digital format sometime in 2000.

APPLYING THE G SCHEDULE

Despite the revision, the basic layout of the cartographic-materials portion of the G schedule has remained the same. The first portion (currently G1000-3122) covers the classification numbers to be assigned to atlases. Next is the portion concerning globes (G3160-3182), and then, the portion covering maps (G3190-9980). There are maps in both the atlas and maps sections that indicate the regional areas for which classification numbers have been established in the schedule. After the maps portion is the section containing special instructions and tables of subdivisions for atlases and maps.

Each section of the cartographic-materials portion of the G schedule starts with class number ranges for celestial cartographic materials, then the world as a whole, and then (with the exception of the globes portion) call number ranges for the various regions of the world. Included in the call number range of world atlases (G1001-1046) are:

- call numbers for atlases of facsimiles (G1025-1026);
- atlases of cities (G1028);
- atlases of islands (G1029); and
- atlases by subject (first historical atlases (G1030-1038) and then other subjects (G1046)).

Then the schedule starts dividing up the world, with call numbers for:

- the Northern Hemisphere (G1050);
- the Southern Hemisphere (G1052);
- the Tropics (G1053);
- the Polar regions (G1054 for general works and G1055 for Arctic regions); and
- maritime atlases (G1059-1061).

The beginning of the maps section starts the same way, although a range for temperate zones is included and the maritime section is dropped.

The major segments of both the atlas and map portions of the G schedule divide up the world. The first division is into the Western and Eastern Hemispheres. The Western Hemisphere is the first portion covered, with the Eastern Hemisphere, Australasia, and the oceans coming later. Each of these sections is further broken down into major regions and then further subdivided into the political entities contained in those regions. For example, the Western Hemisphere in the maps portion starts with a range of numbers for the hemisphere as a whole (G3290-3292), next a range of numbers for North America (G3300-3302 for general works), and then North America is divided into sections. The first section is the Great Lakes aggregation (G3310-3312), then the Atlantic coast and continental shelf (G3320-3321) (further subdivided by the Gulf coast and shelf (G3330-3331)), then the Pacific coast and continental shelf (G3350-3351). After this, the hemisphere is divided by means of the major countries and regions of Greenland, Canada, the United States, the Caribbean area, Mexico, Central America, the West Indies, and South America. Each of these is further subdivided, where appropriate, into smaller political units. Canada is divided into provinces, the United States and Mexico into states, and Central and South America and the West Indies into countries. A similar division is done for the rest of the regions.

Countries (and states in the United States) are given a range of five numbers, ending in 0-4 or 5-9. Other geo-political entities (such as the states of Mexico) are given a range of four numbers, either 0-3 or 5-8. Regions (such as Eurasia and Central America) have a range of three numbers, either 0-2 or 5-7. The sequencing for each range of numbers, which is given in the Schedule in Table III, is as follows:

0 or 5 General

1 or 6 By subject (the subjects are listed in Table IV)

2 or 7 By region, natural feature, etc., when not assigned individual numbers, A-Z

3 or 8 By major political division (Counties, states, provinces, etc.) when not assigned individual numbers, A-Z

4 or 9 By city or town, A-Z

By changing the final digit in the class number, you change the meaning of the number. For example, the range of numbers assigned to the Western Hemisphere is G3290-3292. A general map of the Western Hemisphere would have the base number G3290 (the zero from Table III indicating a general map). If you had a map showing the vegetation types in the Western Hemisphere, you could take the range and give the map the base number G3291 (the one from Table III indicating a subject map). You would not, however, create a call number for a map of Buenos Aires from the range for the Western Hemisphere. This is because the class number range for the Western Hemisphere does not include a number for cities (which would have been G3294, which doesn't exist). Instead, Buenos Aires would be classed in the range for Argentina, which is G5350-5354, and more specifically in G5354, which is the number for cities of that country.

The subject codes, or "Cutters," that are to be used with call numbers ending in either 1 or 6 are given in Table IV. These codes are alphanumeric, with a letter indicating the general form or subject covered and a decimal number to provide for finer analysis. Although they are commonly called subject Cutters (because they are letter-number combinations used to keep items in a given order), they are not mnemonic–that is, they are not derived, as author and geographic Cutters are, from the first letter or letters of a given work–and thus the letter assigned to the subject has no intrinsic connection to the subject. There are seventeen major subject groups, and the number of subtopics under each subject group varies. These codes may be used following either a class number ending in 1 or 6, or any geographic Cutter. The summary of the Form and Subject Subdivision is as follows:

.A Special category of maps and atlases
.B Mathematical geography
.C Physical sciences
.D Biogeography
.E Human and cultural geography. Anthropogeography. Human ecology
.F Political geography
.G Economic geography
.H Mines and mineral resources
.J Agriculture
.K Forests and forestry
.L Aquatic biological resources
.M Manufacturing and processing. Service industries

.N Technology. Engineering. Public works
.P Transportation and communication
.Q Commerce and trade. Finance
.R Military and naval geography
.S Historical geography

Table IV contains the further refinement of each of the sections listed above with the exception of .S, Historical geography. Since each area's history differs, preference is given to chronological subdivisions for each area that are provided in the range for the area for which they have been developed. If a chronology has not been developed for a region, the table includes these general guidelines: .S1 for General history; .S12 for Discovery and exploration; .S65 for World War I; and .S7 for World War II. Any other historical period should fit into these general guidelines. As an example of the usage of the subject codes and going back to our example given earlier, the class number for the vegetation map of the Western Hemisphere would have the base call number G3291.D2.

Following the dictates of Table III, regions and natural features are given base numbers that end in either 2 or 7. Cutters for some regions and natural features have been published in the G schedule as examples from the Library of Congress. For instance, if you had a map of Latin America, it would be given the base number G3292.L3 (G3292 for region in Western Hemisphere, L3 for Latin America). Many of these Cutters have changed over the years, so it's important to verify the Cutter in the publication *Geographic Cutters* (for regions and natural features in the United States) or through online searching. Occasionally, additional Cutters are published in cartographic librarianship journals (e.g., Hoehn, 1992; Sherwood, 1992). Care must be taken in establishing Cutters for regions. For example, one might be tempted to create a Cutter at G3292 for Central America as a region of Latin America. But further exploration of the schedule reveals that a separate call number range has been established for Central America (G4800-4802). It is, therefore, a good idea to check the maps printed in the schedule that show the regions used by LC as well as checking the index to the schedule.

Cutter numbers for main entries are derived from a table that is published in the *Map Cataloging Manual* (p. 1.19). Table A is a simplified version of the standard LC Cutter table.

Table III also contains a fairly detailed list of guidelines to be used in determining where a region should be classed if the region is located in two or more administrative divisions. If a feature is in two administrative divisions, it is classified with the division containing the greater part of the feature. If the feature is evenly distributed within both divisions, the feature is classified with the division that comes first alphabetically. If the feature falls within three or more divisions, it is classified with the next larger geographi-

TABLE A

After the initial letter S, with a second letter of:	a	ch	e	hi	mop	t	u		
use number:	2	3	4	5	6	7 8	9		
After the initial letter Qu, with next letter of:	a	e	i	o	r	y			
use number:	3	4	5	6	7	9			
After other initial consonants, with a next letter of:	a	e	i	o	r	u	y		
use number:	2	4	5	6	7	8	9		
After initial vowels, when the second letter is:	b	d	l m	n	p	r	s t	u y	
use number:	2	3	4	5	6	7	8	9	

cal region that includes the entire feature. The guidelines also cover islands that do not have one distinctive number. Such an island is classified as a regional division of the area of which it is a geographical part; this is done even when the island is also a political unit of the regional division.

The final two sets in Table III allow for the subdivision by major political divisions such as counties, states, and provinces (3 or 8) or by city or town (4 or 9). As elsewhere in the table, guidelines have been established to assist in the creation of class numbers. Cutters for many major political divisions are published in the G schedule itself. Given that the extensive number of cities and towns in the world preclude including Cutters for them in a printed G schedule, the Library of Congress has made available a listing of U.S. Cutters in microfiche format, called *Geographic Cutters*. This publication is available through the Library of Congress' Cataloging Distribution Service. Updates may also appear in the literature (e.g., Hoehn, 1992). One Cutter that does appear for cities in Table III is the Cutter used for collective cities, which is A1. If you had an atlas of the cities in Texas, the base number would be G1374.A1.

As previously noted, subject codes can also be used after a regional Cutter if there is a need for one. The geographic Cutter would always come first, then the subject code. For example, if you had a map of the parks in New York, New York, the call number would be G3804.N4G52 (G3804.N4 for New York City and G52 as the subject code for parks and monuments).

Table III also provides a method for classifying smaller political or administrative divisions within cities, and for some administrative areas. Added to the base number for the larger region is a colon and the number 3, followed by a Cutter for the division. For example, a map of the Sam Hughes Neighborhood in Tucson, Arizona would be classed at G4334.T8:3S35 (G4334.T8 for Tucson, :3 to indicate an administrative subdivision, and S35 for the Sam Hughes Neighborhood). Regions within cities that are not administrative divisions can be classified by adding a colon and the number 2 followed by a Cutter for the region. For example, a map of New York City's Central Park would be classed at G3804.N4:2C4 (G3804.N4 for New York City, :2 to indicate region, and C4 for Central Park).

ADDITIONS TO THE BASE NUMBER

Once the base number for the map has been established, some way to differentiate various maps of the same regions needs to be applied. Although similar components are used in this effort, the way the components are arranged differs for maps and atlases. Atlases follow the standard practices for other monographic and/or serial publications. An item Cutter for the main entry is added to the base number (which can include a regional Cutter and/or a subject code). If the atlas is a monograph, the Cutter number is followed by the date of situation in the call number. If the atlas is a serial, no date is included in the call number although one would be recorded on the item itself. For example an atlas of Santa Barbara County, California, created by the Thomas Brothers Map Company and showing conditions in 1980 would have the call number: G1528.S3 T5 1980 (G1528.S3 for Santa Barbara County, T5 for Thomas Brothers, and 1980 for the date).

A monographic map call number is comprised of the base number followed by the date of situation. If you had a map showing the vegetation types prevalent in Oklahoma in 1930, with the map itself published in 1970, the date added to the base call number of G4021.D2 would be 1930 since the map shows the situation as it existed in that year, not 1970: "The date in a map call number is always the date of situation, *except* when a history (S+) code has been used, in which case the date in the call number is that of publication" (Library of Congress, 1976, p. 206, emphasis added).

Map sets that are on-going, that is, that have maps being added to them over time, are usually cataloged with an open entry. Recording the date of situation in the call number would be difficult, as that changes with each new addition. In these cases, the denominator of the representative fraction of the scale, subtracting the last three digits, is used. A small letter "s" is added in front of the denominator to distinguish it from dates. If the scale of the set is larger than 1:1,000, the denominator is treated as a decimal and preceded by a zero (for example, 1:300 would be given in the class number as s03). Sets that have varying scales have "svar" added to the base number, while sets that have no scale recorded have s000 added. Thus, a set of topographic maps of Georgia at a scale of 1:250,000 issued by the Georgia Department of Transportation would have the call number G3920 s250 G4.

Serial maps have the word "year" added to the date/scale area of the call number (Library of Congress, 1991, p. 1.18). A map of Santa Barbara County created by Thomas Brothers and published annually would have G4363.S3 year .T5 as its call number. As with serial atlases, a date would be recorded on the item itself.

Following the date/scale portion of the map call number is the area for the Cutter number representing the main entry, often the authority responsible for the map or maps but occasionally the title. The Cutter number is derived using the table given above.

PROBLEMS IN USING THE G SCHEDULE

Despite the careful construction of the G Schedule and the modifications that have been made over the years, there remain some problems in using the schedule. One indication of these problems is that there are twenty-three pages of explanation on how to create call numbers in the *Map Cataloging Manual*. Most of the explanations contained in the manual cover situations that could not be fully explained in the schedule itself.

One difficulty in using the G schedule is assigning a class number to a feature or region that is included in three or more political divisions. Following the guidelines in Table III, the feature is classified with the next larger geographical region that contains the entire feature. Unfortunately, the next larger geographical region can include more area that is not related to the feature than area that is, so specificity will often to a greater or lesser extent be lost. For example, an atlas of the Great Plains would be classed in G1382 (regions, natural features, etc., of the West), which includes a considerable area that isn't part of the Great Plains. It is at times difficult to determine what region includes all of the feature or region that is the subject of the piece in hand. Almost all classification schemes have difficulty dealing with non-political areas, and even when an item of political areas is in hand, it may well include states or countries that don't mesh well with the divisions that the G schedule has available.

Another problematic topic is the classification of regions of the oceans. When you look at the portion of the G Schedule that covers oceans (G9095-9794), you might notice that most of the call numbers refer to islands, not the areas in the oceans. As the floors of the oceans are further and further explored and the results of the explorations are mapped, the need to assign classification to ocean regions will increase. Perhaps the best known feature on the ocean floor is the Mariana Trench, located in the South Pacific. There may just be a typographic error in the schedule, but the call number range for the South Pacific Ocean is G9250-9251. G9250 is for general maps of the South Pacific, G9251 is for subject maps. There is no provision for region or natural-feature maps, which given the general regional pattern one would expect to be G9252. Therefore, to be on the safe side, maps of the Mariana Trench would probably be classed at G9232.M3 (G9232 for regions of the Pacific Ocean and M3 for the Mariana Trench). Again, the problem of lack of specificity arises.

Coastal regions have also proved problematic, especially the coastal regions of North America. Maps of these regions have varied greatly on the scale, the content, and the regions included in the maps, especially given the bicoastal nature of countries such as Canada and the United States. Therefore, the Library of Congress has decided that all maps of the North American coast as a whole, the United States and Canadian coasts combined, the United States coast, and the Canadian coast will all be classed at G3320 (Library of Congress, 1991, p. 1.2). Maps that show the Atlantic or Pacific coasts of North America are classed using the base number for the appropriate region, state, or province. All coasts (except where the coast has its own four-digit number) receive the Cutter .C6. This Cutter is used with base numbers ending in either 2 or 7, and treated as any other regional Cutter.

There are some general problems with the schedule as well. A standard criticism leveled at the Library of Congress classification schedules in general and the G Schedule in particular is that they are Anglo-American centric (Frost 1989, p. 50). This is due to the nature of the development of the schedule. The classification schedules have been developed based on the materials collected by the Library of Congress. And, until fairly recently, it has been difficult to obtain maps of some regions. For some areas, obtaining larger-scale maps continues to be problematic. If the maps can't be obtained, they can't be classified.

Another standard criticism is that the schedules are not updated as quickly as countries have name changes. This was particularly true for the atlas portion of the schedule. It could be argued that no classification scheme–except an online one–can keep pace with the changes that occur, as quickly as they occur. The Library of Congress's Geography and Map Division Cataloging Team has as part of its charge to expand and revise the G Schedule. Very few of the other classification schedules that can be used for cartographic materials can say the same (Larsgaard 1987, p. 119).

Despite these problems, the G schedule remains the classification method of choice for many map libraries. Continued work on improving the G schedule will make it even most useful.

BIBLIOGRAPHY

Chan, Lois Mai. *Cataloging and Classification.* 2nd ed. New York: McGraw-Hill, 1994.

Chan, Lois Mai. *Immroth's Guide to the Library of Congress Classification.* 4th ed. Englewood, Colo.: Libraries Unlimited, 1990.

Drazniowsky, Roman, comp. *Map Librarianship: Readings.* Metuchen, N.J.: Scarecrow Press, 1975.

Frost, Carolyn O. *Media Access and Organization.* Englewood, Colo.: Libraries Unlimited, 1989.

Hoehn, Phil. "Cutter List, California Cutter 4362 and 4364." *Information Bulletin,* Western Association of Map Libraries 23(1992): 182-185.

Larsgaard, Mary Lynette. *Map Librarianship: An Introduction.* 2nd ed. Littleton, Colo.: Libraries Unlimited, 1987.

Larsgaard, Mary Lynette. *Map Librarianship: An Introduction.* 3rd ed. Englewood, Colo.: Libraries Unlimited, 1998.

Library of Congress. *Library of Congress Classification: Class G, Geography, Maps, Anthropology, Recreation.* 4th ed. Washington: Library of Congress, 1976.

Library of Congress. Geography and Map Division. *Geographic Cutters.* Washington: Cataloging Distribution Service, Library of Congress, 1987.

Library of Congress. Geography and Map Division. *Map Cataloging Manual.* Washington: Library of Congress, 1991.

Nichols, Harold. *Map Librarianship.* London: Clive Bingley, 1982.

Sherwood, Arlyn. "Cutters, a List." *Information Bulletin,* Western Association of Map Libraries 23(1992): 174-181.

Womble, Kathryn, and Mary Larsgaard. "The Map Cataloging Manual: Autobiography or Leadership Manual?" *Meridian,* no. 7(1992): 33-43.

Subject Analysis
for Cartographic Materials

Katherine H. Weimer

SUMMARY. Cartographic materials portray subject matter, focused on geographical area, with themes and cartographic forms as other facets of interest to users. Subject headings provide access to geographic areas and subject matter, both of which are significant to reference work and organization of map collections. This article focuses on the Library of Congress subject headings system, and its method of application for cartographic materials. Specific formats–including atlases, views, globes, charts and digital maps–and typical problem areas–such as geographic names, coastlines, boundary maps, ancillary maps, facsimiles, topographic quadrangles, and maps accompanying books–are discussed. *[Article copies available for a fee from The Haworth Document Delivery Service: 1-800-342-9678. E-mail address: getinfo@haworthpressinc.com <Website: http://www.haworthpressinc.com>]*

KEYWORDS. Subject analysis, subject cataloging, Library of Congress subject headings (LCSH), cataloging of cartographic materials, map cataloging, geographic subject headings, geographic names

---

Katherine H. Weimer, MLIS, Louisiana State University; BS, Texas A&M University, is Assistant Director of Cataloging, and Head, Monographs Copy Cataloging, Sterling C. Evans Library, Texas A&M University, College Station, TX 77843-5000. She is affiliated with the American Library Association, Association for Library Collections and Technical Services, Map and Geography Round Table, On-Line Audiovisual Catalogers, and Beta Phi Mu.

[Haworth co-indexing entry note]: "Subject Analysis for Cartographic Materials." Weimer, Katherine H. Co-published simultaneously in *Cataloging & Classification Quarterly* (The Haworth Information Press, an imprint of The Haworth Press, Inc.) Vol. 27, No. 3/4, 1999, pp. 385-404; and: *Maps and Related Cartographic Materials: Cataloging, Classification, and Bibliographic Control* (ed: Paige G. Andrew, and Mary Lynette Larsgaard) The Haworth Information Press, an imprint of The Haworth Press, Inc., 1999, pp. 385-404. Single or multiple copies of this article are available for a fee from The Haworth Document Delivery Service [1-800-342-9678, 9:00 a.m. - 5:00 p.m. (EST). E-mail address: getinfo@haworthpressinc.com].

INTRODUCTION

Cartographic materials inherently portray subject matter, focused on the geographical area covered. Secondary features are specific topics or cartographic forms. Geographic and topical subject headings which are assigned provide access to the variety of characteristics of the map. Larsgaard and others have stated that the most frequently asked map reference questions concern geographic area and subject.[1] Subject headings are crucial access components of the map catalog record. This article will focus on the Library of Congress subject headings system, which is widely used by many libraries in the United States.

HISTORY OF SUBJECT ANALYSIS

In 1876, Cutter stated in his "objects" of the catalog the role of subject entries:

> 1. To enable a person to find a book of which . . . the subject is known [and]

> 2. To show what the library has . . . on a given subject [and] in a given kind of literature.[2]

These points guide the formation of the catalog in two areas. The goal of the first objective is the creation of a finding list for locating materials. The second point focuses on the grouping of material by topic or genre. Both objectives are supported by a system of controlled vocabulary.

As thought and discussion on subject analysis evolved in this century, the concepts of specific entry and controlled vocabulary became standard principles in the *Library of Congress Subject Headings* (hereafter referred to as LCSH). The purpose of specific entry is to use an entry term that exactly describes the material, as compared to a broader term or discipline. The use of controlled vocabulary requires that the terms are authorized and maintained. As new knowledge is created, new subject headings must be created. In addition, older subject headings may need revising due to changes in the use of the term.

TOOLS AND MANUALS

A number of manuals and tools are available to guide the cataloger in subject analysis. The Library of Congress publishes a group of materials, headed by the *Subject Cataloging Manual: Subject Headings (SCM).* Additions and changes are noted in their *L.C. Subject Headings Weekly Lists*[3] and *Cataloging Service Bulletin.* Discussion about subdivisions can be found in "SCM Memo H1095 Free-Floating Subdivisions" and *Free Floating Subdivisions: An Alphabetical Index.* An in-depth and heavily used resource is Chan's *Library of Congress Subject Headings: Principles and Application,* which is currently in its third edition. For cartographic material, the Library of Congress, Geography and Map Division's *Map Cataloging Manual* contains a chapter on subject analysis. Its section, "Special Applications and Instructions," details some typical cartographic problems and how they should be resolved.

While manuals and guides concerning subject analysis and headings proliferate, no written code governing these functions exists. Haykin's *Subject Headings: A Practical Guide,*[4] published in 1951, described the Library of Congress' practice, but was not constructed as a code. Over the years, many have called for the creation of a code, charging that the inconsistencies and differing interpretations of the principles are the result of the lack of a code.[5] Nevertheless, persons who assign subject headings must understand the structure of subject headings in order to build internal consistency within the catalog, as well as a predictable access mechanism for the user.

The geographic area is the most important factor in map reference work and organization of map collections. The most common map reference questions include at the very least area and often subject, both of which are represented in the subject headings. Subject analysis, and the assigning of subject headings, serve the collection by gathering together the maps of the same place by subject entries in the catalog. Subject analysis is important not only to the assignment of appropriate subject headings, but because it relates closely to, and is complementary rather than exactly duplicative of, classification.

Determining the subject matter for cartographic materials follows the same principles and strategies used for books and other materials. An overall "read" of the map–including the title, text, indexes, and legend–begins the subject-analysis process. A consideration of publisher's intent, how the item fits into the local collection, as well as the needs of the local user, are also factored into determining what subject headings should be assigned.

Some general instructions from the *Map Cataloging Manual,* Chapter 4, include:

1. There is no limit imposed on the number of subject headings that can be given to a particular item for full-level cataloging, provided they are properly applied.

2. A map of three places that must be classified in the area that contains them all need not receive the tracings for the larger area. The tracing for the three places concerned may be sufficient.
3. A general work may receive a secondary subject tracing for a non-general subject as long as the second subject is not subsumed in the concept of the more general heading.
4. A map of a broad subject receives the subject heading that corresponds to the content of the item as a whole. Only rarely is one component of the broad heading singled out for tracing.
5. All subject headings traced must be justified by something that appears in the record. The justification may appear in the call number, title, series, notes, etc.

As described in the Subject Cataloging Manual, the first subject heading will be "the one that most nearly corresponds to the classification number (i.e., the heading that represents the predominant topic of the work)."[6] For general maps, the geographic area is the first element in the subject string, and a form subdivision is added (e.g., *Texas–Maps*). For most topical maps, the first subject heading is the topical heading to which the geographic location is attached (e.g., *Mines and mineral resources–Colorado–Maps*). Other topical maps may have as first subject entry a heading for the geographic area with a topical or form subdivision attached (e.g., *Texas–Road maps*). Often, cartographic materials contain subject matter covering many topics. In Chapter 4 of the *Map Cataloging Manual,* a "Subject Headings Decision" table provides formulas for determining when a map with numerous subjects is assigned specific headings, broad headings, or both. For instance, "three distinct subjects in a work usually receive the broader subject heading, unless the heading is very much broader than the combination of the 3 individual headings. Four subjects always receive the broader subject heading, regardless of the extra scope."[7]

Experienced monograph catalogers will recognize this as different from guidelines for text materials, which are that: "for a work covering two or three topics treated separately, a heading representing precisely each of the topics is assigned. The two or three specific headings are assigned in favor of a general heading if the latter includes in its scope more than three subtopics."[8] Another difference from subject cataloging for books is the fifth point stated above concerning justifying subject headings. Subject headings for books need not be justified in notes,[9] or any other manner.

GEOGRAPHIC COVERAGE

Understanding the geographic aspect is crucial to proper subject analysis. Both name and subject headings, their construction and their authority records, are important in the subject analysis of maps since geographic names are the predominant heading for maps. Geographic subject headings consist of two types. The first is the political jurisdiction, e.g., a country, state or city, such as France, Ohio, or New Orleans. These often change over time, due to change in political structures. Jurisdictional names are constructed following descriptive cataloging rules in the *Anglo-American Cataloging Rules, second edition, revised* (AACR2R).

The other type of geographic subject heading is the geographic or topographic feature, which includes named rivers, lakes, or mountains, or named places (e.g., battlefields, National parks or forests), as well as established regions, such as *Canada, Western* or *Southwest, Old*. Non-jurisdictional names are constructed using the SCM Memo H690. Guidelines for qualifying geographic names are given in SCM Memo H810.

Comprehending when a specific heading is constructed under which guidelines is sometimes unclear. While headings for highways and parks are understandably subjects, and colleges and airports are clearly names, the headings for refugee camps are subjects, and headings for concentration camps are names. The SCM Memo H405 clarifies the two types of headings with their respective USMARC tagging under two lists. The first is the Name Authority Group Headings, the second is the Subject Authority Group Headings. The SCM Memo H405 is also commonly referred to as the 'Division of the World.'

One common difficulty with determining the place name(s) for a map is the name change of the geographic area shown on the map. Wars and political unrest have led to changes in names of many locations. The SCM specifies in Memo H708 that catalogers assign subject headings under the latest name of political jurisdiction. SCM Memo H710 details what should be done with jurisdictions which have merged or split. An example is the heading *Russia (Federation),* which is valid for the current jurisdiction. The cataloger is directed in the authority record to do the following: "DESCRIPTIVE USAGE: Name heading for the government after 1991. Use 'Russia' as qualifier for places within the country."[10] By comparison, a note in the authority record for *Russia* directs the cataloger in its proper use: "DESCRIPTIVE USAGE: This heading is to be used for corporate bodies and jurisdictions which represent Tsarist Russia from ca. 15th cent. to the 1917 Revolution; corporate bodies and jurisdictions which existed into the Soviet period, i.e., 1917-1991, should use name headings for either Russian S.F.S.R. (qualifier R.S.F.S.R.) or Soviet Union depending on jurisdictional level."[11] The following countries have specific memos addressing the headings use for each jurisdictional change: Germany, see H 945; Soviet Union, see H 1023; and Yugoslavia, see H 1055.

Catalogers should not confuse the use of a place name as a qualifier with its use as a subdivision. State names, when used as a qualifier to a place, are abbreviated (as specified in AACR2R[12]), but are not abbreviated when used as subject subdivisions. For example, a general map of Lincoln, Nebraska, will be found under the subject heading, *Lincoln (Neb.)–Maps,* while a thematic map may have a subject heading formulated as *[Topic]–Nebraska–Lincoln–Maps.* The impact that this has on keyword searching is momentous and unfortunately not often understood by library users. Unless the online catalog software has software that brings up for the user all occurrences of both the abbreviation and the full name whenever either is input, both the abbreviated version of the state name and full name must be searched to find all maps concerning that place.[13]

Catalogers should be aware of the inversion with the form of name when the natural feature begins with a generic term, such as "Mount" or "Lake," as we see with *Michigan, Lake* and *Mexico, Gulf of.* The exception is when the generic term is an integral part of the name, as is the case with *Mount Morris Reservoir (N.Y.).* The SCM describes both of these situations in Memo H690.

ROLE AND USE OF AUTHORITY RECORDS

The authority record has a valuable role in a system which is based on a controlled vocabulary. The authority record shows the source of the heading, cross references, classification numbers or ranges, and the rules used in its construction, and may include scope notes. In the USMARC Format for Authority Data, the field 4XX gives "see from" references from unauthorized headings. The 451 field contains a tracing for a geographic name "see from" reference. As shown in the example below, the heading, "Amon Carter Lake (Tex.)," is not established, but directs the user to the authorized heading, *Amon G. Carter, Lake (Tex.).* In the USMARC Authority Format, the field 5XX gives "see also from" references. The 551 field contains a tracing for a geographic name "see also from" reference. These are headings which are established and valid under certain conditions–typically earlier, later, or related established jurisdictional names. The subfield $w in the 4XX and 5XX fields denotes the relationship of the heading, such as earlier/later names ($a, $b) or broader/narrower headings ($g, $h).

Further on in the USMARC Format for Authority Data, the field 670 contains a citation and information about the heading from a consulted source. The first source listed is the item being cataloged. Further sources, such as dictionaries or gazetteers, are contained in a separate 670 field. The primary source to use when establishing a name in the United States is the form of name found in the U.S. Board on Geographic Names (BGN), or

Geographic Names Information System (GNIS) (http://www-nmd.usgs.gov/www/gnis/gnisform.html). A secondary source for United States place names is the annual, *Rand McNally Commercial Atlas and Marketing Guide.* The most current source for foreign geographic names is the GEOnet Names Server (GNS) of the United States National Imagery and Mapping Agency, which is available at http://164.214.2.59/gns/html/index.html; the BGN foreign-country gazetteers in hardcopy form (paper or microfiche) may also be consulted, but many of these have not been reissued in a good many years and are thus out of date. Sources which are consulted where no information was found are cited in field 675 (Source Data Not Found).[14]

An example of an authority record from OCLC, which shows tags, references and sources, follows:

1 010   sh 87006532

2 040   DLC |c DLC |d DLC

3 005   19941227064448.4

4 151 0   Amon G. Carter, Lake (Tex.)

5 451 0   Amon Carter Lake (Tex.)

6 451 0   Lake Amon G. Carter (Tex.)

7 550 0   Lakes |z Texas |w g

8 550 0   Reservoirs |z Texas |w g

9 670   Work cat.: Walters, P. Amon Carter Lake extension, c1986.

10 670   BGN, 7/14/87 |b (Lake Amon G. Carter, tank, 33° 28' 09" N, 97° 51' 56" W)

11 675   Lippincott; |a Rand McNally; |a Web. geog.

SUBDIVISIONS

Topical subject headings can be further broken down by geographic subdivisions. Geographic subdivisions–in the form of jurisdiction, region or feature–are used regularly for thematic maps and cartographic materials. Within the subject authority record, or printed list, is a phrase or tag which authorizes use of a geographic subdivision. In print form, the term, "(May Subd Geog)," appears beside the heading. Online, a coded value of i in field 008/06 in the subject authority record, and in OCLC's fixed field, "Geo subd: i" indicates whether the subject heading may be subdivided geographically. Geographic subdivisions are constructed in the bibliographic record as a $z subfield of the 650 field, such as, *[Topic] $z Nebraska $z Lincoln $v Maps*.

Chronological subdivisions further define a subject heading, or subject heading/subdivision combination, by breaking it down into time periods. Most are associated with wars, or historical, economic and political periods, such as the heading, *Great Britain–History–Anglo Saxon period, 449-1066–Maps*. Those of particular interest to cartographic materials specify geological periods, such as, *Geology, Stratigraphic–Paleozoic–Maps*. Chronological subdivisions can be used only when established in combination with a subject heading. The only exceptions are the somewhat restricted free-floating century subdivisions, *–16th century,–17th century,–18th century*, etc. The subdivision,*–Early works to 1800*, is assigned for items which were written or issued before 1800. This includes facsimiles of those items.

Lists of subject subdivisions can be found in the *Subject Cataloging Manual*. All catalogers should be aware of the "Free-Floating Subdivisions," found in "Memo H1095." These form and topical subdivisions may be assigned to any subject, without the usage being established in an authority record. Form subdivisions are denoted with a diamond. Some subdivisions relate closely to cartographic materials. Sections "H1140 (Names of Places)" and "H1145.5 (Bodies of Water)" list subdivisions which can be used after place names. Some common subdivisions under place are the topical subdivision, *–Buildings, structures, etc.*, and the form subdivision, *–Maps, Topographic*. The order of subdivisions in the subject string was one topic of the Subject Subdivisions Conference, held in 1991, which, "recommended that the standard order of subdivisions be [topic]–[place]–[chronology]–[form]. . . . Toward achieving the recommended order, new subdivisions for which geographic orientation is logical are now established with authorization for further subdivision by place. On a case-by-case basis, subdivisions not previously divided by place are being authorized for geographic subdivision."[15] Memos H860 and H870 give further instructions. The USMARC format designates special subfield tags to each subdivision type as follows: $x subject; $y chronological; $z geographic; and, recently implemented, $v, form.

Many form subdivisions relate closely to the specific material designation (SMD). The SMD is prescribed in the physical description area of AACR2R, 3.5B1.[16] Also, a number of the subject Cutters are forms. See the Appendix for a list of cartographic SMDs and their most commonly associated form subject subdivisions and subject Cutters.

The Library of Congress recently implemented subdivision authority records. The subdivision data is in a 18X field, with codes in the 073 field which refer to the memo number from the *Subject Cataloging Manual*. Specific guidelines are in the 1998 *Update Number 2* of the *Subject Cataloging Manual: Subject Headings*. Also, the Library of Congress is including the 781 linking field for subdivision forms of geographic names.

THEMATIC MAPS

As mentioned earlier, reference queries for cartographic materials frequently include questions about topics. Thematic or topical cartographic materials often require one or more topical subject headings or subdivisions. The Library of Congress' controlled list of subject headings is the primary source for most libraries. Their subject headings are available in print, on-line, on microfiche and CD-ROM. The most familiar source may be the annually printed *Library of Congress Subject Headings* (LCSH, also often called the "red books") which is a hefty multi-volume (currently four) set. The range of topics represented in a map collection may be surprisingly varied for those unfamiliar with this medium. Some often-seen subject headings for cartographic materials are *Petroleum industry and trade, Mines and mineral resources,* and *Earthquakes*. In addition, almost any topic can be represented cartographically. Some examples of *Library of Congress Subject Headings* used for maps in the author's library are *Public radio–United States–Maps* and *Pink bollworm–Southwestern States–Maps,* which point the searcher respectively to the titles, *WHIMPUR Map (WHere Is My PUblic Radio)* and *Pink Bollworm Quarantine.*

ORDER OF SUBJECT HEADINGS

The *Map Cataloging Manual* instructs catalogers on the order of subject headings:

> The first subject heading generally should be the one that most nearly corresponds to the classification number (i.e., the heading that represents the predominant topic of the work). If a work has two equally important major topics, assign heading(s) for the second of these topics immediately after the heading(s) for the first, and before any headings for secondary topics. For items including more than one geographic

area tracing as well as at least one topical heading for each place, the subject headings are arranged according to the topic treated, rather than grouped by area. . . . There is no set order for the presentation of the topics. They may be alphabetical, in the order they appear in the record, in order of importance, etc. Subject headings constructed to include form subdivisions and form headings are usually the last subject headings given.[17]

The *Subject Cataloging Manual* gives similarly worded instructions in Memo H80.

CLASSIFICATION

The following is a very brief summary of classification of cartographic materials; for details, see the previous article in this volume by Susan Moore, "Navigating the G Schedule." Classification of cartographic materials is quite different from classification for other materials. The primary organizing component is geographic, and the subject matter is secondary–the exact opposite of almost all other portions of the Library of Congress classification. A portion of the map classification number is the subject Cutter. It ranges from cartographic form (.A35 for Panoramas), to topical (.H8 for Petroleum and natural gas), and chronological (.S1+ for Historical Geography). The subject Cutters do not have a direct correlation to a specific subject heading; there are some Cutter and subject subdivision combinations that are often used together, which are compiled in the Appendix. Although very broad in nature, the subject Cutters should not be dismissed; they function well to group maps within a geographic area by topic.

OTHER FORMS OF CARTOGRAPHIC MATERIALS

Atlases

Thematic atlases cover a range of topics, such as agriculture or transportation, or specialized information such as mortality rates or religion. Subject analysis for atlases is fairly straightforward, but it is important to recognize the limitations of the subject heading *Atlases*. Only general world atlases published in the United States receive the subject heading *Atlases*. Those showing the world, but published in other countries, receive the subject heading *Atlases* qualified by the country of publication (e.g., *Atlases, Austrian*). Atlases showing a particular country or topic receive the appropriate subject heading, with the subdivision–*Maps* (e.g., *California–Maps,* and *Food supply–Maps*).

Views and Panoramas, Photo Maps and Satellite Image Maps

Bird's eye views and panoramas depict places observed from the air. The subdivisions –*Aerial views,* and/or –*Maps, Pictorial,* are assigned. The *Map Cataloging Manual,* Chapter 9, gives detailed information on the description and subject analysis for specific types of views and panoramas.

Similarly, other form subdivisions describe specific methods used to obtain pictures of the Earth's surface. The subdivision, –*Remote-sensing images,* is used for those images obtained through a remote-sensing process. A more specific subdivision is, –*Remote-sensing maps,* which is used for remote-sensing images, including photos, upon which grids or other mapping data elements have been added. In the past, the subdivision used in this situation was –*Photo map* and it is often seen on older records.

Globes

Globes of the earth are simply assigned the subject heading *Globes.* Globes of the solar system or planets are assigned *Celestial globes.* The subdivision, –*Globes,* is used only when established in conjunction with an authorized heading, such as an individual planet (e.g., *Mars (Planet)–Globes*). Raised-relief globes typically are assigned both *Globes,* and *Relief models,* in that order. Another specific type of globe, which is intended to be used by the blind, is assigned the heading *Globes for the blind.*

Bathymetry Maps and Nautical Charts

Two major categories of maps of the waters are the bathymetric map and the nautical chart. The bathymetry map contains detailed relief of the floor of an ocean, lake or other body of water. The form subdivision, –*Bathymetric map,* is assigned, for example, *Pacific Ocean–Bathymetric maps.* Nautical charts also depict relief information, but their principle function is to chart directions and hazards to navigation for vessels. They typically contain nautical measurements, and information about shorelines, tides, currents and harbors. The heading, *Nautical charts,* is assigned, which is subdivided geographically, for example, *Nautical charts–South Atlantic Ocean.* Often a second subject string is assigned, with the geographic heading as the first element, and the subdivision, –*Navigation,* added, e.g., *South Atlantic Ocean–Navigation.*

Digital Maps and Geographic Information Systems

Digital maps and remote-sensing images, and other geographic information in digital form, are distributed on disks and CD-ROM or accessed through the Internet. They typically receive subject headings based on the topic and place that they describe, with only a general form subdivision added (e.g., *–Maps*), such as, *Geology–Appalachian Region–Maps*. No further subdivision for the computer file aspect is assigned, with one exception. The genre heading, *Maps–Digital* (expanded to *Maps–Digital–Raster* and *Maps–Digital–Vector* when appropriate) is increasingly assigned in the 653 field (Uncontrolled index term). Older catalog records for digital maps may contain the form subdivision, *–Databases*. The use of this subdivision, however, became limited in 1996, to only those materials which fit the definition of a database, as described in the SCM Memo H1520. Catalogers should remember that in subject analysis for any type of software, the Library of Congress instructs catalogers not to assign headings for a named computer program (e.g., *ARC/INFO (Computer program)*) as a subject heading to the program itself. The name heading is only assigned for items *about* the software.[18] Additionally, the subdivision, *–Computer programs,* can only be used with works *about* computer software or programs.[19]

PROBLEM AREAS

There are some problem areas in subject analysis of cartographic materials, and the assigning of headings, which lend themselves to discussion and clarification.

Coastlines

Assigning subject headings to maps of coastlines can be confusing. The *Map Cataloging Manual* goes into some detail in Chapter 4, "Subject Access." Headings are assigned by body of water, and follow the principles of indirect geographic subdivision. Those which are qualified with an area larger than a country, or qualified by a country not used in indirect subdivision, are divided directly (e.g., *Botany–Pacific Coast (U.S.)–Maps*. Those which are qualified by states of the U.S., provinces of Canada, parts of the U.K. or republics of the Soviet Union are divided indirectly (e.g., *Geology–Oregon–Pacific Coast–Maps*).[20]

Coastline names are established and qualified by the land mass to which it relates (e.g., *Pacific Coast (Chile), Pacific Coast (B.C.), Pacific Coast (Japan), Pacific Coast (Wash.),* or *Gulf Coast (Miss.), Gulf Coast (Ala.), Gulf Coast (Fla.), Gulf Coast (U.S.),* etc. Names of international bodies of water, however, are not qualified unless there is a conflict. Examples of headings for bathymetry maps and nautical charts are, *Pacific Coast (Wash.)–Bathymetric maps,* and *Nautical charts-Gulf Coast (U.S.).*

Boundary Maps

Boundary maps need not be confusing to work with, even though they do not typically include a whole political or geographic area. Guidance can be found in the SCM "Memo H1333.5." For maps showing international boundaries, use the form, *[Location]–Boundaries–Maps.* With maps showing boundaries between two or more countries or states, the subdivision *–Boundaries* will be further subdivided by place, i.e., *[1st Location]–Boundaries–[2nd location]–Maps.* A second subject string is also created which reverses the order of the locations, thus, *China–Boundaries–India–Maps* and *India–Boundaries–China–Maps.* Named boundaries are established under guidelines found in the *Subject Cataloging Manual* and are assigned when appropriate, for example, *Mason-Dixon Line–Maps.* City maps may have need for boundary-type subject headings. Commonly used subject strings are, *Landowners–[Place]–Maps,* and, *Real property–[Place]–Maps.*

Ancillary Maps, Insets, and Subsidiary Maps

Chapter 4 of the *Map Cataloging Manual* gives detailed instructions for when inset maps receive subject headings. In general, specific categories of insets are deserving of subject headings, particularly when they have place-name indexes. Subsidiary maps that are not indexed may still receive a subject heading when it is deemed valuable to the user to have that access through the catalog. Condensed from the *Map Cataloging Manual,* the guidelines are as follows: A subsidiary map of a town, military fort, airport, or college or university, receives a tracing if it is indexed. Other types of subsidiary maps may receive a tracing, particularly those that are indexed, and would be of value to the user, such as a map of a ruin or archaeological site, or park. If an archaeological site or ruin itself is not traced, consider a secondary subject heading for archaeology or excavations. A subsidiary map mentioned in a title, subtitle, cover, etc., even if not indexed, frequently receives a tracing depending on prominence, detail, typography, etc. A subsidiary map that receives a tracing does not automatically receive the same topical treatment as the main map. Any topical treatment must be justified by the content of the subsidiary map alone. Subsidiary maps, especially insets, which portray a specific subject frequently receive subject tracings, regardless of the presence or absence of an index. Each subsidiary subject map on CIA maps receives a separate subject tracing, regardless of the quality or extent of subject information.[21]

Peripheral Areas

Geographic areas, especially jurisdictional areas, are not known for their square or rectangular shapes. They often are bounded by rivers, lakes or coastlines which zigzag across the landscape. Maps, therefore, do not show only tidy, square or rectangular plots of land. Cartographers may include extraneous areas just to create a more attractive layout. Sometimes these nearby areas are mentioned in the title, or notes, although often they are ignored. Catalogers are instructed in the *Map Cataloging Manual* to, "eliminate peripheral material from consideration,"[22] in subject analysis.

Unidentified Areas and Imaginary Places

The *Map Cataloging Manual* instructs catalogers that "every map of an identifiable place must have a subject heading for the area. Maps of places that cannot be identified receive no geographic subject headings."[23] They are classed at the end of the G range, in G9980. Maps of imaginary places are assigned an authorized name heading, such as, *Fountain of youth (Legendary place),* or, *Narnia (Imaginary place),* and are classed near the end of the G range, in G9930. Maps which show multiple imaginary places are assigned a general heading, such as, *Imaginary places–Maps,* or *Geographical myths–Maps.*

Thematic Maps Lacking Direct Geographic Entry

While this author has attempted to present the methods of assigning subject headings according to agreed-upon methods promoted by the Library of Congress, there is one aspect of the guidelines that may inhibit access, and should be discussed. The guidelines do not prescribe direct geographic access under every circumstance. For instance, with a topical map, the first subject heading may be topical, and will be subdivided geographically in most cases. Studwell and Schreiber suggest automatically adding a subject heading with the geographic area whenever the area involved is not already the initial element in a subject headings string.[24] So, a topical map would then always receive at least one subject entry with the first element being the geographic area, such as a secondary subject string, [Area]–[form]. While this is not a prescribed practice, catalogers may determine that their users have a need for the additional access point, and may choose to add headings. While online catalogs with keyword searching mean that users will find a thematic map of Europe with just the topic-divided-by-area subject heading, that same keyword searching in effect "tears apart" the pre-coordinated term that the cataloger has worked so hard to put together, and this will lead to some false drops.

Facsimiles and Early Works

Facsimile maps are more broadly defined than the definition of non-map facsimiles found in AACR2R. The map "facsimile" reproduction *may* differ from the original in some bibliographic elements (e.g., scale, coloration, type of paper, etc.), as well as additions, changes, or modifications of bibliographic details (e.g., title, publication details, etc.).[25] Early works "produced prior to 1800, or facsimiles of those items receive the chronological subdivision –*Early works to 1800* following each topical or geographic subject heading."[26] The *Subject Cataloging Manual* "Memo H1595, Facsimiles, #3," instructs catalogers, "For facsimile collections of maps, or facsimile editions of a single map, use the subdivision –*Facsimiles* under any heading or subdivision that designates maps or specific types of maps." "Memo H1576, Early Works," clarifies the positioning of the subdivisions: "3. As a general rule, locate the subdivision as the last element in the string. If it is necessary to bring out the form of the original work, add the appropriate form subdivision immediately before –*Early works to 1800,* for example, *[topic]–Hand-books, manuals, etc.–Early works to 1800.* To bring out the form in which the modern publisher issued the work, add it after –*Early works to 1800.*" The following examples show the use of these subdivisions: *Railroads–West (U.S.)–Maps–Facsimiles*; *Washington (D.C.)–Maps–Early works to 1800; Coatlinchan (Mexico)–Maps, Manuscript–Early works to 1800–Facsimiles;* and, *Macao–Aerial views–Early works to 1800–Facsimiles.*

Topographic Quadrangles

Topographic maps, produced by the U.S. Geological Survey, are among the most common maps found in most general collections. These are cataloged by the U.S. Government Printing Office. The large series are cataloged collectively, organized by state and type of map. In addition, each sheet is cataloged individually and appears in the *Monthly Catalog.*[27] Each sheet in the *7.5 Minute Series (Topographic)* is assigned a subject heading for the name of the state and subdivision –*Maps, Topographic,* (e.g., *Illinois–Maps, Topographic*). Sheets in another quadrangle series, the *Geologic Quadrangle Map (GQ-) Series,* are assigned the subject heading *Geology,* which is then subdivided geographically by the state and county represented, then further subdivided with the form heading –*Maps* (e.g., *Geology–Nevada–Lincoln County–Maps*). Neither the *7.5 Minute Series,* nor the *Geological Quadrangle Map (GQ-) Series* sheet maps is assigned more specific subject headings for the feature name, although added-title entries may be made for the quadrangle name. Studwell suggests the quadrangle name be used as a geographical entry, or that a new subject heading, such as "Quadrangles (Geology)" be established and subdividable geographically.[28] Libraries should determine if the current practice of assigning general headings provides enough access for their users, or if more detailed subject headings for topographical maps are called for.[29]

Maps Accompanying Books

The following is a very brief summary of this topic; for further information, see in this collection of issues on cataloging cartographic materials the article by Dorothy Prescott. Maps are often produced to support a larger work, namely books or other texts. While these maps are considered secondary to the larger text, there is often a need to bring out the cartographic content by assigning subject headings. The SCM Memo H1865 instructs catalogers to assign map headings if at least 20 percent of the text of a work consists of maps, or discusses maps, even though the primary emphasis of the work is on the textual content. These map headings are assigned in addition to the appropriate headings assigned to the text. Map headings are not assigned, however, if the maps are routine or incidental to the text.

FUTURE

Cartographic information is increasingly being produced and distributed in electronic format. At the same time, online catalogs are moving to web-based interfaces. This provides unique challenges and opportunities for increasing methods of access to digital libraries of cartographic materials. Searching with a map-based, hypertext, graphical interface has many advantages, particularly given the spatial aspect of cartographic inquiries.

We currently have some examples of spatial search methods on the Internet. The U.S. Geological Survey's (USGS) Global Land Information System provides Websites for selecting and ordering USGS National Aerial Photography Program (NAPP) Photos (PhotoFinder) and USGS *7.5 Minute* printed paper maps (MapFinder). These provide hypertext maps which allow searchers to click on the site of interest, as well as textual indexes, searchable by name or zip code. The central web site is http://edcwww.cr.usgs.gov/webglis/. The Microsoft® TerraServer (http://terraserver.microsoft.com/) is a popular site which uses spatial or textual access to digital orthophoto quadrangles from the U.S. Geological Survey (USGS), and the SPIN-2 database of Russian satellite images. Users can access the DOQs by place name, coordinates, or a visual imaging map search, which includes zoom capabilities. The MIT Digital Orthophoto Browser is another good example of a graphical search mechanism. The site provides orthophotos of the Boston area, and access to associated metadata files. Searchers can click on the large-area grid, and zoom into the specific location of interest. This Web site is: http://ortho.mit. edu/. It also provides a textual list of the orthophotos available at the site.

A good example of an in-house subject authority file for digital cartographic information can be found on the Texas State Archives Historical Map Collection Webpage. Their Map Archive database (http://www.tsl.state.tx.us/ lobby/maps/index.html) is searchable by place name, cartographer name, year, subject or map type, and map number. Its subject search uses a controlled list of subject headings, which includes some cross references, to locate maps in their collection. Among the headings and descriptive information in their subject authority file are: accident sites (industrial accidents, oil rig accidents, EPA accident sites); aerial photographed areas; agriculture laboratories, USDA (sites); crop production; airline routes, use also the proper name of airline. A number of the maps are zipped JPEG map images which are downloadable through the appropriate viewer software.

Early research comparing the text-based search to the graphical interface reveals user preference for the graphical interface.[30] In web-based catalogs, the MARC 21 field 856 provides a direct link to hypermaps. The joining of the graphical search method with controlled vocabulary forms a very powerful information access structure. While this concept is valid, Larsgaard points out many weaknesses in our current subject headings for digital geospatial data, such as no approved LCSH headings for raster data, and no such heading strictly for digital geospatial data.[31] As Internet resources for cartographic materials increase, form and genre headings will become more meaningful collocation devices, which will require more specific headings be established.

NOTES

1. Larsgaard, Mary Lynette. *Map Librarianship*, 3rd ed. (Englewood, Colo.: Libraries Unlimited, 1998), 276.

2. Cutter, Charles A. *Rules for a Dictionary Catalog*, 4th ed., rewritten (Washington, DC: GPO, 1904), 12.

3. Online at http://lcweb.loc.gov/catdir/cpso/wls.html.

4. Haykin, David Judson. *Subject Headings: A Practical Guide* (Washington, DC: GPO, 1951).

5. For further reading, see: Chan, Lois Mai. *Library of Congress Subject Headings: Principles and Application*, 3rd ed. (Englewood, Colo.: Libraries Unlimited, 1995), 12; Sheila S. Intner. *Interfaces: Relationships Between Library Technical and Public Services* (Englewood, Colo.: Libraries Unlimited, 1993), 61-66; and William E. Studwell, "Why Not an 'AACR' for Subject Headings?," *Cataloging & Classification Quarterly* 6(1; 1985): 3-9 and also his, "Ten Years after the Question: Has There Been an Answer?,"*Cataloging & Classification Quarterly* 20(3; 1995): 95-98.

6. Library of Congress. Geography and Map Division. *Map Cataloging Manual* (Washington, D.C.: Library of Congress, 1991), 4.1.

7. Ibid, p. 4.4.

8. Chan, 174.

9. *Subject Cataloging Manual,* Memo H180, #2.

10. OCLC database, 12/15/98, Library of Congress Control Number 92056007.

11. OCLC database, 12/15/98, Library of Congress Control Number 80001203.

12. *AACR2R,* Appendix B (Abbreviations), p. 607-608.

13. Tenner, Elka and Katherine H. Weimer. "Reference Service for Maps: Access and the Catalog Record," *Reference & User Services Quarterly* (Winter 1998) (forthcoming)

14. Library of Congress. Network Development and MARC Standards Office. *USMARC Format for Authority Data,* 1993 ed. (Washington, DC: Library of Congress, 1993)

15. *Subject Cataloging Manual,* Memo H860.

16. Gorman, Michael and Paul W. Winkler, eds. *Anglo-American Cataloging Rules,* 2nd ed., 1988 revision (Chicago: American Library Association, 1988): 108.

17. *Map Cataloging Manual,* 4.1.

18. *Subject Cataloging Manual: Subject Headings.* Memo H2070.

19. Chan, 228.

20. *Map Cataloging Manual,* 4.8.

21. Ibid, Ch. 4.

22. *Map Cataloging Manual,* 4.2.

23. Ibid.

24. Studwell, William E. & Robert E. Schreiber. "Sounding Board: A Simple and Successful method to Enhance Map Subject Access," *Information Bulletin,* Western Association of Map Libraries, 16(3; June 1985): 357-8.

25. *Map Cataloging Manual,* 8.1.

26. Ibid, Chapter 4.

27. *Government Printing Office Cataloging Guidelines,* 3rd ed. (Washington, DC: Cataloging Branch, Library Programs Service, U.S. GPO, 1990): 31.

28. Studwell, William. "Cataloging Column: The Invisible Geologic Quadrangle," *Information Bulletin,* Western Association of Map Libraries, 21(2; March 1990): 98.

29. Tenner.

30. Morris, Barbara. "CARTO-NET: Graphic Retrieval and Management in an Automated Map Library," *Bulletin,* Special Libraries Association, Geography and Map Division, 152 (1988):19-35.

31. Larsgaard, Mary Lynette. "Cataloging Planetospatial Data in Digital Form: Old Wine, New Bottles–New Wine, Old Bottles," *Geographic Information Systems and Libraries: Patrons, Maps, and Spatial Information,* ed. by Linda C. Smith and Myke Gluck (Urbana-Champaign: University of Illinois, 1996): 27-28.

SUGGESTED READING

Miksa, Francis. *The Subject in the Dictionary Catalog from Cutter to the Present.* Chicago: ALA, 1983.

Studwell, William E. *Library of Congress Subject Headings: Philosophy, Practice, and Prospects.* New York: The Haworth Press, Inc. 1990.

TOOLS

Chan, Lois Mai. *Library of Congress Subject Headings: Principles and Application.* 3rd ed. Englewood, Colo.: Libraries Unlimited, 1995.

Library of Congress. Cataloging Policy and Support Office. *Subject Cataloging Manual: Subject Headings.* 5th ed. Washington, DC: Library of Congress, Cataloging Distribution Service, 1996- .

Library of Congress. Geography and Map Division. *Map Cataloging Manual.* Washington, D.C.: Cataloging Distribution Service, Library of Congress, 1991- .

APPENDIX

SMDs, Common Form Subdivisions and Related Subject Cutters

SMD and 007/01 (Maps)	*Form Subdivision*	*Subject Cutter*
atlas = d or map = j	–Aerial views	.A3/.B52
	–Bathymetric maps	.C2
	–Historical geography –Maps	.S
	–Index maps	.C1 (except topo, .A2)
	–Maps	
	–Maps–Early works to 1800	
	–Maps–Facsimiles	
	–Maps for children	
	–Maps for the blind	.A7
	–Maps for the visually handicapped	.A7
	–Maps, Comparative	.B8

SMD and 007/01 (Maps)	*Form Subdivision*	*Subject Cutter*
	–Maps, Digital	.A25
	–Maps, Manuscript	
	–Maps, Mental	.A67
	–Maps, Outline and base	.A1
	–Maps, Physical	.C2
	–Maps, Pictorial	A5
	–Maps, Topographic	.C2
	–Maps, Tourist	.E635
	–Remote-sensing maps	.A47
	–Road maps	.P2
	–Zoning maps	.G44
map section = s	–Maps	.C57
profile = k	–Maps	.C2
relief model = q	–Relief models	.C18
remote-sensing image = r	–Remote-sensing images	.A47
view = y	–Aerial views	.A3/.B52

RETROSPECTIVE CONVERSION
OF COLLECTIONS
AND QUALITY CONTROL

A Survey Technique for Map Collection Retrospective Conversion Projects

Paige G. Andrew

SUMMARY. Although much has been written about the need for, methodologies, costs, and other aspects of retrospective conversion little exists in the literature regarding retrospective conversion of cartographic materials, and map collections specifically. Reference is usually made to the need to survey the collection for conversion, but the

Paige Andrew, Assistant Professor, is Maps Cataloger Librarian for the Pennsylvania State University Libraries, University Park, Pennsylvania. Formerly he was Maps Cataloger, from 1986-January 1995, for the University of Georgia Libraries, Athens, Georgia. He holds a BA degree with a major in Geography from Western Washington University, Bellingham, Washington and a MLS from the University of Washington, Seattle, Washington.

Address correspondence to: Paige G. Andrew, E506 Pattee Library, The Pennsylvania State University Libraries, University Park, PA 16802 (Email: pga@psulias.psu.edu).

[Haworth co-indexing entry note]: "A Survey Technique for Map Collection Retrospective Conversion Projects." Andrew, Paige G. Co-published simultaneously in *Cataloging & Classification Quarterly* (The Haworth Information Press, an imprint of The Haworth Press, Inc.) Vol. 27, No. 3/4, 1999, pp. 405-412; and: *Maps and Related Cartographic Materials: Cataloging, Classification, and Bibliographic Control* (ed: Paige G. Andrew, and Mary Lynette Larsgaard) The Haworth Information Press, an imprint of The Haworth Press, Inc., 1999, pp. 405-412. Single or multiple copies of this article are available for a fee from The Haworth Document Delivery Service [1-800-342-9678, 9:00 a.m. - 5:00 p.m. (EST). E-mail address: getinfo@haworth pressinc.com].

author was unable to locate a description of a random sampling technique that explains how it is applied and what the outcome was.

This article introduces the use of a random sampling technique with a major university map collection. The University of Georgia's Maps Collection was surveyed to ascertain how much of the existing maps card catalog needed to be converted to an electronic form for use in the local online public access catalog. In addition, the samples pulled from the survey were searched against the OCLC union catalog to determine the proportions of records that could be found in OCLC and loaded into the Georgia Libraries Information Network (GALIN), the online catalog, with no cataloging intervention versus the degree to which the maps cataloger would have to either adjust existing records available or create original records for the online catalog. *[Article copies available for a fee from The Haworth Document Delivery Service: 1-800-342-9678. E-mail address: getinfo@haworthpressinc.com <Website: http://www.haworthpressinc.com>]*

KEYWORDS. Retrospective conversion, maps, cartographic materials, bibliographic records, cataloging, random sample, OCLC, survey

INTRODUCTION

Data conversion for libraries is the process of turning non-computer-readable library records into a computer-readable format.[1] Libraries have been establishing and completing retrospective data conversion projects for thirty years now, the vast majority for collections of monographs and serials. Library literature is replete with details and descriptions of retrospective conversion projects. Most "of the literature on retrospective conversion focuses on the choice of a vendor or the means for conversion."[2]

It has only been within the last ten years that, with some exceptions, cartographic collections, and in particular sheet maps, have received the same level of attention towards migrating from manual to online access for the patron. Collections of maps, atlases, globes, and most recently electronic forms of cartographic material typically are the last, or one of the last, kinds of library material given the necessary attention and funding by library administrators. "Most libraries involved with retrospective conversion projects are concentrating on monographs and serials, leaving maps, media, and other less traditional formats for the future."[3] Even in the area of retrospective conversion maps and cartographic collections remain the proverbial "stepchild" as noted by several map librarians over the years.[4,5]

Thankfully, map collections of all sizes and types have been successfully making retrospective conversions from the card format to online bibliographic records–through in-house cataloging projects or more prevalently via outsourcing to agencies such as OCLC's RetroCon Service–and sharing their data by means of the major bibliographic utilities. Bibliographic records for maps and other cartographic materials can also now be shared by providing entry to the online public access catalog (OPAC) via the Internet and by selling bibliographic records such as in the form of MARCIVE tapes. One of the questions, however, that seems not to have been answered in the literature has to do with determining how much of the map collection needs to be converted when a portion of it already exists online, a question that needs serious consideration before time and money is spent on the conversion process itself. Another question begging an answer in the preparation stage of a retrospective conversion project is, "What kind of a hit rate can I expect when searching for matching records in the bibliographic utility I am using?" This paper will describe a particular survey technique used to prepare a report[6] for the administrators at the University of Georgia Libraries in 1994 that answered the two questions posed above. Based in part on this report, the Maps Collection at the University of Georgia has completed an 11,500-title retrospective conversion project.

BACKGROUND

The Maps Collection at the University of Georgia Libraries, one of the largest in the country, was considered "completely cataloged" (with the exception of an approximately 2,500-title backlog that had grown over a three year period when the Libraries lacked a staff member with map cataloging skills), when a full-time professional librarian was hired as maps cataloger in the Fall of 1986. The Maps Collection was founded in the early 1960s when it was transferred to the Libraries from the Geography Department and has been continuously cataloged in a variety of ways using different cataloging standards over the years. Individual bibliographic records in the maps card catalog and shelflist housed in the Maps Room included "brief" or "minimal-level" information all the way through to "full-level" records with multiple subject headings, a detailed physical description, and multiple access points when needed. In short, the quality of records available in the card catalog (composed of separate author, title, and subject units housed together) and the shelflist ran the gamut. However, the Libraries administration decided to make the commitment to provide full-level OCLC/US-MARC-standard records for cartographic materials in the Maps Collection by hiring a full-time professional to catalog the maps in the collection.

Simultaneously, the decision was made that it was time to create online bibliographic records for the Maps Collection, dropping manual card production. The process of "mainstreaming" or integrating maps into its OPAC, then called MARVEL and later re-named GALIN after a system conversion, along with monographs, serials, microforms, audiovisual material, music, and other forms of information, began in April 1987. While manual production of catalog and shelflist cards was halted, the Libraries continued to purchase OCLC-produced cards for both the maps card catalog and its shelflist until October 1995.

From April 1987 through January 1995 the Maps Cataloger created online records for the Maps Collection using existing OCLC copy that matched titles in the collection and by creating original bibliographic records in OCLC. Two outcomes of cataloging online are sharing bibliographic records for unique items held by the University of Georgia Libraries and providing records for non-unique items for which no record had yet been created in OCLC. In this manner the 2500-title backlog was eliminated while at the same time new receipts were cataloged as they were received in the Maps Collection.

Therefore, at the time when a survey of the collection was called for in 1994 the card catalog and shelflist for the Maps Collection consisted primarily of catalog cards created in-house over decades, but also included an ever growing amount of OCLC-produced cards.

THE SAMPLING METHOD EMPLOYED

Library managers need accurate information about patrons, staff, files, and collections in order to make informed decisions. Yet often the population to be examined is so large as to make consideration of each member infeasible. Sampling comes to the rescue in such situations. It allows librarians to make useful generalizations about a large number of patrons, files, books, or whatever, by examining only a small fraction of the population. It is in essence a shortcut that extends the practical range of a librarian's data-gathering activities.[7]

The "population" in the case of the Maps Collection was the maps shelflist. It was chosen as the benchmark for surveying the maps collection to determine how much needed to be retrospectively converted. Not surprisingly, "a shelflist . . . is the most common bibliographic file to be selected for data conversion."[8] An additional outcome of the survey, as will be seen below, was the ability to determine how much and what kinds of "copy" would be available for cataloging use on OCLC versus how much would require creating original bibliographic records.

In reality, the Maps Cataloger "borrowed" the survey technique from a colleague who had used the same technique to successfully sample the serials collection several years earlier. The technique can be used on any type of shelflist because it relies on physically "measuring," and identifying, cards to be pulled from shelflist drawers using a random sample method. Butler, Aveney and Scholz provide instructions for "choosing a random sample from a shelflist," in Appendix A in their article. They point out that "the initial consideration for determining the size of the sample is the number of volumes in the collection that are represented in the catalog. For example, a university catalog with a shelflist of 800 to 900 trays, having 800 to 900 cards per tray, would require a sample of only one or two cards per tray . . . In general, the smaller the file, the higher the portion of it that must be checked in order to have a statistically reliable sample."[9]

The above-mentioned appendix to the article, "The Conversion of Manual Catalogs to Collection Data Bases" contains: "a page divided into 10 'rulers'. Each ruler has five points, numbered 1 through 5. The numbers appear at different intervals and sequences on the 10 rulers. By cutting a copy of the page into 10 sections, numbered 0 through 9, 10 rulers are created. These can be copied multiple times and numbered 10-19, 20-29, 30-39, or one for each drawer in the shelflist. By then using the numbers on each ruler, the necessary number of cards can be selected from each tray and photocopies can be made to provide an analyzable sample of the shelflist or other file."[10] The Serials Cataloger had done just this, copying 10 rulers onto thick pressboard for use during the sampling part of his project. The Maps Cataloger borrowed these same 10 rulers for the maps shelflist sampling project.

TAKING THE SAMPLE FROM THE MAPS SHELFLIST

The maps shelflist at the time of sampling contained 30 drawers, or "trays." The cards in all 30 drawers were first measured, one drawer at a time, and collectively the outcome was 236.5 inches. Using the 10 "rulers" mentioned above each drawer was then randomly sampled, rotating the set of rulers as the sampling progressed. Wherever a line on a ruler struck a card in the drawer being sampled, that card (or cards if it happened to be a second or third card in a set representing one title) was pulled and photocopied. If the card selected turned out to be a part of a set of holdings cards for a series title, the card representing the title itself was pulled for photocopying. Anywhere from 1 to 7 titles were pulled from each shelflist drawer, most drawers producing six or seven titles, with the average number of titles pulled per drawer being 6.13 cards.

The result of this random sample obtained 184 titles, of which 48 were already cataloged on OCLC (the titles had been cataloged since 1987 and thus 26.1% of the sample would not need converting since they were also represented in GALIN).

The remaining 73.9% of the sample, or 136 titles, was subsequently searched against the OCLC's WorldCat. This was done to determine how many matching records existed for use in a conversion project, the level of quality and completeness of those matching records found, and how many matches would not be found, representing the percentage needing original records created to complete the conversion project.

OUTCOME OF THE OCLC SEARCHES

Below is a breakdown of the types of records found when the group of photocopied sample titles were searched against the OCLC union catalog, along with the number of shelflist records where a match was not found.

A.	Library of Congress copy found	47
B.	Library of Congress copy for a different edition found	8
C.	Member Institution Full-Level copy found (I-level)	25
D.	Member Institution copy for a different edition found (I-level)	5
E.	Member Institution copy tape-loaded into OCLC (M-level)	5
F.	Member Institution Minimal-Level copy (K-level)	2
G.	No matching copy found in OCLC	44
	Total of Titles searched against OCLC	136

Percentages of Types of Copy Found

A.	Percent of Titles where no copy was found on OCLC	32.4%
B.	Percent of Titles with Library of Congress copy available	34.5%
C.	Percent of Titles with Member Copy (all levels) and LC copy for a different edition combined	33.1%
	Total of Titles needing retrospective conversion	100%

Of the 136 titles searched the majority had either Library of Congress (DLC $c DLC) matching copy, 47 titles, or no matching copy found, 44 titles. This indicated that a portion of the Maps Collection would be relatively easy to convert to online records on the one hand, but additionally a portion of the collection would require the time and expertise of the Maps Cataloger to provide originally created online records. The remaining 45 titles in the sample where some type of matching copy was available on OCLC represent a range of manual intervention necessary to have full-level (i.e., AACR2R standard or "high quality") records loaded into GALIN.

In addition, if one combines the number of titles where no copy was found with both Library of Congress and OCLC member-institution matching copy for a different edition, the percentage of titles needing original records created (no copy found) jumps from 32.4% to 41.9%. (Local practice at the University of Georgia Libraries for edition copy was to "borrow" matching portions of an edition copy record in order to create a new bibliographic record for the edition in hand.)

The final outcome determined that 73.9%, or nearly three-fourths, of the Maps Collection would require conversion from the existing card catalog to online bibliographic records. These online bibliographic records for maps would reside side-by-side with other forms of information represented in GALIN. Of this percentage it was also determined that a significant portion of the work would reside in creating original bibliographic records to represent those titles in the Maps Collection where records did not exist, whether the work was done in-house or outsourced.

CONCLUSION

Although other sampling techniques have probably been used elsewhere to assist in determining the scope of a retrospective conversion project for a map collection the author was unable to locate articles explaining how the sampling was done. What *was* found in a few instances are articles describing retrospective conversion and/or retrospective cataloging projects for map collections as a whole.[11,12,13]

By using a random sampling technique established and proven, both within the University of Georgia Libraries and at other institutions, the Maps Cataloger was able to determine the proportion of the Maps Collection that would require retrospective conversion. In addition, by carefully monitoring the type of matching copy found as the searching part of the project was done an accurate picture was painted as to how much of the Maps Collection would need to be cataloged from scratch versus what portion could be converted using existing bibliographic records found on OCLC.

The resulting report drafted by the Maps Cataloger recommended that "The results of the sampling and searching should be used as a basis for setting up procedures to complete a full conversion project for the University of Georgia Libraries' Map Collection as well as assist in making decisions regarding doing such a project in-house versus contracting out or 'outsourcing' the work needing to be accomplished."[14] Using OCLC's RetroCon Service[15] unit, the University of Georgia Libraries successfully completed a retrospective conversion for the Maps Collection in July 1998.

NOTES

1. Ruth C. Carter and Scott Bruntjen, *Data Conversion* (White Plains, NY and London: Knowledge Industry Publications, Inc., c1983), 1.

2. Howard Pasternack, "Online Catalogs and the Retrospective Conversion of Special Collections," *Rare Books and Manuscripts Librarianship* 5 (1990): 73.

3. Rebecca Elsea Dinkins, "Map Retro on a Shoestring," *Special Libraries Association, Geography and Map Division* Bulletin, No. 153 (Sept. 1988): 3.

4. Walter W. Ristow, "What About Maps?" In *The Emergence of Maps in Libraries* (Hamden, Conn.: The Shoe String Press, 1980), 32.

5. Mary L. Larsgaard, *Map Librarianship: an Introduction* (Littleton, Colo.: Libraries Unlimited, Inc., 1978), 226.

6. Paige G. Andrew, unpublished report, *Preliminary Investigation for Establishing a Conversion Project for the Maps Collection at the University of Georgia Libraries,* submitted February 27, 1995.

7. Richard M. Dougherty and Fred J. Heinritz, *Scientific Management of Library Operations.* 2nd ed. (Metuchen, N.J. & London: The Scarecrow Press, 1982), 210.

8. Carter and Bruntjen, *Data Conversion,* 24.

9. Carter and Bruntjen, *Data Conversion,* 24.

10. Brett Butler, Brian Aveney, and William Scholz, "The Conversion of Manual Catalogs to Collection Data Bases," *Library Technology Reports* 14 (1978): 201.

11. Anita T. Sprankle, "Map Retrospective Conversion: the Latest Development in One University's Map Collection," *Special Libraries Association, Geography and Map Division Bulletin* No. 165 (Sept. 1991): 15-18.

12. Dinkins, *SLA G&M Division Bulletin,* 3-10.

13. Pierre-Yves Duchemin, ed., "Conversion of Map Catalogues [special issue]," *INSPEL* 28 (1994): 7-199.

14. Andrew, *Preliminary Investigation . . . ,* 3. For an excellent starting place see Daphne C. Hsueh, "Recon Road Maps, Retrospective Conversion Literature, 1980-1990," *Cataloging and Classification Quarterly* 14 (3/4) (1992): 5-22.

15. *OCLC RetroCon Service,* see the Web site: http://www.oclc.org/oclc/promo/7307crs/7307.htm

Retrospective Conversion and Cataloging of a Major Academic Map Collection: The University of Washington Story

Kathryn Womble

SUMMARY. This article is intended to provide information to the person faced with an uncataloged map collection. The article will discuss how various projects to catalog and classify a large existing map collection were completed at the University of Washington Libraries (UW). Project planning, standards, personnel issues and costs will be discussed. Information will be presented about outsourcing map cataloging, utilizing MARCIVE/U.S. Government Printing Office cataloging records and completing a shelflist conversion project. This article deals with the cataloging and classification of print maps and aerial photographs; atlases and electronic mapping products were not included in these projects. *[Article copies available for a fee from The Haworth Document Delivery Service: 1-800-342-9678. E-mail address: getinfo@haworthpressinc.com <Website: http://www.haworth pressinc.com>]*

KEYWORDS. Map cataloging, outsourcing, MARCIVE, retrospective conversion

INTRODUCTION

Many librarians have inherited large or small map collections at some point in their careers. Some of us intentionally sought positions in large map

Kathryn Womble is Head, Map Collection and Cartographic Information Services, University of Washington Libraries, P.O. Box 352900, Seattle, WA 98195-2900 (E-mail: kwomble@u.washington.edu).

[Haworth co-indexing entry note]: "Retrospective Conversion and Cataloging of a Major Academic Map Collection: The University of Washington Story." Womble, Kathryn. Co-published simultaneously in *Cataloging & Classification Quarterly* (The Haworth Information Press, an imprint of The Haworth Press, Inc.) Vol. 27, No. 3/4, 1999, pp. 413-428; and: *Maps and Related Cartographic Materials: Cataloging, Classification, and Bibliographic Control* (ed: Paige G. Andrew, and Mary Lynette Larsgaard) The Haworth Information Press, an imprint of The Haworth Press, Inc., 1999, pp. 413-428. Single or multiple copies of this article are available for a fee from The Haworth Document Delivery Service [1-800-342-9678, 9:00 a.m. - 5:00 p.m. (EST). E-mail address: getinfo@haworthpressinc.com].

413

collections loaded with "underprocessed" materials and occasionally wonder what we were thinking! Map collections tend to be in various states of bibliographic control, from fully cataloged and classified[1] to fairly chaotic. The first piece of advice to the person who has been assigned to bring an unruly map collection under control is, don't panic! There is considerable help available in the form of committed map librarians and catalogers, as well as a myriad of map cataloging tools and existing cataloging records. In addition, electronic discussion lists and professional organizations are good places to get advice on map cataloging. Lists such as MAPS-L and AUTO-CAT are good online forums for asking questions and discussing map cataloging topics.[2]

This article describes projects completed at the University of Washington Libraries where institutional support for the Map Collection is strong. The UW Libraries is also a large institution where standard library practices and systems have been in place for many years. The ideas presented here may need to be adjusted for smaller collections or collections where staff and funding are not readily available. To garner support from within and outside your institution for map cataloging projects, you must be visible and every chance you have, you must talk about your users needs and how your map collection serves them and what you need to serve them better. Money is available to support you, but it may take some effort to go after it. Geographic information is critical for researchers, planners and students, and getting it into the catalog in order to be more accessible for those users requires support from your institution.

THE DECISION TO CATALOG MAPS

Map collections represent large investments of time and money in the creation of the collection and in the space and labor needed to maintain them. Like other library materials, maps provide a wealth of information, both current and historic. And, like other library materials, maps must be cataloged in order to give them heightened visibility and use. User access is the number one reason maps should be cataloged. If you are committed to maintaining a map collection for people to use, you should commit to cataloging maps. Mainstreaming maps by cataloging them in a centralized online catalog along with books, journals and other library materials will provide the avenue of access that the users of these materials deserve.

Cataloging maps helps users outside of your institution know of their existence for resource sharing purposes. Cataloging simplifies the creation of cartobibliographies and is useful for inventory and circulation purposes.[3] Collection development is also simplified when records appear online. It is easier for the selector, and much easier on the maps as well, to check the catalog for what is already in the collection rather than to rifle through the drawers to determine what needs to be acquired. Although map collections vary in size and current state of control, this author strongly suggests that national standards be used to catalog and classify the collection, regardless of size. Cataloging is expensive, but it must only be done one time for the records to be available indefinitely for reference, circulation, and collection development. And, if cataloging is done to accepted standards, then integration of records with other systems will be possible in the future if desired.

DEFINE THE MAP COLLECTION'S CURRENT STATE

Start planning by defining the current state of the map collection as well as possible. How many map titles and map sheets do you estimate you have? Keep in mind that one title may include hundreds of sheets in a set. The number of titles is important for estimating cataloging time, and the number of sheets is important for labor and time estimates if you are creating individual item records and barcoding individual items. For example, at UW, we circulate maps published 1960 and after; earlier maps generally do not circulate. We began barcoding individual maps and air photos in 1994 for automated circulation and it has greatly decreased the staff time needed for manual overdues and billing. We began with the most frequently circulated materials: aerial photographs and U.S. Geological Survey (USGS) topographic maps of Washington State. Barcoding and creating item records for Washington topos and air photos is an ongoing student project.

If manual cataloging records already exist count titles from those records. If a card catalog exists, you can make a title estimate by counting how many cards make up several inches, then measuring the entire shelflist to estimate the number of titles. If no records exist, embark on a counting project and actually count map sheets and titles. It is helpful to have an idea of the size of the collection to make estimates of time and money involved in cataloging. For more details on how to estimate the size of your map collection and what might need to be accomplished see the article prior to this one, *A Survey Technique for Map Collection Retrospective Conversion Projects.*

Are there existing manual or online catalog records? What information has been recorded? Title, author, scale, publishing information, subject headings? Are the maps classified? What classification system or systems have been used? Estimate the percentage of the collection that already has some form of cataloging and classification recorded. Again, this information will help you decide how much work there remains to be done. You will want to establish a minimum level of cataloging record content that will assist your users and compare what you already have accomplished toward this goal.

Describe how the maps are physically organized. For example, are they filed by class number, filed alphabetically by country name or filed by accession number? If you want to do any physical reorganization of the maps you will need to account for additional labor costs.

What standards and systems, if any (classification, cataloging standards, manual vs. online catalog, bibliographic utility), are already being used in your organization for organizing other library materials? Try to make your map cataloging plans fit into what the rest of your institution is doing or has done for cataloging. You may receive better institutional support if you do not try to reinvent the wheel.

Note the general condition of the collection. If there is a need for preservation measures cataloging will take longer due to repairs or special handling. Also, pay attention to the breakdown of languages and general date ranges. Language expertise may be important when looking at personnel issues. If your collection contains many western government agency and commercially produced map sets (American, Canadian, Western European) from the mid-twentieth century forward you will find that many catalog records already exist. If you have a collection of manuscript or rare maps, or maps not from major government agencies or commercial mapping companies, you will need to create a good deal of original cataloging records.

SUGGESTED STANDARDS

There are many ways to organize, catalog and classify maps. Will you use national standards or non-standard local schemes, or some combination of both? This decision may be influenced by cataloging and classification systems already being used in your organization, but it is highly recommended that you use Library of Congress Classification (G schedule), Library of Congress Subject Headings, and AACR2R descriptive cataloging standards (Anglo-American Cataloguing Rules, 2nd ed. revised).[4] Use of the LC G-classification schedule will physically organize the map collection in a geographic order and descriptive cataloging standards will allow you to take advantage of hundreds of thousands of existing MARC 21-based map cataloging records already created using these standards. As of July 1, 1998, the OCLC database contained 505,518 map records;[5] the Research Libraries Network (RLIN) as of September 1998 contained 393,352 map records;[6] the Washington Library Network (WLN) as of August 1998 contained 232,858 map records.[7]

By choosing international cataloging standards, you will be able to incorporate map cataloging records into a centralized catalog with other library materials. It is inevitable that your catalog will migrate from system to system in future years and adhering to standards will make those transitions less difficult.

You may choose from several levels of cataloging. You may choose to create full-level cataloging for local, regional, and state materials, or materials depicting a particular geographic area or theme of most importance for your users. At the same time you may create less detailed records for other items. Descriptions of category level standards for maps may be found in AACR2R, sections 1.0D and 3.0D. In addition, the Library of Congress' *Map Cataloging Manual*[8] has an appendix on less-than-full level cataloging guidelines that are used at Library of Congress' Geography and Map Division.

A bookshelf containing standard cataloging volumes, with a few specialized books and tools added for maps, is needed to perform online map cataloging to the national standard level. (See Figure 1.) In addition, the map cataloger will need adequate space to work. A large table is essential.

FIGURE 1. Suggested Map Cataloging Texts and Tools

To catalog maps, have on hand the standard texts and tools used for cataloging and classification, plus:

Anglo-American Cataloguing Committee for Cartographic Materials, 1982. *Cartographic Materials: A Manual of Interpretation for AACR2*. Chicago: American Library Association; Ottawa, Canadian Library Association; London The Library Association. (A revision of this work is currently under way.)

Geographic Cutters [microfiche], 1987. Washington, D.C.: Geography and Map Division, Library of Congress.

Larsgaard, Mary Lynette, 1998. *Map Librarianship: An Introduction*. 3rd ed. Englewood, Colorado: Libraries Unlimited.

Library of Congress. Subject Cataloging Division, 1992. *Super LCCS, Gale's Library of Congress Classification Schedules: Class G, Geography, Maps, Anthropology, Recreation, combined with additions and changes through 1998*. Farmington Hills, Mich.: Gale Research Inc., 1999.

Library of Congress, Geography and Map Division, 1991. *Map Cataloging Manual*. Washington, D.C.: Cataloging Distribution Service.

Wood, Alberta Auringer and James A. Coombs, 1996. *Index to the Library of Congress "G" Schedule: a Map and Atlas Classification Aid*. Chicago, Illinois: Map and Geography Round Table of the American Library Association.

A meter stick or tape measure with centimeters marked.

A Map Scale Indicator for measuring scale. Available from: Dr. Clifford H. Wood, Dept. of Geography, Memorial University of Newfoundland, St. Johns A1B 3Y1 Canada, *http://www.mun.ca/geog/muncl/msi.htm*   e-mail: chwood@morgan.ucs.mun.ca

PERSONNEL AND COSTS

Chances are good that a number of people with different skills and pay levels will work on various facets of a map cataloging/conversion project. For example, at UW librarians have done original and copy cataloging, as well as classification work. A librarian planned, wrote requests for proposals, and oversaw an OCLC RetroCon shelflist conversion project, a cataloging outsourcing project and the implementation of using MARCIVE Government Printing Office map cataloging records. Library technicians contributed to copy cataloging and classification and students have assisted with map classification and barcoding projects.

Do you have in-house personnel who will be assigned to catalog and classify maps? Will it be a new hire–full or part-time–permanent or temporary? Are you willing to explore outsourcing options? Map cataloging requires good basic cataloging knowledge plus learning to apply some unique fields to the record such as mathematical data, usually called the "scale field" (which includes projection and coordinates), the physical description fixed field for maps (field 007), and three fixed fields (008 field). Commonly used subject headings, notes, map measuring techniques and classification applications need to be learned as well. Above all, map cataloging requires a desire to work with maps and make their existence known to library users. If you have catalogers in-house who need additional training specifically for map cataloging contact your regional office of OCLC. They have been sponsoring map cataloging workshops across the country, bringing in an experienced map cataloger to teach them. Or, post a message on MAPS-L and see if someone is willing to come to your institution for a daylong workshop.

Costs for in-house map cataloging will vary depending upon salary levels, makeup of the collection, and your access to existing records via a bibliographic utility. Map cataloging at UW over the last twelve years using OCLC plus utilizing MARCIVE map records from 1976 onward has resulted in 79% copy cataloging and 21% original cataloging. In coming years the percentage of original cataloging will rise because we have now utilized most of the copy cataloging that is available for our collection.

Time estimates for cataloging based on UW map cataloging experience and Mary Larsgaard's new 3rd edition of *Map Librarianship*[9] can be based on:

- 5 to 6 titles per hour for copy cataloging
- 2 to 3 titles per hour for original cataloging

depending, of course, on the impact of other possible related activities such as barcoding, call number marking and/or multiple sheet titles or sets.

UNIVERSITY OF WASHINGTON'S MAP COLLECTION CATALOGING PROJECTS

The UW Map Collection contains over 257,500 maps, 70,000 aerial photograph prints of Washington State, 5,800 microforms, 4,000 books, and 630 computer files. The Map Collection is responsible for cataloging maps and air photos and is just taking on cartographic computer files. The Central Monographic Services Division catalogs atlases and cartographic computer files, as well as maps in locations other than the Map Collection. The Map Collection serves as the regional Federal Depository Library for maps and has large collections of USGS topographic and geologic maps, monographic maps from other agencies and nautical and aeronautical charts. The Collection includes worldwide topographic map sets, primarily from the 1940s onward, as well as many gifts and purchased monographic maps and map sets. The Map Collection is committed to cataloging the collection and has approached the task in a variety of ways.

Figure 2, roughly in chronological order, lists map cataloging or processing projects that have taken place at the UW Map Collection since 1982 when the first Department of Education Title II-C Cataloging Grant was begun. The projects are each discussed below.

LC CLASSIFICATION AND BRIEF SHELFLIST CARDS

In 1982/83, as part of a Libraries'-wide Department of Education Title II-C Cataloging Grant, maps covering the Pacific Northwest published before 1940 were identified and a brief shelflist card was created. About 600 titles were recorded in the Map Collection shelflist.

By the mid 1980s most of the maps in the UW Map Collection were housed in flat map cases and organized geographically as outlined by LC classification. However, a frustrating problem encountered was that many map sheets did not have call numbers marked on them, they were simply filed in the appropriate geographic area with other general, thematic, regional and city maps. The Map Collection provides open access to its materials and all users may remove materials from the map drawers. Students who refiled the maps often had difficulty figuring out where to put them since the map drawers were organized by LC call number ranges but many maps were not marked with LC call numbers.

In the early 1990s the Map Collection began a major push (while simultaneously cataloging newly acquired materials and Pacific Northwest materials) to classify every map in the Collection and create a brief handwritten shelflist card for each title. This project's aim was to assist staff and users in retrieving and refiling maps, and to build an inventory of what the Map Collection owned.

FIGURE 2. Map Cataloging and Classification Projects at the UW Libraries' Map Collection

Project	Level of Staff Involved	Projects Dates	Number of Titles	Cost*
LC Classification/ Brief Shelflist Cards	Librarian Technician Student	1982-1997	19,400 classification numbers assigned	Estimate for well-trained person: 12-15 per hour
In-house Copy Cataloging using OCLC	Librarian Technician	1986-present	11,850	Estimate for well-trained person: 5-6 per hour
In-house Original Cataloging using OCLC	Librarian	1988-present	2,630	Estimate for well-trained person; 2-3 per hour
MARCIVE Retrospective Map Records	Librarian Technician Student	1995-1997	11,700	Estimate for well-trained person to locate map, barcode, and create item record: 6-8 per hour
MARCIVE Ongoing Map Records	Technician	1995-present	505	Estimate for well-trained person: 8-10 per hour
Outsourcing Map Cataloging	Rickert & Associates	1996-present	257 copy 1140 original	$6300
Shelflist Conversion	Librarian Student	1996	--	$750 for student prep work
	OCLC RETROCON	1996	2506 copy 3330 original	$40,000
In-house Air Photo Minimal Level Records	Library School Graduate Student	1998	150	$800

*Estimated costs will vary depending upon staff experience, skills and salary levels.

By 1997 over 19,000 call numbers were assigned and written on map sheets, and on brief shelflist cards. Librarians, technicians, and selected students helped with this project. They literally stood at a map drawer with the LC G-Schedule, a shelflist drawer and a pile of blank catalog cards and classified for hours. A call number was written on the lower right hand corner of the map in pencil. (Adhesive labels were used in the 1980s, but many are now falling off. We now write call numbers in pencil on the map sheet without a label.) A shelflist card containing the LC call number, along with a date, the map title, author and scale, and number of sheets in a set (no subject headings) was completed and filed.

This project built a classified map collection and shelflist that gives staff information about the collection's scope and depth. Collection development is now much more efficient. The librarian can check holdings in the online catalog and the shelflist rather than looking through map drawers. For multiple-sheet map sets, however, looking at graphic indexes in the drawers is still necessary for determining holes in the set. For every map in the collection there should now be a shelflist card or a record in the online catalog.

IN-HOUSE MAP CATALOGING

In 1986 a half-time librarian in the Map Collection began copy cataloging maps on an OCLC terminal located in the unit. (The Map Collection is one of three units outside of the Monographic Services Division and the Serials Services Division approved to do its own cataloging. The others are the Music Listening Center and the East Asia Library.) All records for maps added to the collection since 1980 were searched and holdings were added to existing OCLC records. Then pre-1980 acquisitions were searched. In 1986 and 1987, close to 7,400 map titles were searched in OCLC and about 55% already had records in OCLC. In addition, approximately 800 map titles for newly published maps were searched in OCLC and only 35% of the newer titles had existing records.

In 1988 a second Title II-C Grant application was submitted to the Department of Education requesting funds to catalog maps of the Pacific Northwest at the universities of Washington and Oregon.[10] The grant application was successful and the project ran from October 1, 1988 through June 30, 1990. Only one of the two cataloging positions advertised for this grant at the University of Washington was filled and time spent training staff during the grant project was significant. The grant resulted in the UW adding over 1,000 original cataloging records to OCLC plus tagging copy (finding a match with existing OCLC bibliographic records) for another 205 records. Staff funded by the grant also codified map cataloging procedures and policies, including adding many cutters for Washington State places to LC's *Geographic Cutters.*

Beginning in the mid-1990s map cataloging slowed significantly while the classification and shelflist project was emphasized. During this time we also loaded MARCIVE records and did a shelflist recon project. The cataloging emphasis continued to be on Pacific Northwest maps and newly acquired materials. From the late 1980s to the present in-house map cataloging has added about 2,630 original map records to OCLC, and over 11,850 OCLC map records were tagged for copy.

As new maps continue to arrive they are cataloged using OCLC and given LC classification numbers and subject headings. An exception is if newly acquired materials are in non-Roman characters and no language expertise is immediately available to help the map cataloger. In this case, the map gets an LC class number and is filed in the public map drawers; additionally a brief shelflist card is written and filed. These materials will be cataloged in the future.

The Map Collection shelflist still contains about 9,000 titles that need original cataloging. In addition, there is a backlog of totally unprocessed maps and aerial photographs that needs cataloging and classification. This backlog will add another estimated 3,000 titles to the shelflist bringing the number of titles still to be cataloged to about 12,000. This includes a large number of map sets, foreign geologic map series, and Washington State aerial photography.

In 1998 a library school graduate student completed a small but important in-house minimal level cataloging project. The student created bibliographic records directly in the online catalog for about 150 aerial photograph sets for Washington State. In retrospect, these records should have been entered into OCLC. Minimal level cataloging will continue to be utilized with some materials as we work our way through the remaining maps and photos needing to be cataloged.

LOADING MARCIVE RECORDS

Purchasing existing bibliographic records is another option for some materials. In 1994 the UW Libraries purchased bibliographic records from MARCIVE, Inc. The Libraries loaded about 350,000 records created by the Government Printing Office for federal documents cataloged from 1976 onward. Companies such as MARCIVE sell GPO bibliographic records along with linked barcodes, and their database includes many map cataloging records. The UW Libraries also purchased "ongoing" GPO bibliographic records from MARCIVE and continues to load them into the online catalog monthly for recently published federal documents and maps.

The result of the March 1995 load of retrospective GPO records from MARCIVE was about 15,300 map records loaded directly into the local online catalog (suppressed from public view until a code was changed in the record). We did not purchase linked barcodes with these records. We chose not to purchase individual records for USGS topographic maps but rather to catalog those as sets (as there is copy in OCLC by state). The MARCIVE records were loaded into our Innovative Interfaces Inc. (III) catalog as "bare bibs"–that is, bibliographic records without attached item records.

Of the 15,300 map records loaded, about 3,200 overlaid map records already existing in III by matching on the OCLC record number. From mid-1995 through 1997 a librarian coordinated the work of three students to create item records and barcode the remaining 12,100 map records. This involved assigning a LC call number if the map did not already have one, barcoding the map and creating an item record in III. In the end, a number of bibliographic records were deleted because we decided not to use individual MARCIVE records for nautical or aeronautical charts, but rather to catalog those titles as sets.

Because the Libraries purchased ongoing MARCIVE records and barcodes, every month we get a number of barcodes and records that we match with the maps. Most of these are for monographic maps, including USGS geologic maps.

Dr. HelenJane Armstrong published a thorough article about loading GPO map records into a local online catalog that goes into greater depth about what to expect during such a project.[11] Her article also includes a brief history of map cataloging at the University of Florida.

The UW Map Collection chose to class selected U.S. government map sets using non-LC classification. Included are USGS topographic maps and geologic maps, as well as nautical and aeronautical charts. We did this because they are filed by map or chart numbers in map cases or file cabinets separate from the LC classified maps. Figure 3 shows the non-standard call numbers that have been used for these sets. This is a dangerous practice that I do not recommend as it relies on agency or map set titles in some cases, and, as we all know, these are subject to change!

RETROSPECTIVE CONVERSION

Another project that the UW Map Collection embarked on during six months in 1996 was the use of retrospective conversion (recon) services from OCLC RetroCon. This happened as part of a larger UW Libraries' recon project.

We sent 5,800 map catalog cards to OCLC RetroCon (about 1,200 had OCLC numbers written on the cards, so these were easy to match). RetroCon tagged copy for about 2,500 titles (43%) and transcribed the card information into OCLC as K-level records for the other 3,300. No subject headings were included since headings on the cards were very incomplete. RetroCon's cost for this project was about $40,000 and resulted in most of the 5,800 cataloging records being added to UW's catalog.

Costs for these types of projects will vary widely based on what kind of bibliographic information the map collection starts with. The Map Collection's shelflist as we embarked on this project contained different types of

FIGURE 3. Non-Standard Call Numbers Used for U.S. Depository Map Sets

Map Set	Call Number
USGS 7.5' topographic	[State abbreviation] 7.5 [sheet name date] example: WA 7.5 Aberdeen 1994
USGS 1:100,000 topographic and planimetric	US 100 [sheet name date map type] (map type = t for topographic, p for planimetric, tb for topo/bathymetric) example: US 100 Colville 1984t
USGS 1:250,000 topographic	US 250 [sheet name date] example: US 250 Spokane 1980
USGS geologic maps	USGS [series designation and number] example: USGS I-1481
Nautical charts	Nautical US [chart number edition date] or Nautical foreign [chart number edition date] example: Nautical US 11307 35 ed. 1998
Aeronautical charts	[Series designation chart number] example: TPC C-6D

cards from various time periods. A student was hired to go through the shelflist and separate the cards into the following categories:

1. Shelflist cards duplicating records already in the online catalog. These cards were recycled. About 9,600 cards.
2. Typed shelflist cards with an OCLC record number on the card, but no record in the online catalog. These cards represented records that were tagged in OCLC in the 1980s, but many were never loaded into the on-line catalog. Sent to RetroCon to search and tag OCLC record. About 1,200 cards.
3. Typed shelflist cards without an OCLC record number on the card and no record in the online catalog. Sent to RetroCon to search and create K-level record if no record is found. About 4,600 cards.
4. Handwritten very brief shelflist cards including call number, title, and scale and date if they were prominent on the map. These cards were not sent to RetroCon due to lack of information for a match or to create an original record. These cards remain in the shelflist and represent maps requiring original cataloging.

The OCLC RetroCon staff were wonderful to work with, especially Mr. David Van Dyk, Senior Conversion Specialist and our project liaison. Retro-

Con presented project specifications that estimated the number and cost of cards to be processed and detailed the technical specifications for the project. We worked with RetroCon staff to finalize those specifications that included over 25 pages of definitions to identify a matching record! This sounds like a lot, but if you think about the variety of information that may exist in your shelflist, you will realize that many situations for many different MARC fields need to be addressed. The author chose not to include these twenty-five pages of specifications in this article, but will be happy to discuss them if contacted.

OUTSOURCING MAP CATALOGING

In 1995, due to an academic leave, the Map Collection's staff of librarians dropped from two to one. To continue map cataloging while losing a full-time librarian a contract cataloging project was identified and requests for proposals were sent to OCLC's TechPro Services and to Rickert & Associates of Clarksville, Maryland. Rickert was chosen to catalog a project of about 200 newly acquired maps. Rickert's costs and turnaround time were much better than those proposed by OCLC. Since November 1995 Rickert has cataloged close to 1,400 maps (over 80% is original cataloging) for the UW Map Collection. Rickert & Associates provided top-quality map cataloging for a very reasonable price.

The Libraries provided Rickert with a UW OCLC password to use from Maryland. The Libraries supplied Rickert with PRISM on a diskette to be used only for UW cataloging. Rickert connects to OCLC using a modem and all cataloging appears in the OCLC database with the UW Libraries' OCLC code.

Through 1998 Rickert completed seven cataloging projects for the Map Collection. Each project began with work done by Map Collection staff. Newly acquired materials were gathered and counted and a request for cost proposal was prepared and sent to Rickert.[12] Rickert made an estimate based on the number of map titles and sheets per title, languages and estimate of original versus copy cataloging, and how much classing was already done.

Maps were mailed to Rickert for cataloging using insured UPS. Before mailing a student photocopied panel titles to keep as a record of what was sent. The student also barcoded each sheet with an unlinked barcode. Rickert cataloged the maps and emailed OCLC "save" numbers to the Map Collection. The UW map librarian reviewed the records and staff checked the shelflist for call number duplication. Once approved via email, Rickert updated the records in OCLC and mailed the maps back to Seattle.

The outsourcing option assisted greatly with map cataloging during a time of reduced staffing. The Map Collection staff is back to 100% with the addition of a GIS Librarian and more original cataloging will be done again in the unit. The timing for more cataloging done within the unit could not be better as most maps remaining to be cataloged require original cataloging and include older or fragile materials that should not leave the collection.

CONCLUSION

The Map Collection at the University of Washington Libraries will continue to acquire and catalog new maps, continue to work through the shelflist to catalog the collection retrospectively, and line up assistance to catalog non-Roman alphabet materials. In addition we have begun to catalog cartographic computer files in the unit.

The projects discussed in this article have greatly increased user access and use of maps. Just in the past three years the number of users who come to the Map Collection with a catalog printout in hand has greatly increased. If the map they have found a record for is not *exactly* what they are looking for, we have at least led them to the Map Collection through the catalog and can assist them in finding other relevant maps.

The map classification project was useful especially in helping student employees refile maps in their proper places, but cataloging and barcoding maps are the real goals. Cataloging maps gets them into the database with all other library materials where they are visible to the users who need them. And that, after all, is why we're here!

NOTES

1. "UC Berkeley Library Completes Map Conversion," Western Association of Map Libraries *Information Bulletin*, 29, no. 1 (Nov. 1997): 33.

2. To subscribe to MAPS-L (Maps and Air Photo Systems Forum), send an e-mail message with the words "subscribe maps-l" in the body of the message to: listserv@uga.cc.uga.edu
To subscribe to AUTOCAT (Library Cataloging and Authorities Discussion Group), send an e-mail message with the words "sub AUTOCAT your name" to: listserv@listserv.acsu.buffalo.edu

3. Mary L. Larsgaard and Katherine Rankin, "Helpful Hints for Small Map Collections," <http://www.sunysb.edu/libmap/larsg.htm> American Library Association, (Map and Geography Roundtable Electronic Publication no. 1). Rev. by Stephen Rogers, 1997.

4. Excellent overviews of general map cataloging issues, resources and tools may be found in Mary L. Larsgaard, *Map Librarianship: An Introduction* 3rd ed. (Englewood, Colo.: Libraries Unlimited, 1998). See also, Larsgaard, Mary L., "Cataloguing and Classification" in *Information Sources in Cartography,* ed. C.R. Perkins and Robert B. Parry (London; New York: Bowker-Saur, 1990).

5. "WorldCat Statistics by Format, July 1, 1998," OCLC *Pacific News Update & Calendar* 136 (August 1998): 5. For more information about OCLC Inc., see <www.oclc.org>

6. E-mail from Anne Stanton, RLG Information Center, Sept. 11, 1998. For more information about RLIN, the Research Libraries Information Network or RLG (Research Libraries Group), see <www.rlg.org>

7. E-mail from Elizabeth R. Cowart, WLN, Sept. 14, 1998. For more information about WLN, see <www.wln.org>

8. Library of Congress, Geography and Map Division. Map Cataloging Manual (Washington, D.C.: Cataloging Distribution Service, 1991).

9. Mary L. Larsgaard, *Map Librarianship: An Introduction* (Englewood, Colo.: Libraries Unlimited, 1998).

10. Department of Education Title II-C grant opportunities ended in September 1997 for "Strengthening Research Library Resources–Discretionary Grants to Major Research Libraries" (CFDA No. 84.091). For grant opportunities in your area, look at the current volume of *The Big Book of Library Grant Money*, prepared by the Taft Group for the American Library Association, Chicago.

11. HelenJane Armstrong, "An Academic Map Library Loads GPO Cataloging Tapes: A Case Study of Plans and Impacts," Special Libraries Association, Geography and Map Division *Bulletin*, 177 (September 1994): 2-34.

12. Sample request for proposal letter sent to off-site map cataloger:

Ms. Frances M. Rickert

Rickert and Associates
14009 Brighton Dam Road
Clarksville, MD 21029

Subject: Request for Cost Proposal

Dear Ms. Rickert:

This letter is to request a cost proposal from you for a project to catalog approximately 298 map titles (413 sheets and 8 accompanying texts) for the University of Washington Libraries' Map Collection utilizing OCLC. Most of the titles are in English, except for 12 Spanish, 3 Russian, 2 French, 1 Vietnamese (copy cat), 1 Danish, 1 Norwegian, 1 Finnish and 1 German. 202 of the titles (311 sheets and 4 texts) already have LC call numbers written on them, so they do not need classification.

I estimate about 85% of these materials will require original cataloging although some related copy exists.

Specifications for original and copy cataloging are listed below. This proposal may not exceed $xxx (including labor, postage, telecommunications, tax and any other charges). Also, please estimate the number of weeks you think you will need to have the maps to complete this cataloging project.

Please feel free to call or send me an e-mail if you have any questions about this proposal.

Sincerely, Kathryn Womble

APPENDIX

MARC Tags Required if applicable for Original Cataloging

Fixed fields:
Type, RecG, Relief, Desc, Bib lvl, Enc lvl, Indx, Source, Govt pub, Base, Dat typ, Lang, Ctry, Dates
040
007 Should be the same for most maps we send on this project
020
034 Include coordinates only if they appear on the map
041 Only if non-English or multiple languages. Not included if only one
    language or if it would just say "eng"
043
052
090 Fully-formulated LC call number
1xx
245
246
255 If no scale on map, use "Scale not given." We do use a Map Scale
    Indicator to convert bar scales to approximate representative fractions.
260
300
4xx
5xx
6xx
7xx
8xx
945 for local holdings
  For example:
    945 1_ G7031.E635 1962 .S6 |l mpgen |t 16 |b 39352036580413
    or
    945 1_ G7031.E635 1962 .S6 |d text |l mpatl |t 3 |b 39352038583942
    (where delimiter "d" is volume information, delimiter "l" is location
    code, delimiter "t" is item type code and delimiter "b" is barcode.
    Specific information about codes would be provided when contract
    begins.)
948 for local overlay information
  For example:
    948 .b30858293 (where .b30858293 is Innovative record number)

Required for copy cataloging once record is verified to be correct record

Add 945 for local holding information (see examples above)
Add 948 for local system overlaying (see example above)
Add 052 if not in record
Add LC call number and LC subject headings if not in record

Enhancing in OCLC's Maps Format:
A Participant's View

Arlyn Sherwood

SUMMARY. This article explores one of OCLC's cooperative quality control efforts, the Enhance Program, specifically in the Maps Format. Various aspects of participation in the program, such as the application process, reasons for participation, training, typical experiences, types of changes made to records, and the benefits of participation are discussed. One cataloger's twelve-year experience with Maps Format records forms the basis for a list of the most common changes made to map records in the Enhance Program. The list is offered in the hope of further improving the quality of map cataloging in the creation of OCLC records. *[Article copies available for a fee from The Haworth Document Delivery Service: 1-800-342-9678. E-mail address: getinfo@haworthpressinc.com <Website: http://www.haworthpressinc.com>]*

KEYWORDS. OCLC, Enhance Program, Maps Format

INTRODUCTION AND BACKGROUND

Although e-mail messages looking for simple ways to "index" or "catalog" map collections in libraries still appear in cyberspace, the majority of

Arlyn Sherwood (MLS, MA in Literature) has served as Map Librarian at the Illinois State Library since 1974, working with the OCLC Maps Format since 1980 and Enhancing since 1986.

Address correspondence to: Arlyn Sherwood, Illinois State Library, 300 South Second Street, Springfield, IL 62701.

[Haworth co-indexing entry note]: "Enhancing in OCLC's Maps Format: A Participant's View." Sherwood, Arlyn. Co-published simultaneously in *Cataloging & Classification Quarterly* (The Haworth Information Press, an imprint of The Haworth Press, Inc.) Vol. 27, No. 3/4, 1999, pp. 429-441; and: *Maps and Related Cartographic Materials: Cataloging, Classification, and Bibliographic Control* (ed: Paige G. Andrew, and Mary Lynette Larsgaard) The Haworth Information Press, an imprint of The Haworth Press, Inc., 1999, pp. 429-441. Single or multiple copies of this article are available for a fee from The Haworth Document Delivery Service [1-800-342-9678, 9:00 a.m. - 5:00 p.m. (EST). E-mail address: getinfo@haworth pressinc.com].

those map librarians who want their cartographic collection to attract the widest possible audience have come to realize that the bibliographic records for the map collection must dwell with the bibliographic records of the whole library's collection. For all practical purposes this means the cartographic records have to be in MARC 21 standard form*–and the most prominent utility for map records using the MARC 21 standard is OCLC.*

OCLC's WorldCat is a 40 million record bibliographic database, used by catalogers all over the world to share cataloging records for all types of library materials. The database was created in 1971 by a few Ohio librarians using the original book format but has since grown into a database of world-wide coverage with eight formats. Catalogers began contributing map format records to WorldCat in 1976. Map records currently constitute about .01% of WorldCat.

In 1984, the OCLC Users Council adopted an 8-point "Code of Responsible Use," (now "Principles of Cooperation"), 6 points of which deal with the issue of quality. The code of ethics, which those institutions contributing original records to the database have voluntarily agreed to live by, include:

1. provide training to enable staff to use the database effectively,
2. avoid creating duplicate records,
3. report errors promptly, thereby improving the database quality,
4. input current cataloging promptly,
5. input original cataloging according to national standards and as promulgated in OCLC's *Bibliographic Formats and Standards,*[1] and
6. use the appropriate MARC formats.

Although every library that utilizes OCLC has made the promises above, in reality duplicate records, records that do not reflect national standards in all respects, and an unwillingness to report errors promptly in a paper or email process continues to be a problem.

Quality is not only an ideal, but an economic goal. Productivity decreases when catalogers have to evaluate duplicate records and choose between them or when catalogers must take time to correct records, e.g., for typographical errors. Time is also wasted when interlibrary loan staff have to combine holdings from multiple records to create the best lender string. Thoroughness is sacrificed when frontline reference staff miss relevant citations due to a subject heading misspelling.[2]

Summarizing from Calhoun, Campbell, and Gehring's article, errors that appear in the database have been categorized in the following way: too much (duplicates); too little (records with insufficient information); and too many versions (incorrect name and subject headings and MARC 21 coding errors and changes). In the first category human and computer methods were employed to solve the "too much" problem, to report and merge duplicate records. OCLC staff corrected the errors they found. Additionally, member libraries mailed in change requests to notify OCLC staff to correct errors, but until 1991 duplicate records could only be merged manually by OCLC staff. Thereafter, a duplicate detection and resolution algorithm was employed to merge duplicates automatically for the books format. For other formats, including maps, separate programs run automatic comparison algorithms against incoming records to the WorldCat database to help prevent duplicates from being loaded.

Four methods were utilized to solve the "too little" problem: (a) maintain standards for minimal level (Level K) records, (b) encourage full cataloging (Level I) for as many materials as possible, (c) allow authorized users to upgrade less-than-full records, and (d) provide new ways to add call numbers or subject headings to minimal-level records.

To assist in resolving the "too many versions" problem,[3] "as early as 1972 OCLC installed error detection capabilities within the system–checks for illegal indicators, subfield codes and text characters,"[4] and since then automated authority control (run on specific occasions from 1990 to 1993), record validation, and database scanning have been applied to this problem as well.[5]

EMERGENCE OF COOPERATIVE QUALITY CONTROL EFFORTS

OCLC has been concerned about database quality from day one. At first, quality control of the database was concentrated at OCLC headquarters. Though OCLC's regional network affiliates had always stressed training and education for those individuals beginning to work in the database and those advancing their skills, errors reported were outpacing the staff's ability to correct them in a timely fashion, so cooperative programs were born. Libraries in the CONSER* Program began working on the quality of serial records in 1975. The ability for member libraries with full-mode cataloging authorization to upgrade minimal-level records (encoding level K, M, 2, 5, or 7) was authorized in 1985. Authorization to "enrich" records by adding call numbers, 006 and 007 fields, subject headings, genre terms, or contents notes was also granted to full-mode libraries in later years. But the program highlighted in this article–the Enhance Program–was created in 1984.

ENHANCE EXPLAINED

The Enhance Program, like all OCLC cooperative enrichment programs, was designed to distribute the responsibility of maintaining the quality of the

OCLC database.[6] OCLC created Enhance in response to the expressed need to speed up the error-correcting process.[7] Enhance was to provide a cost-effective mechanism for authorized users to add and correct data in the master bibliographic records online as well as to improve the usefulness of records and the quality of cataloging while additionally reducing the number of paper change requests sent to OCLC staff.[8,9] Enhance would be for monographs what CONSER was for serials.[10,11]

OCLC installed Enhance in December 1983 in response to the need for a database of high enough quality so that paraprofessionals could take care of member-input records[12] and a way for those libraries already correcting records for their own use to share this valuable time.[13] Application instructions were published and applications to the Enhance Program were encouraged for the first time. Initially, twenty institutions were selected by OCLC staff to Enhance the books, scores, and sound recordings formats. OCLC staff reviewed 10 records input by each institution and subtracted points from a total of 100 for each error found. Error values ranged from 1 for a minor typographical error to 25 for a duplicate record or non-use of the Name Authority File. Applicants who scored above 90, and had a high volume of records, were selected. The remainder of the first 20 were chosen from the 70-90 scores, supplementing the geographic coverage of the first group. Based on estimates supplied by the chosen applicants 50,000 records could be "enhanced" annually.[14]

In June 1984, the first 20 institutions chosen received Enhance training. Training, then as now, consists of both technical and philosophical elements. The "rules" of Enhance are set forth in the *Cataloging User Guide*,[15] (second edition), especially Chapter 6 on replacing records, and *Bibliographic Formats and Standards* (second edition), especially Chapters 4 and 5 on when to input a new record and quality assurance. The technical elements include how to lock and replace a record, what database activities should and should not be done while in the Enhance mode, and the credit process. The philosophical rules include: (1) Enhance applies only to editing member records, (2) Enhance applies only to editing I-level (full-level) member records, (3) the Enhance participant must follow national practices in correcting errors, not simply changing a field because s/he would not make the same judgment call as the cataloger who input the record as an original, and finally (4) an Enhance participant can only correct records for which s/he is cataloging the same piece.

A second round of applications brought 18 more institutions on board by April 1985. A new format, the media (currently visual materials) format, now had Enhance participants in addition to the previous three formats.[16] One of these new Enhance participants, the University of Wyoming, joined the force to Enhance in the books format. Their decision to apply for Enhance status was based upon four factors: the ability to make permanent changes directly online, the ability to use less experienced paraprofessionals with an improved database, pride in the department, and the credits received.[17] They published their thoughts on their first year's positive experience as Enhance participants. They found participation worth the effort because it enabled the library to improve the quality of the database, it helped train staff, it contributed to Cataloging Department morale, it allowed them to update records online, and it offered a billing credit. Although Enhance participation necessitated new workflow patterns within their cataloging department it did not hamper production. And the money wasn't a major factor![18] On the other hand, the University of Maryland Health Sciences Library found that initially it slowed their workflow by about 20%. They liked the billing credit but the primary motivation was improvement of the database while simultaneously meeting their own cataloging needs.[19] At the University of Pittsburgh, reasons for seeking Enhance status included: the Catalog Department was already committed to cataloging in accordance with full national standards and much effort was being expended editing member records only for local use, a desire to contribute to the cause of cooperative cataloging, and their long association with OCLC.[20]

ENHANCE FOR MAPS

Not until the third round did the map format have an Enhance participant. Those selected in late 1985 and trained in January 1986 were Mary Nell Maule (Georgia State University) and myself (Illinois State Library). Authorization to Enhance is granted to an institution but that library is responsible for someone to perform at the Enhance level in the specific formats for which they have been granted authorization and, additionally, for a minimum number of Enhance transactions per year. Unfortunately, Georgia State University withdrew from Enhance participation in the map format in 1987. Thankfully, that same year the University of Illinois at Urbana-Champaign (Nancy Vick Edstrom) was added, as was the National Agricultural Library. In 1990 the GPO* was granted Enhance authority in the map format, but the University of Illinois withdrew as Nancy Edstrom left campus. In 1991, the National Agricultural Library withdrew, but the University of Georgia (Paige Andrew) was added. In 1992, the USGS* library was authorized, but subsequently dropped. When Paige left the University of Georgia campus in 1995, their authorization was withdrawn, but Paige re-emerged at Pennsylvania State University where he applied for, and was granted, map format Enhance authorization once again later that same year. So, Penn State, the Illinois State Library, and the GPO remain the current map format Enhance participants.

However, the New Mexico Institute of Mining and Technology currently has temporary authorization for the duration of a special project.

Training for Enhance authorization is no longer on site at OCLC in Dublin but is accomplished through the "Enhance Training Outline" (ETO), which documents all policies and procedures for Enhance institutions and supplements other OCLC documentation. The March 1998 revision of the ETO is available online at http://www.oclc.org/oclc/cataloging/enhance/outline.htm "Enhance Requirements and Application Instructions" and "Procedure Used to Evaluate Records Submitted with an Enhance Application" are also on the OCLC web site. As of 1989, the Enhance application process was opened year round, regardless of format.

Additionally, the BIBCO Program, a National Level Enhance Program run by OCLC and the Library of Congress exists, but there are no map participants at this point in time.

EXAMPLE OF PRODUCTION LEVELS IN ENHANCE

As of 1997, over 140 institutions participate in OCLC's Enhance Program for all formats and they have performed over 1.2 million Enhance transactions. Enhance transactions are not profiled by format at OCLC, but I have kept my own statistics from 1986 to date. As of July 1998, there were 505,518 total map records and 236,978 I-level map records on OCLC.[21] I have enhanced 1.4% of the total map database and 3% of the eligible I-level map database. Over the last 12 1/2 years, I have enhanced 47% of the member records I have edited, out of a total of 14,983 records. The statistics don't fall evenly from month to month, for several reasons. I am only a part-time map cataloger and interruptions, such as moving the library in 1990, learning to live with a new local system in 1991 and again in 1998, along with other duties as assigned, have a definite impact on production.

WHY GET INVOLVED?

The reason I applied for Enhance (at the urging of two other State Library staffers at the time–Arlene Schwartz and Janet Lyons Dickinson) sounds very similar to the ones mentioned above. My library's policy was to perfect all fields on a record for our own use, so why not just "punch a couple more keys" and share the improvements with everybody! Arlene and Janet didn't lead me far astray. Enhance participation is intended to fit within a library's regular workflow and it has worked this way for me.

My method for performing Enhance work is as follows. I review the

OCLC record previously found and printed for the particular map. Obviously I am not truly cataloging online for several personal and institutional reasons. I review the entire record, fixed field by fixed field, variable field by variable field, subfield by subfield, etc., making sure that every field completely matches the piece in front of me, making sure that every subject, series, and name heading is correct and up to date, and making sure that every field corresponds to I-level standards, the AACCCM* manual, and *Bibliographic Formats and Standards.* If everything is technically perfect, i.e., doesn't need Enhancing, I only add our local information to an edit form, put the form in the pile to be produced in the future, and send the map on to the stacks. If there is a subject heading, for instance, that I didn't want to use locally, I cannot remove it from the master record if it is technically correct. I can note on the edit form to delete it from the local record when I go back to produce the record for us, but it must remain in the master record.

If I find a problem with the member record that does need Enhancing, such as a missing required field (e.g., 034), an incomplete field (e.g., an 052 of class number 4102 lacking subfield "b" information and/or the subfield indicator itself), or conflicting information between fields (e.g., the relief statement in the fixed field doesn't indicate the same types of relief that the 500 note does), I make a note on the edit form about how it should be corrected. From there I add my local information to that same edit form and then put it in a separate pile of edit forms that will be Enhanced before they are produced. When I have a good-sized pile of edit forms batched "to be Enhanced," I sit down at the terminal, bring up the first record, lock it, input only the changes made for Enhance purposes, and then replace the record. The replaced master record will appear back on the screen, at which point I input the local data and any local changes I want to make to our record but am obligated not to make to the master record (such as taking out any 650 _4's), and produce the record.

TYPES OF CORRECTIONS
AN ENHANCE PARTICIPANT MAKES

So, for those of you who have just been through your first OCLC map format workshop and now feel prepared to catalog that entire unique map collection that has never been cataloged before, here are a few of the common problems I find (and which you can help prevent in the future in your original cataloging):

1. The biggest problem is poor proofreading or quality control. For instance, perhaps there is a record for an earlier edition of your map already in the database. You have pulled up the record for that earlier edition, used the

"new" command on OCLC to create an original record with most of the earlier data intending to change all of the fields appropriate for the new edition, but actually forgot to go back and change one field, and produced the new record. Now the fixed field data doesn't agree with information in the body of the record! (North Carolina State University listed this as a common problem back in 1988!)[22] There are several instances in an OCLC record where the same information is recorded in more than one field, once in English and once in coded form, and the information given must match. The scale in the 034 subfield "b" should be the same scale as that recorded in the 255 subfield "a," or, a "bg" code in the relief fixed field has to translate to "Relief shown by shading and spot heights" in a 500 note. The best solutions toward avoiding this type of problem are: to put your work in save and come back to it a day later, re-read it carefully, compare it to your edit form if there was one; have another cataloger check over your cataloging; or both, and then produce it.

2. Another field in which errors commonly occur in the map format is the Index fixed field. The meaning of the field is not well understood. There are only two coding choices to input, 0 or 1. Zero means no index and 1 means there is an index. But it's the term itself that is misunderstood. Additional terms given in OCLC's documentation that help to explain this fixed field's use are location index and gazetteer. If the map contains numbers in red circles on tourist destinations which are listed and elaborated upon in numbered order along a margin of the map, or if the map is bordered by a grid matrix and an alphabetical list of streets in that town with grid locations given is printed on the back (or anywhere on the map), then that map is considered to be indexed and the field must be coded with a "1." An index is not an index map to part of a set, nor a location map, nor a legend. Conversely, if no "one-to-one" correlation between information given as part of the map and a list exists then the field is coded "0."

3. A third common error is the lack of an 052 field for every geographic area named in a 65X field. Every 650 field in the record with a subfield "z" has to have a corresponding 052 field that represents the same specific geographic area. Every subfield "a" in a 651 field has to have a corresponding 052 field that represents the same specific geographic area. There *can* be more 052s than 65Xs but not vice versa. If the record has:

650 _0 Harbors |z Louisiana |z New Orleans |v Maps.

651 _0 Gulf Coast (La.)

then the record needs:

052 4014 |b N4

052 4012 |b C6

If the record has:

650 _0 Outdoor recreation |z New Mexico |v Maps.

651 _0 Cibola National Forest (N.M.) |v Maps.

then the record needs:

052 4321

052 4322 |b C5

For example, in OCLC reccord number 40852625 which is a record for a map concerning sea-floor disposal sites off of Boston Harbor, there are three 052s for Cape Cod Bay, Martha's Vineyard, and Massachusetts Bay, but only one subject subdivision, for Massachusetts Bay. To access all area and sub-area cutters already created, the map cataloger must have copies of the Library of Congress G-classification schedule (kept updated) and the Library of Congress, Geography and Map Division's *Geographic Cutters.*[23] The NUC-CM* microfiche and LOCIS* are also very useful tools. Even with all these tools, not all potential cutters necessarily exist, especially because the microfiche version of *Geographic Cutters* only covers areas for the United States. This will be rectified in the near future once the electronic version of the G-classification schedule is made available to the public, as it will contain links to geographic cutter tables for places worldwide. For the time being I often request that LC G&M* cataloging staff create geographic cutters when needed.

4. A fourth common error occurs when cataloging one sheet of a possible series as a separate monographic item. The "CrTp" fixed field for "type of cartographic material" should be coded "a," for single map, rather than "b" for a map series. The code should reflect how the cataloger is cataloging this map, not whether it was published as part of a series. If a single 1:100,000-scale topographic quadrangle from USGS is the total subject of a specific record, then the "CrTp" fixed field should be coded "a." If the entire series is cataloged as one record, then the "CrTp" fixed field should be coded "b."

5. Other common errors include: not following prescribed punctuation, not including the vertical scale in the 034/255 fields when cataloging a geologic section or cartographic diagram in the map format, not understanding the difference between 5 maps, 5 maps on 1 sheet, and 1 map on 5 sheets (and the corresponding "CrTp" fixed field), not understanding when to use the term "remote-sensing image" versus "map" in the 300 subfield "a" and the 007 subfield "b," not capitalizing the first word in the 255 subfield "b" and many others too numerous to mention individually.

6. While not "errors" originally, rule modifications over the years have led to necessary changes and revisions to bibliographic records for maps residing in WorldCat. Since I am a typical map librarian, with a typical map library's acquisitions budget that necessitates accepting all donations, a substantial percentage of my current cataloging is of older maps that therefore have older records, if any. Even though the Enhance authorization does not require updating pre-AACR2* records, I do. In addition, updating is needed even for records created after the implementation of AACR2 as some of the map format procedures and policies have changed since AACR2's adoption. For instance, the 007 field was implemented afterwards, so AACR2-based records lacking the 007 field exist in the database. There are also records with out-of-date name and series headings even though they were correct when input. Numerous recent changes in the Library of Congress Subject Headings have affected common map subject headings, especially those relating to remote sensing images and digital/electronic maps. All of the above lead to necessary changes and revisions to bibliographic records for maps residing in WorldCat, and warrant the skills of an Enhance participant to make these changes.

CATEGORIES OF CHANGES PERFORMED
ON BIBLIOGRAPHIC RECORDS

OCLC has done some studies on the numbers of changes actually made to an Enhanced record (in all formats combined) and the types of changes made. The number of changes made to a record was found to average 2.6 in a 1992 report and 4.5 in a 1994 report. The 1992 report found that: (1) a large number of Enhance transactions represented only cosmetic changes that neither increased the usefulness of a record nor added access points, (2) some records had been replaced more than once, and (3) Enhance, instead of the Database Enrichment Program, had been used only to add call numbers in many cases. Despite these three disappointing findings, the report found that overall Enhance libraries were doing valuable and highly accurate work.[24]

The 1994 report found: (1) an increase in the number of subject headings added over the 1992 report, (2) the number of editorial changes had doubled since 1992, accounting for 62.9% of the total types of changes made to master records, and (3) a substantial increase in the number of cosmetic changes. (Another interesting finding is that the publication dates for over 50% of the Enhanced records were post-1990 and 5.2% were items published before 1900).[25] OCLC continues to urge that Enhance work concentrate on substantive rather than minor cosmetic changes, and that the Enhancing library review the entire record, make all changes needed, and replace the record only once. In fact, a second replace on the record by the same institution is no longer credited. And since software changes were made in 1991 the system can differentiate between a Minimal Level Upgrade (MLU) replace versus an Enhance replace, even when done in Enhance authorization. With the advent of Format Integration in 1996, instructions to libraries include the message *not* to Enhance a record if all that it needs are corrections caused by Format Integration.[26]

PERSONAL REFLECTIONS
ABOUT THE ENHANCE PROGRAM

1. It would be very helpful if we could Enhance DLC*'s pre-AACR2 records as there are a substantial number in the database that I have updated for our use but, since I can't change a master record with a blank encoding level, I can't share the benefit. However, BIBCO participants can change national libraries' records.

2. Enhance only applies to I-level records. K-level records are eligible only for MLUs. In theory then, K-level records would need more work to bring them up to perfection than would I-level records, but that is often not the case. Some libraries are apparently timid about how well they understand the map format and label their work K-level when in fact, it is better work than some I-level records. (A study of replaces during fiscal year 1988-89 found that Enhance libraries were responsible for the majority of MLUs, even though any library can perform an MLU without any special authorization).[27] Following the standards more closely is a desirable goal.

3. We are a GPO regional depository library so we receive every map coming through the depository program. A lot of depository libraries are purchasing the MARCIVE* products for the depository map collections. MARCIVE takes the record straight from GPO; all libraries receiving MARCIVE map records do not get the benefit of any Enhance work I do on GPO records.

4. Since the Enhance authorization is an acknowledgement of expertise, you get calls for cataloging advice and requests to do workshops! (You can decide whether that's a plus or minus!)

CONCLUSION

My estimation of the benefits of Enhance authorization echoes what others have put in print:

a. If you improve the record for your library, you might as well share it–it isn't much more work!

b. There's a quality master record on OCLC for everyone that follows you. It's too bad that a lot of libraries have used records before I get to Enhance them, even on non-federal maps.

c. If you're still working on OCLC instead of your local database and want to take the time to improve the WorldCat, changing the master record is easier than sending, even by e-mail or via OCLC's Web site, the change request forms.
d. There's always that billing credit, which has increased over the years–but I don't ever see my library's OCLC bill!

There's no way a map format workshop leader could ever cover all of the circumstances you will run across while cataloging maps. If you've been through all the workshops offered and you've digested all the appropriate rules in all the appropriate tools and are still stumped, try posting your cataloging question on MAPS-L. You'll probably get several responses. If not, give it your best shot. If you're wrong, Paige or I can fix it–if we find it while working on the same title!

**ACRONYMS*

MARC 21 – machine-readable cataloging (U.S. standard, formerly USMARC)
OCLC – formerly Online Computer Library Center, Inc., 6565 Frantz Rd., Dublin, OH 43017-3395
CONSER – Cooperative Online Serials Program
GPO – Government Printing Office
USGS – United States Geological Survey
AACCCM – Anglo-American Cataloguing Committee for Cartographic Materials. Published *Cartographic Materials: A Manual of Interpretation for AACR2.* Chicago: American Library Association, 1982. (Revision in progress).
NUC-CM – *National Union Catalog: Cartographic Materials.* Washington, D.C.: Library of Congress, 1993- .
LOCIS – Library of Congress Information System. telnet: locis.loc.gov
LC G&M – Library of Congress Geography and Map Division
AACR2 – *Anglo-American Cataloguing Rules,* Second Edition, 1988 Revision. Chicago: American Library Association, 1988, with amendments.
DLC – OCLC's symbol for the Library of Congress
MARCIVE – MARCIVE, Inc., P.O. Box 47508, San Antonio, TX 78265-7508

NOTES

1. OCLC, *Bibliographic Formats and Standards,* 2nd ed. (Dublin, OH: OCLC, 1996).
2. Karen Calhoun, Nancy Campbell, and Donna Gehring, "How OCLC and Member Libraries are Improving the Online Union Catalog." *OCLC Newsletter*, No. 204 (July/August 1993): 16.

3. Ibid., 17.

4. Rowland C.W. Brown, "Quality and OCLC." *OCLC Newsletter*, No. 169 (July/August 1987): 3.

5. Calhoun, op. Cit., p. 17.

6. Carol Davis, "Enhance Applications Received; More Libraries Sought." *OCLC Newsletter*, No. 159 (October 1985): 7.

7. Judith J. Johnson and Clair S. Josel, "Quality Control and the OCLC Data Base: A Report on Error Reporting." *Library Resources & Technical Services*, 25:1 (1981): 46.

8. Richard O. Greene, "Enhancement of Records in the Online Union Catalog." *OCLC Newsletter*, No. 153 (June 1984): 2.

9. Mary K. Roach, "Report of the Cataloging Advisory Committee Meeting." *Action For Libraries*, No. 153 (June 1981): 6.

10. "OCLC to Strengthen Online Quality Control for Database." *OCLC Newsletter*. No. 153 (June 1984): 1.

11. Carol Davis, "Enhance Training Completed; Libraries Gear Up for Autumn." *OCLC Newsletter*, No. 154 (September 1984): 8.

12. Johnson, op. cit., p. 47.

13. John M. Sluk, "Enhancing a National Database." *Cataloging & Classification Quarterly*, 6:1 (Fall 1985): 34.

14. Greene, op. cit.

15. OCLC, *Cataloging User Guide, 2nd ed.* (Dublin, OH: OCLC, 1983-).

16. Carol Davis, "New Enhance Libraries Chosen." *OCLC Newsletter*, No. 157 (April 1985): 1.

17. Carol White and Martha Hanscom, "OCLC Enhance: An Experience." *Action for Libraries*, 11:9 (September 1985): 1.

18. Martha J. Hanscom, Carol J. White, and Carol C. Davis, "The OCLC Enhance Program: Some Practical Considerations." *Technical Services Quarterly*, 4:2 (Winter 1986): 21.

19. Barbara G. Smith, "Everything You Always Wanted to Know About Enhance, but. . ." *Start of Message*, 33 (October 1984): 8.

20. Sluk, op. cit.

21. "WorldCat Statistics." *Information Bulletin* (ILLINET/OCLC Services), No. 305 (July 1998): 10-11.

22. Walter M. High, "The Dilemma of Enhance." *Cataloging & Classification Quarterly*, 9:1 (1988): 135.

23. Library of Congress, Geography and Map Division, *Geographic Cutters*, 2nd ed. (Washington, D.C.: Cataloging Distribution Services, 1989).

24. Jay Weitz to Enhance participants, Dublin, OH, 14 December 1992, memorandum.

25. Jay Weitz to Enhance participants, Dublin, OH, 1 September 1994, memorandum.

26. Jay Weitz to Enhance participants, Dublin, OH, 20 May 1996, memorandum.

27. Ellen Caplan to Enhance participants, Dublin, OH, 15 August 1990, memorandum.

CARTOGRAPHIC MATERIALS IN AN ARCHIVAL SETTING

Cataloguing Cartographic Materials in Archives

Hugo L. P. Stibbe

SUMMARY. After a brief review of the history of the development of cataloguing rules for cartographic materials, both within the Anglo-American Cataloguing Rules community and internationally, the author

After obtaining his MSc in Geography, Hugo L. P. Stibbe started his career as the first University Map Curator at the University of Alberta in Edmonton. He obtained a doctorate from the University of Utrecht in the Netherlands and accepted a position at the National Archives of Canada in Ottawa in the National Map Collection in 1973. He worked in the National Archives for the next 26 years in various capacities throughout the organization. From 1987 until 1998 he was the National Archives' Senior Archival Standards Officer. He was Chair of IFLA's Section of Geography and Map Libraries, chaired the initial Working Group on the *ISBD(CM)* (1977), was General Editor of *Cartographic Materials: A Manual of Interpretation for AACR2* (1982) and, in the last 10 years, was active in the International Council on Archives where he is Project Director and Secretary of the ICA Committee on Descriptive Standards, the body which developed the *ISAD(G)* (1994) and *ISAAR(CPF)* (1996).

[Haworth co-indexing entry note]: "Cataloguing Cartographic Materials in Archives." Stibbe, Hugo L. P. Co-published simultaneously in *Cataloging & Classification Quarterly* (The Haworth Information Press, an imprint of The Haworth Press, Inc.) Vol. 27, No. 3/4, 1999, pp. 443-463; and: *Maps and Related Cartographic Materials: Cataloging, Classification, and Bibliographic Control* (ed: Paige G. Andrew, and Mary Lynette Larsgaard) The Haworth Information Press, an imprint of The Haworth Press, Inc., 1999, pp. 443-463. Single or multiple copies of this article are available for a fee from The Haworth Document Delivery Service [1-800-342-9678, 9:00 a.m. - 5:00 p.m. (EST). E-mail address: getinfo@haworthpressinc.com].

discusses the differences of the cataloguing of this material in archives compared with that in libraries. He states that the cataloguing of this material in archives relates to how the material was created or produced in the context of how it was used. Archival cataloguing (or archival description as archivists prefer to call the process) emphasizes methodology over the material itself. The paper attempts to give a short overview of the methodologies, their development and application. It uses diagrams and an example to illustrate the archival methodology and concludes that cataloguing records and archival description records produced by these methods may co-exist on the same database but that current computer systems developed for library applications have, in general, not developed the capability to display open hierarchically structured linked multilevel descriptions such as required by the archival methodology of multilevel description. *[Article copies available for a fee from The Haworth Document Delivery Service: 1-800-342-9678. E-mail address: getinfo@haworthpressinc.com <Website: http://www.haworthpressinc.com>]*

KEYWORDS. Multilevel descriptions, cartographic material in archives, Dutch Project CCK, provenance, original order, archives standards, archival description, fonds, maps in archives, ISAD (G), ISAAR (CPF)

INTRODUCTION

What is cataloging of cartographic material in archives and how does it differ from cataloging cartographic materials in a library setting?

Why have a separate paper on cataloging cartographic materials[1] in archives? Isn't all cartographic material and the cataloging of it the same? The short answer is yes, all cartographic material is the same; the definitions employed for it certainly are, but the cataloging of this material depends on how the material is created or produced on the one hand and how it is used on the other. Therein lies the difference between the library and the archives approach. Archival cataloging (or archival description as archivists prefer to call the process) emphasizes methodology over the material itself. This paper attempts to give a short overview of the methodologies, their development and application. The reader is cautioned that, in order to understand this paper completely, reading the cited archival literature is highly recommended.

The recognition of cartographic materials as a separate entity from books within libraries, needing its own guidelines for cataloging, goes back quite some time. *Notes on the Cataloging, Care, and Classification of Maps and Atlases* (1921) contains that famous (or infamous, depending on one's point of view), often-quoted phrase of Philip Lee Phillips, "The cataloging of maps and atlases differs very little from the cataloging of ordinary books."[2] Map librarians have had to struggle for decades with their superiors, supervisors, administrations, and library cataloging rule-makers to overcome this notion. Strangely enough, even though this sentence purports that there does not need to be any difference in the cataloging of maps from that of books, the very fact that the booklet was written and produced acknowledged the need for such guidelines and, therefore, that there may be a difference in the cataloging between the medium of books and cartographic materials.

In the cataloging of maps, the basic goals of the library catalog were followed in the past–that is, to produce catalog cards with title, author and subject entries, and descriptions which follow the book layout of an entry by recording the title, edition, imprint, and collation, followed by notes. Scale was recorded in a note and the matter of series was problematic because the term has a different meaning for books than it does for maps. Information on map series was provided in a note. This approach to the cataloging of cartographic materials was maintained for many years at the Library of Congress and followed by most other institutions with map collections in libraries, which had librarians in their employ.

In archives, the general approach to making information about archival documents available was by means of finding aids in a great variety of forms, such as guides and inventories, lists, and collection specific descriptions. There was very little standardization in the production of such finding aids, and such standards as there were were institution-specific, or were left to be developed by the individual archivist who compiled and produced the finding aid. The term "catalog" or the act of creating it, "cataloging," does not appear in the International Council of Archives publication, *Dictionary of Archival Terminology*.[3] Rather, the concept of a catalog or cataloging is closest to the terms employed in archives, "calendar" and "calendaring." One of the two meanings of the term "calendar" the *Dictionary* defines is, "A list, usually in chronological order, of précis of individual documents in the same [archival] series or of a specified kind from a variety of sources, giving all content and material information essential to the user."[4] This does not mean that archives did, or do not, have catalogues in the library sense of the word. But such catalogues are rare, and the practice of describing individual documents in archives is not common. For cartographic materials (and other non-textual archival records, such as motion pictures, photographs, etc.), the situation is different. Maps are, and were, very often described individually in lists, inventories and on card systems. Generally, the card systems were not of the library type and the standardization according to rules, if present, were thus also of the local type.

The practice in mainstream libraries of treating the cataloging of cartographic materials the same as books continued with minor changes through the appearance of *Rules for Descriptive Cataloging in the Library of Congress,* which evolved into the *Catalog Rules for Author and Title Entries* compiled by committees of the American Library Association and the (British) Library Association (1908) and the *A.L.A. Cataloging Rules for Author and Title Entries,* the second edition of which appeared in 1949. These, in turn, evolved into the *Anglo-American Cataloging Rules* (1967) and finally, in 1978, the second edition of those rules and its revision, the *Anglo-American Cataloguing Rules, 2nd Edition, 1988 Revision.* The latest emanation of these rules appeared in 1998 and includes all revisions approved by the Joint Steering Committee for revision of AACR (JSC), but does not contain any substantial changes in the approach to cataloguing since 1978. It is referred to as the *Anglo-American Cataloguing Rules, 2nd Edition, 1998 Revision.* The rules are now also available on CD-ROM, as part of *Cataloger's Desktop* software.

Only with the second edition of the rules in 1978 was there any substantial change in the cataloging treatment of cartographic materials. It is this author's perception that this occurred mainly because of the international efforts by the map librarian community in the Section of Geography and Map Libraries of the International Federation of Library Associations (IFLA)[5] where the development of the *ISBD(CM): International Standard Bibliographic Description for Cartographic Materials* took place. At this time the JSC, working on a major revision of the 1967 edition, stated its intent to adhere to the ISBD standards which had already been developed in IFLA–the ISBD(M) for monographic publications, and the ISBD(S) for serial publications. The IFLA Office for Universal Bibliographic Control, together with the JSC, initiated the working group on the *ISBD(G): International Standard Bibliographic Description.* General ISBD(G) was developed to enable the revision of AACR2 to proceed coherently to a single international standard rather than to diverse ISBDs. All existing ISBDs and those under development–including the ISBD(CM)–were subsequently revised to conform to the ISBD(G). Thus AACR2, and its subsequent revision, conform to the ISBD(G) and nominally to all the specific media ISBDs. The first editions of the ISBD(G) and ISBD(CM) were published in 1977.

ISBD(CM) introduced substantial changes. The main changes were in:

- the naming of the materials, formerly under the heading of "maps," to "cartographic materials," which had the effect of expanding this category of materials to include the types of material defined by the ISBD(CM) (see note 1);
- the creation of a special area of description for recording mathematical data (such as scale), much closer to the beginning of a bibliographic description rather than in a note; and

- perhaps the most important change, the introduction of multilevel description as an option to describe parts of a cartographic work in separate and linked descriptions.

Unfortunately, this last feature was either not properly or clearly understood by the editors of AACR2, or purposely not brought into the second edition in an unambiguous fashion. More about this later.

With the introduction of AACR2 and the substantial changes mentioned, the need for rule interpretations for cartographic materials became evident. *Cartographic Materials: A Manual of Interpretation for AACR2*[6] originated from work by the Association of Canadian Map Libraries (ACML). Its National Union Cataloguing Committee had been working on Canadian rules for the cataloguing of maps for many years, with the goal of alleviating the difficulties of applying the rules in the 1967 edition of the AACR. When the second edition of AACR appeared, the Committee decided to internationalize the work and asked American colleagues to look into the possibility of producing an interpretation manual based on the Canadian rules. The Committee called a meeting in Ottawa in 1979 and also invited British and Australian colleagues. This meeting resulted in the establishment of the Anglo-American Cataloguing Committee for Cartographic Materials (AACCCM) which, in a meeting in Washington, D.C. in April-May 1981 made the final decisions on the rule interpretations to chapter 3 of AACR2, resulting in the manual. *Cartographic Materials* was published in 1982.[7]

The publication of the ISBD(CM) and *Cartographic Materials* may surely be considered the most significant development for the cataloguing of cartographic materials. *Cartographic Materials* is now used and accepted in all of the English speaking world and has been influential in other parts of the world. It also has been the basis for many of the enhancements and improvements to the Machine Readable Cataloguing (MARC) formats, used extensively in libraries and institutions adhering to library standards.[8] One important aspect directly related to the MARC format must be mentioned. In 1972, the MARC format was adapted in the Netherlands as the basis for the "data structure" (as we called it in these days) of the system used by the Dutch Project CCK (Central Catalogue Cartography) at the University of Utrecht. Since the Dutch do not use the *Anglo-American Cataloguing Rules,* but have their own rules, they were adapting the format to satisfy their rules and what they saw as essential requirements for map documentation. It is in the Dutch Project CCK that multilevel description originated.[9] The technique, as a descriptive technique encoded in rules, was subsequently incorporated in the ISBD(CM). The cataloguing of cartographic materials thus reached full maturity.

Meanwhile, in archives much of the cartographic materials acquired as single items, not related to a record-keeping system, was catalogued in the same way, using AACR2 and *Cartographic Materials*. However, the approach in archives to describe its holdings[10] is based on some very important and basic principles and precepts–provenance and original order–which have evolved over many years. The evolution of these principles and of archival theories and practices, and the encoding of them in manuals and other writings, were covered exhaustively and comprehensively in a paper by Terry Cook which was given at the XIIIth International Congress on Archives in Beijing, China in 1996 as the key paper of Session III and published in its proceedings.[11] The first articulation of these principles was in the famous Dutch Manual of 1898, which celebrates its centenary this year.[12] As Cook remarks:

> The Dutch authors' chief contribution was to articulate the most important principles (or 'rules' as they called them) concerning both the nature and the treatment of archives. The trio stated in their very first rule, which to them was, 'the foundation upon which everything must rest,' that archives are 'the whole of the written documents, drawings, and printed matter, officially received or produced by an administrative body or one of its officials. . . .' Rules 8 and 16 enunciated the twin pillars of classic archival theory: archives so defined, 'must be kept carefully separate' and not mixed with the archives of other creators, or placed into artificial arrangements based on chronology, geography, or subject: and the arrangement of such archives 'must be based on the original organization of the archival collection, which in the main corresponds to the organization of the administrative body that produced it.' There, simply stated, are the concepts of provenance and original order. . . . [The authors] believed that by so respecting the arrangement of original record-keeping systems, the administrative context in which the records were originally created could be elucidated.[13]

This approach clashes with library cataloguing only in that it emphasizes the whole in the approach to archival description before the individual "record" (which in the library context becomes individual "work") and that the description has to be in the "administrative context" in which the records were originally created. The principles are based on an orderly transfer of government or corporate records to archival repositories to preserve their original order and classification; this may give some practical problems with modern records transfer where creators change frequently, and where the "origin" is not always recognizable or clear, and it gives particular problems with electronic records.[14] In general, though, the principles can be maintained for modern records if sufficiently elucidated and modified or adapted in their modern application. They may also be applied to archives of individuals and families.[15]

ARCHIVAL DESCRIPTION/CATALOGUING:
TWO DIFFERENT TRADITIONS

Archivists proceeded to develop their own descriptive guidelines, manuals and standards based on the principles of provenance and original order. In Canada work began on *Rules for Archival Description* in 1969.[16] In the United Kingdom there was the publication of the *Manual of Archival Description* (MAD) which reached its second edition in 1989 and was accompanied by the *MAD User Guide,* published in the same year.[17] To maintain the principle that an archive or archival collection is, "the whole of the written documents, drawings, and printed matter, officially received or produced by an administrative body or one of its officials. . . . ," both these publications advocated multilevel description where one proceeds by first describing the whole and thereafter the parts, and linking them in a hierarchy. Thus the precept that archival description proceeds from the general to the particular was born. The multilevel technique of description advocated in ISBD(CM) was adopted for this purpose.

In the United States, Steve Hensen's *Archives, Personal Papers, and Manuscripts: A Cataloging Manual for Archival Repositories, Historical Societies, and Manuscript Libraries* appeared first in 1983, and later, in 1989, as a book in second edition published and endorsed by the American Society of Archivists.[18] The Manual is generally referred to as APPM. The title indicates to whom the Manual was directed. Since it is an interpretation of Chapter 4 AACR2, Manuscripts (Including Manuscript Collections), it is in the same vein as *Cartographic Materials,* which interprets Chapter 3, AACR2. The Manual pays scant attention to multilevel description and does so only in the context of AACR2, rule 13.6, where multilevel description is regulated only within a single record, not multiple (linked) records. In this, the AACR2 technique of multilevel description does not differ in any substantial way from the *In:* analytic advocated as a separate analytical approach in the same chapter. Thus, in the United States, the two traditions or approaches to archival description unfolded, where archives which follow the record-keeping practice of describing collections is one tradition, and the library approach is the other. Outstanding examples of the former are the National Archives and Records Administration (NARA) and some (probably most) State records centers, while of the latter, the Library of Congress and state historical societies come most quickly to mind.

THE INTERNATIONAL COUNCIL ON ARCHIVES STANDARDS:
ISAD(G) AND ISAAR(CPF)

The work of the International Council on Archives (ICA)[19] Ad Hoc Commission on Descriptive Standards had its origin in an Invitational Meeting of

Experts on Descriptive Standards held in Ottawa, Canada, 4-7 October 1988. The meeting was hosted and sponsored by the National Archives of Canada in cooperation with the International Council on Archives.[20]

A number of papers presented at the meeting dealt with what the state of affairs regarding archival descriptive standardization efforts were in the various countries from which speakers came. It became clear that such efforts were at most at the very beginning stages of development or not developed at all. To at least some of those who were present, it also became painfully evident that some of the delegates were neither sure what the meeting was about nor cognizant of the problem and, therefore, of the need for standards of archival description.

Nevertheless, the meeting came to a unanimous conclusion that there was a need for international archival descriptive standards, and passed a resolution asking the ICA to establish a working group. This working group became the ICA Ad Hoc Commission on Descriptive Standards (ICA/DDS). The Commission, with UNESCO funding, developed ISAD(G), published in 1994 and ISAAR(CPF) published in 1996.[21] These two standards, for the first time in archival history, encode the principles enunciated a century ago in the Dutch Manual. The acronyms were purposely chosen to be analogous to the ISBDs. The ISAD(G) was developed to describe the archival materials themselves and the ISAAR(CPF) to describe and document the creators of that material and the context of creation. Before these standards were produced, there was a *Statement of Principles*,[22] again modeled after the document of the same title put together by international experts on cataloguing under the auspices of IFLA before its Office of Universal Bibliographic Control proceeded with its work on the ISBDs.

An important aspect of the archival *Statement of Principles* is the definitions of the terms used. The definitions embody the assumptions the Commission made and in the context of which the standards, the ISAD(G) and the ISAAR(CPF), were formulated and ought to be read. In particular, it is important to discern what the Commission adopted as its definition of archival description which it defined in the glossary as:

> ***Archival Description***. Creation of an accurate representation of the fonds and its component parts by the process of capturing, collating, analysing, and organizing any information that serves to identify archival material and explain the context and records systems which produced it.

This definition uses the word *fonds*, which the *Statement* defines as:

> ***Fonds***. All of the documents, regardless of form or medium, naturally generated and/or accumulated and used by a particular person, family, or corporate body in the conduct of personal or corporate activity.

ISAD(G) specifies its purpose to be to:

a. ensure the creation of consistent, appropriate, and self explanatory descriptions;
b. facilitate the retrieval and exchange of information about archival material;
c. enable the sharing of authority data; and
d. make possible the integration of descriptions from different repositories into a unified information system.

It further states that, as general rules, these are intended to be broadly applicable to descriptions of archives regardless of the nature or extent of the unit of description. The rules guide the formulation of information in each of twenty-six elements that may be combined to constitute the description of an archival entity.

This latter point acknowledges and emphasizes that *all* archival description is fundamentally description of "collectivities" of material and that these collectivities may be organized in sub-collectivities which may be further subdivided, etc. Such organization is called "arrangement" in English and "*classement*" in French, and is done on the basis of the principles of provenance and original order. Each collectivity or unit of arrangement becomes a unit of description. Thus, there may be many "units of description" in a collectivity of archival material. These units of description, being divisions and subdivisions of the whole collectivity called a fonds or collection, naturally group themselves into hierarchical levels characteristic of a tree structure.

Before it enumerates and gives rules of each of the 26 elements of description that may be combined to constitute an archival description, the ISAD(G) first sets out rules for the units of description of these levels which it calls multilevel description and hence, multilevel description rules. These are fundamental because these rules set up the structure of an archival description, as illustrated in the appendix, which is like a tree. This schema is the same as in the *Statement of Principles*. (See Figure 1.)

The rules of multilevel description do not prescribe a fixed set of levels in the hierarchy, nor do they make it compulsory to describe all levels even if they do occur in particular cases. Thus, one may add or subtract levels as each case demands. The appendix shows one possible occurrence as an example. Between the dotted lines in the schema it tries to indicate the most commonly occurring levels (most typical situations), even though there may be sub-levels at any level but the lowest. The names given to each level (e.g., series, files, items) are not necessarily current in different archival traditions. The labels in the schema are examples and may be valid only for the English language.

FIGURE 1

APPENDIX

A1 The model shows some typical situations and does not include all possible combinations of levels.

A2 Any number of intermediate levels are possible between any shown in the model.

MODEL OF THE LEVELS OF ARRANGEMENT OF A FONDS

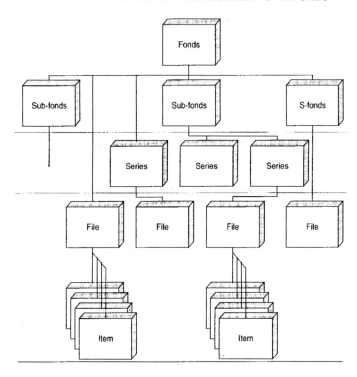

The ISAD(G) enumerates the fundamental principles of multilevel description as follows:

2.1 Description from the General to the Specific

PURPOSE: To represent the context and the hierarchical structure of the fonds and its parts.

RULE: At the fonds level give information for the fonds as a whole. At the next and subsequent levels give information for the parts being described. Present the resulting descriptions in a hierarchical part-to-whole relationship proceeding from the broadest (fonds) to the more specific.

2.2 Information Relevant to the Level of Description

PURPOSE: To represent accurately the context and content of the unit of description.

RULE: Provide only such information as is appropriate to the level being described. For example, do not provide detailed file content information if the unit of description is a fonds; do not provide an administrative history for an entire department if the creator of a unit of description is a division or a branch.

2.3 Linking of Descriptions

PURPOSE: To make explicit the position of the unit of description in the hierarchy.

RULE: Link each description to its next higher unit of description, if applicable, and identify the level of description.

2.4 Non-Repetition of Information

PURPOSE: To avoid redundancy of information in hierarchically related archival descriptions.

RULE: At the highest appropriate level, give information that is common to the component parts. Do not repeat information at a lower level of description that has already been given at a higher level.

These four rules were designed to put forth a logical construct of descriptions (an open hierarchical tree structure) which is universally applicable to any archival fonds or collection, and both economical to implement and easy to adopt or adapt in the construction of finding aids.

Cartographic materials in archives, described using the archival method, have to conform to these standards. In Canada, the Canadian archival community developed *Rules for Archival Description,* mentioned earlier (see note 16). It has a chapter on cartographic materials which conforms with the archival methodology of description. The best way to show what the consequences are of these archival standards of description, and the differences between the archival approach and library cataloging, is by means of illustrations.

EXAMPLES:
THE SAME ITEM IN AN ARCHIVAL RECORD
AND A LIBRARY CATALOG RECORD

The following example shows an archival multilevel description where, at the item level of description, there occur two of the items in the file, the deed of the purchase of a property and the map of the property. These examples were prepared according to the Canadian *Rules for Archival Description,* cited earlier, and compatible with ISAD(G).

Methodist Church (Canada). Missionary Society

14: Methodist Church (Canada) Missionary Society fonds. [ca. 1851-ca. 1930], predominant 1884-1925. 15.34 m.

The Missionary Society of the Methodist Episcopal Church Canada Conference, Methodist Church in Canada was established in 1824. When this Church joined with the British Wesleyans to establish the Wesleyan Methodist Church in Canada in 1833, the Society evolved into an Auxiliary of the Wesleyan Missionary Society (Great Britain) to support the growth of domestic missions, including missions to Aboriginal People. This union was ended in 1840, but resumed in 1847. In 1854 the British Hudson's Bay Territory missions were transferred to the Missionary Society in Canada, which gradually took over the responsibility of all mission work from Britain beginning in Central Canada and the Northwest. The Society, with some changes in administrative structure, existed as part of the Methodist Church of Canada and the Methodist Church (Canada). The object of the Society came to be the support and enlargement of the aboriginal, French, domestic, foreign and other missions, carried on under the direction of the central committee and board, and later also under the Conferences. In 1906, the missions were divided between two new Departments–Foreign and Home.

Fonds consists of the following series: General Board of Missions, 1865-1925; correspondence of the General Secretaries, 1868-1923; foreign mission records, 1888-1950; home mission records, 1906-1927; financial records, 1899-1930; quarterly returns of aboriginal institutes and day schools, 1902-1923; printed ephemera; and constitution and financial records of the Superannuation Fund for Lay Missionaries of Foreign Fields, 1919-1929.

Finding aid: See series descriptions.

See also United Church of Canada Board of Overseas Missions fonds (502) for records of missions continued by the United Church after the 1925 Union.

Location Number: See series descriptions.

I. Wesleyan Methodist Church in Canada. Missionary Society. II. Methodist Church of Canada. Missionary Society.

14/3: Records re foreign missions. 1888-1950, predominant 1888-1925. 4 m.
Series consists of records re the following missions: West China, 1891-1931; West China Union University, 1896-1950; and Japan, 1873-1925.
Finding aid: See subseries descriptions.
Location Number: 78.084C, 78.096C-78.098C.

14/3/1: West China Mission collection. 1891-1931, predominant 1891-1925. 2.2 m.
The Canadian Methodist Mission in West China was established in 1891.
Subseries consists of correspondence of the General Secretaries of the Methodist Church (Canada) Missionary Society; copybook of W.J. Mortimore; minutes of the West China Mission Council; reports, financial records, property registers, manuscripts of historical and biographical studies, and other material relating to the evangelistic, pastoral, educational and medical work of the West China Mission.
Finding aid: 19.
Location Number: 78.096C.

I. Canadian Methodist Mission of West China. Mission Council.

14/3/1/1: Canadian Methodist Mission Property Register, West China. 1899-1923. 1 cm.
File consists of Canadian Methodist Mission Property Register pages, West China for Chengtu College University and Chengtu City.

14/3/1/1/1: Chengtu, College University, No. 1, University Site, East of Administration Building skirting east and west road to Silk School with some breaks, 1914: [Land deed]. 1922. 1 p.
"Date of purchase 1914"
"Date of Registration . . . Oct.1922"

14/3/1/1/2: Plan of Chengtu, College University, No. 1, University Site, East of Administration Building skirting east and west road to Silk School with some breaks, 1914 : [Cartographic material]. Scale [1:1250]. 1 map : ms ; 10 x 30 cm.

NOTE: Even though the finding aid from which the fonds to sub-series examples were copied has (topical) subject added entries, these have been left off in the example.

Like the display in the finding aid from which these examples were extracted, the above display of descriptive records shows the part-to-whole hierarchy by indentation as well as by the numbering scheme given to the levels of description and the parts of the fonds by the archives. Each level of description is indented one unit of indentation from its immediate parent. Thus, if there were more than one series description, it would be at the same indentation as the series description shown.

The numbering scheme—fonds numbers, series numbers, sub-series numbers, etc.—also show the hierarchy. The fonds example here is fonds number 14. The series in this fonds are numbered 14/1, 14/2, 14/3, etc. The sub-series is a further division of this number, e.g., 14/1/1, 14/3/2, meaning sub-series 1 of series 1 of fonds 14; sub-series 2 of series 3 of fonds 14, etc. This numbering subdivision continues to the lowest level of description given in the example, the item level description of which there are two, shown at the same level of indentation.

Of note in these examples is that there is only one main entry and that each description may have its own added entries because they are separate records. The records are linked, each to its respective parent record, according to the ISAD(G) multilevel rule 2.3, "Linking of descriptions," and rule 2.4, "Non-repetition of information," insures economy of description, but also requires that certain information is inherited from parents in the hierarchy above the relevant record, such as the main entry. The main entry of the map in the example is:

Methodist Church (Canada). Missionary Society.

If there had been an author or cartographer mentioned on the ms. map, corporate or personal, it would have appeared as an added entry with the appropriate qualifier, such as *"(cartographer)"* or *"(surveyor)."* In this respect, it should be mentioned that both RAD and the ISAD(G) make a distinction between author and an archival creator. In ISAD(G)'s glossary the term "Creator" is referred to "Provenance," which is defined as, "The organization or individual that created, accumulated and/or maintained and used documents in the conduct of personal or corporate activity."[23] The definition of "Author" in RAD retains its AACR2 definition of "Personal author."[24] In an archival description there are no main entries for any of the records that are linked to a parent. In this way, the whole set of records form one integrated whole linked to the provenance of the fonds. The context of that single map, its significance as a record of evidential value as well as its significance as an instrument in the administration of the Missionary Society of which it forms a part, is maintained and protected by this method of description. This is the archival approach to the description of materials, and if the materials have cartographic records (or any other media type record), they follow the same archival rules.

The schema in Figure 2 models the hierarchy of descriptions of an archival fonds, using the above set as an example. It also shows the relationship of these descriptions to authority records that accompany the descriptions and all the links that will have to be made:

    a. between descriptive records to link the records in the hierarchy, linking child records to their parent;

    b. between the main entry and its authority record;

    c. between added entries and their authority records;

    d. between authority records (*see* and *see also* linkages).

The authority records are shown as a separate set of records at the right hand side of the diagram, to the right of the dotted line. This implies a separate but linked authority system. Not all library systems work this way.

The same map as a library catalog record would likely take the following form:

> Plan of Chengtu, College University, No. 1, University Site, East of Administration Building skirting east and west road to Silk School with some breaks, 1914 [cartographic material]. Scale [1:1250]. 1 map : ms ; 10 x 30 cm.

Since the map does not have an author, be it a cartographer, legal surveyor, or the like, it is difficult to know what its main entry would be or which added entries to make. It probably would get the same entry, "Methodist Church (Canada). Missionary Society," as main or added entry and a title added entry. Perhaps there would also be an added entry for the Chengtu College University as a topical subject heading. Compared with the archival record, this catalog record would have lost its significance as a record of a transaction–its evidential value–for it has lost its connection to the function it served as a legal instrument, and its connection to the deed that accompanied it. It would also have lost the relationship related to the other records in the file, the (archival) sub-series, the (archival) series and the fonds of which it is a part. These other records may reveal why the Mission Society bought and maintained property in China and how it went about obtaining it. There may be many other pieces of information of a historical nature in these records. That information may be of interest to patrons, researchers or users of the map archives and may shed additional light on the map and what it depicts or signifies. But when the map is individually catalogued without its contextual information and links, it has interest only as a geographical depiction of a site in China where a college was built. Not very impressive, in comparison with

FIGURE 2

ARCHIVAL DESCRIPTION

RELATIONSHIPS BETWEEN AUTHORITY AND DESCRIPTIVE RECORDS

Illustrated with an example

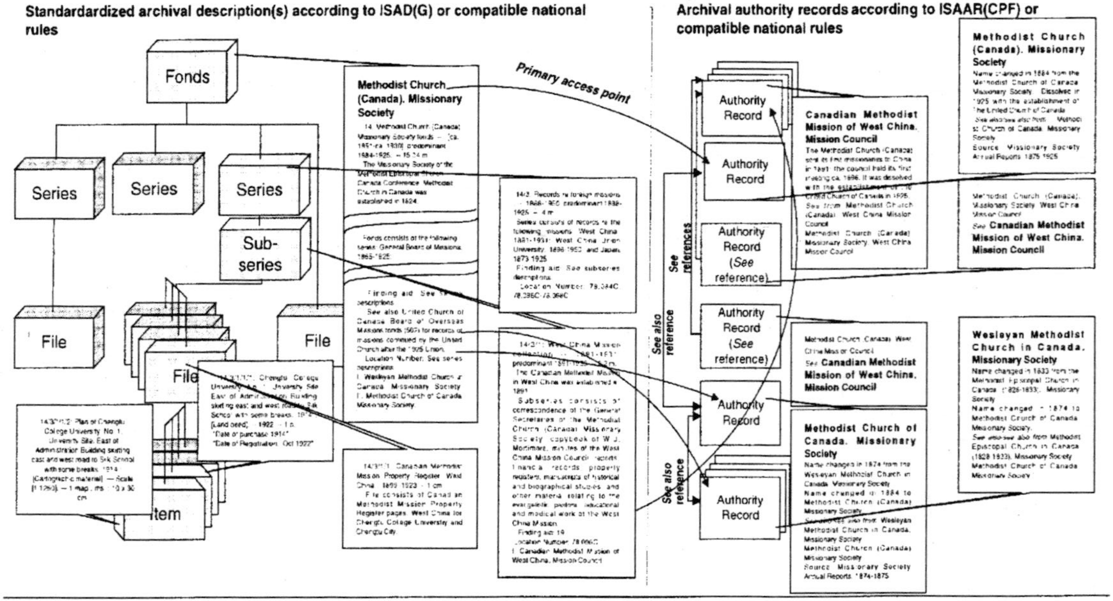

Chart designed and produced by Hugo L.P. Stibbe, c1998

its archival-record presence! A cataloger would have to make an inordinate number of notes to bring out all that its position in the hierarchy of descriptions in the fonds shows, a hierarchy which, in the example above, only depicts one lineage, the one which links the immediate parents in the hierarchy. The total fonds and all of its parts are, of course, much larger than is indicated in the example, as it has many series, and files.

TOWARD AN INTEGRATED WAY OF ARCHIVAL DESCRIPTION OF CARTOGRAPHIC MATERIALS IN ARCHIVES

Archival descriptions and bibliographic cataloging records may exist side by side in the same database. A question that is often asked about cartographic records that are part of a fonds description is whether these records can be found easily when they are "buried" inside a set of descriptions that represent a fonds and its parts. The answer depends on the design of the system on which the records reside. Most current library systems cannot accommodate multilevel descriptions that are linked in an open hierarchical structure such as that illustrated in Figure 2. By means of the linkage fields, the MARC 21 formats allow, and fully provide for, linkages of the kind described to link child records to parent records, no matter how many levels there are in a particular case of a hierarchy. But most systems do not do anything with these links in terms of display and other output products, no matter how diligently archivists and map librarians may fill out and encode these linkage fields. In terms of searching, each of the descriptions in the hierarchy of an archival description is a separate record. Thus, they may have their own added entries in addition to the main entry for the fonds level. Searching would bring up any relevant records which have such access points.

The problem occurs in the subsequent display. Library systems only bring up the hit record(s) and not the records that are the parents in the hierarchy. When a hit displays a record at a lower level than the top (fonds/collection) level, that record depends on all of its linked parents to display the complete information.[25] If these parent records are not recovered and displayed, the system cannot be used for an archival application, nor, I might add, for multilevel cartographic cataloging records such as sheets of a map series. Some systems allow for searching the links and then bringing to the screen, one by one, the linked records. This is hardly adequate for studying the integrated record. The MARC 21 formats now also provide descriptions created by the archival method to be identified in the Leader CP08, Type of Control. It provides the code "a" for "Archival control." In Leader CP06, Type of Record, there are codes for all types, including "p–Mixed material." Thus it is possible to have MARC 21 records for archival controlled descriptions (the methodology, rather than the material) co-exist with bibliographic records.

CONCLUSION

There are many more aspects regarding the differences and similarities of archival description and library or bibliographic description one could write about and discuss in detail, but space available does not permit doing so. In conclusion, there are no obstacles to the integration of archival and bibliographic descriptions for cartographic materials, not in the rules and rule interpretation manuals now available, nor in the MARC 21 formats. At the time of this writing, what seems to be lacking are systems that can make use of these possibilities. There are hopeful signs. Some vendors or organizations are appearing on the market with systems which are capable of dealing with multilevel description or were specifically designed to satisfy ISAD(G) and ISAAR(CPF) requirements. For example, a partner of IBM in Spain, Informática el Corte Inglés, S.A.,[26] seems to have developed such a system. Also, the new MINISIS, developed by the International Development Research Centre (IDRC) in Canada, is working on an archival package with the desired capabilities.[27] Others are the Advanced Revelation DBMS, a Vancouver firm, Eloquent Systems Inc., which has an archival package,[28] and CUADRA STAR/Archives.[29] Map librarians who want to fully implement multilevel descriptions, as they are meant to function according to ISBD(CM) and/or ISAD(G), should look at such systems.

NOTES

1. The term "cartographic materials" only became prevalent with the publication in 1977 of the *ISBD(CM): International Standard Bibliographic Description for Cartographic Materials* (London: IFLA International Office for UBC). It employed the term in its title and defined it as, "all materials representing, in whole or in part, the earth or any celestial body at any scale, such as two- and three-dimensional maps and plans; aeronautical, navigational and celestial charts; globes; block-diagrams; sections; aerial, satellite and space photographs; atlases; bird's-eye views; etc." (p. 1, 0.1.1 Scope).

2. Philip Lee Phillips, *Notes on the cataloging, care, and classification of maps and atlases, including a list of publications compiled in the Division of Maps* (Washington, Government Printing Office, Library Branch, 1921), 7.

3. International Council on Archives. *Dictionary of Archival Terminology = Dictionnaire de terminologie archivistique: English and French with Equivalents in Dutch, German, Italian, Russian and Spanish,* edited by Peter Walne. 2nd Revised Edition. München, etc.: KG Saur, 1988. It should be noted that the forthcoming 3rd

edition of this work (DAT III) will include the term "Catalogue" and that this is defined similarly, although not word-for-word identically, to the library definition in the *ALA Glossary of Library and Information Science* (Chicago: ALA, 1983).

4. *Idem*, p. 29. Calendaring is hardly ever used these days in archives due to the considerable labor this technique requires.

5. IFLA later changed its name by adding, "and Institutions" at the end, but leaving the acronym IFLA as is.

6. Anglo-American Cataloguing Committee for Cartographic Materials. *Cartographic Materials: A Manual of Interpretation for AACR2.* Hugo L.P. Stibbe, General Ed., Vivien Cartmell and Velma Parker, Ed. Chicago, etc.: American Library Association, etc., 1982.

7. The manual is currently being revised.

8. This paper does not deal with aspects of automation. Therefore, a discussion on the impact of the rules on the MARC formats falls outside its scope. However, the reverse may also be true–the impact the MARC formats had on rules. One such aspect is discussed in the next paragraph.

9. This often-forgotten origin of multilevel description in the world of cataloguing cartographic materials is documented in my doctoral dissertation, *MARC-Maps: The History of its Development and Current Assessment* (The Hague: NOBIN, 1976, Chapter IV (especially note 13 of this chapter); and appendix IV, tag 090. More precisely, the technique was introduced in the CCK by Evert Hans van de Waal, the Project Director of the CCK and should be attributed to him. After having been introduced in the ISBD(CM) the technique was later adopted for archival description and incorporated in the *General International Standard Archival Description: ISAD(G)* developed by the Ad Hoc Commission on Descriptive Standards (now a Committee) of the International Council on Archives. (See note 21 for the full citation of this standard.) See particularly: Waal, E.H. van de "Over titelbeschrijving van kaarten: Ervaringen, opdedaan tijdens de eerste fase van het proefproject CCK," *Open Vaktijdschrift voor bibliothecarissen, literatuuronderzoekers, bedrijfsarchivarissen en documentalisten* 4, dec. 1972, pp. 1-9 [in off-print]. (Title of article in translation: "About the cataloguing of maps: Experiences gained from the first phase of the pilot project Central Catalogue Cartography.")

10. As mentioned before, archives prefer the word, "description," to indicate the act of making holdings accessible to users, rather than "cataloging." I shall thus distinguish the archival use and library way of making holdings accessible by means of using the words, "description" and "cataloging," respectively. The "holdings" referred to here form the bulk of materials in archives. These relate to records acquired through records keeping systems, and kept together on the principles of provenance and original order, the two most important archival principles, explained further in the text.

11. Cook, Terry. "Archives in the Post-Custodial World: Interaction of Archival Theory and Practice Since the Publication of the Dutch Manual in 1898." *Archivum* XLIII (1997): 191-214.

12. Muller, Samuel, Robert Fruin and Johan A. Feith. *Handleiding voor het ordenen en beschrijven van archieven,* translated into English by Arthur H. Leavitt as the *Manual for the Arrangement and Description of Archives,* drawn up by direction of

the Netherlands Association of Archivists by S. Muller, J.A. Feith and R. Fruin, 2nd ed. With a new foreword by Ken Munden. New York: H.W. Wilson, 1968.

13. Cook, "Archives in the Post-Custodial World," p. 193.

14. See, for example: Mulè, Antonella. "The Principle of Provenance: Should it Remain the Bedrock of the Profession?" *Archivum* XLIII (1997): 233-256.

15. For this "elucidation" see, among other writings, my article, "Implementing the Concept of Fonds: Primary Access Point, Multilevel Description and Authority Control," *Archivaria* 34 (Summer 1992): 109-137.

16. Planning Committee on Descriptive Standards. *Rules for Archival Description.* Ottawa: Bureau of Canadian Archivists, 1969. This work, which was published in English and French at the same time, took some 10 years to complete, is in loose leaf form, and is continuously updated. Its current "publication date" is 1990 and updates are posted at the URL:
http://www.cdncouncilarchives.ca/radltr_e.html
For a retrospective of the approach taken and the history of its development, see my article, "Archival Descriptive Standards and the Archival Community: A Retrospective, 1996," *Archivaria* 41 (Spring 1996): 259-274.

17. Cook, Michael and Margaret Procter. *A Manual of Archival Description.* 2nd ed. Brookfield, Vt.: Gower Publishing Co., 1989.
Cook, Michael and Margaret Procter. *MAD user guide.* Brookfield, Vt.: Gower Publishing Co., 1989.

18. Hensen, Steven. *Archives, Personal Papers, and Manuscripts: A Cataloging Manual for Archival Repositories, Historical Societies, and Manuscript Libraries.* 2nd ed. Chicago: Society of American Archivists, 1989.

19. For map librarians, curators and archivists, the acronym ICA stands for International Cartographic Association. The other major non-governmental organization that carries the same acronym (ICA), is the International Council on Archives. This will be the only one occurring in this paper.

20. The proceedings of this meeting were published as, *Toward International Descriptive Standards: Papers Presented at the ICA Invitational Meeting of Experts on Descriptive Standards, National Archives of Canada, Ottawa 4-7 October 1988 /* Compiled and edited with the financial assistance of the Toronto Area Archivists Group Education Foundation = *Projet de normes internationales de description en archivistique: Communications présentées à la réunion restreinte d'experts en normes de description, Archives nationales du Canada, Ottawa, du 4 au 7 octobre 1988 /* Receuil rassablé et publié avec l'aide financière de la Toronto Area Archivists Group Education Foundation. München; New Providence; London; Paris: Saur, 1993. (ISBN 3-598-11163-0)

21. *ISAD(G): General International Standard Archival Description /* adopted by the Ad Hoc Commission on Descriptive Standards, Stockholm, Sweden, 21-23 January 1993: final ICA approved version. Ottawa: 1994. (ISBN 0-9696035-1-7). Christopher J Kitching *(chair);* Hugo LP Stibbe *(project director and secretary).* At head of title: Conseil international des archives, International Council on Archives. Issued also in French under the title: *ISAD(G): Norme générale et internationale de description archivistique.* (ISBN 0-9696035-2-5). (This standard is currently being reviewed as part of its 5-year review cycle. The next edition is expected to become available in late 1999 or early 2000.)

ISAAR(CPF): International Standard Archival Authority Record for Corporate Bodies, Persons and Families / prepared by the Ad Hoc Commission on Descriptive Standards, Paris, France, 15-20 November 1995: final ICA approved version. Ottawa: International Council on Archives, 1996. (ISBN 0-9696035-3-3). Christopher J Kitching (*chair*); Hugo LP Stibbe (*project director and secretary*). At head of title: Conseil international des archives, International Council on Archives. Issued also in French under the title: ISAAR(CPF): Norme internationale sur les notices d'autorité archivistiques relatives aux collectivités. Aux personnes et aux familles. (ISBN 0-9696035-4-1)

Both standards may be viewed and downloaded from the ICA Word Wide Web site at URL:http://www.archives.ca/ica/cgi-bin/ica?04_e

22. *Statement of Principles Regarding Archival Description, First Version Revised* / adopted by the Ad Hoc Commission on Descriptive Standards, Madrid, Spain, January 1992. Ottawa: The Commission, Feb. 1992.

23. *ISAD(G)*, p. 4.

24. AACR2-1988R, p. 620.

25. This is the consequence of the ISAD(G) multilevel rule 2.4, "NON-REPETITION OF INFORMATION," which states that information already given at a higher level need not be repeated at a lower level. This same rule also appears in the Canadian *Rules for Archival Description* (RAD) cited earlier.

26. Informática el Corte Inglés, S.A. is located at Travesía de Costa Brava 4, planta 3, 28034 Madrid, Spain. E-mail: archbib@ibm.net

27. Information about IDRC and MINISIS may be viewed on the Internet at the URLs respectively: http://www.idrc.ca/index_e.html; and http://www.idrc.ca/nayudamma/minisis_96e.html

28. Information about Eloquent Systems and their archival package, The GENCAT Archives System, may be obtained from URL: http://www.eloquent-systems.com/

29. Information about CUADRA STAR/Archives may be obtained from its contact person: Ted Wolfe, voice telephone (310) 478-0066, and from CUADRA STAR's Internet site at URL: http://www.cuadra.com

Index